Agricultural Science and Management

Agricultural Science and Management

Editor: Jude Boucher

www.callistoreference.com

Callisto Reference,
118-35 Queens Blvd., Suite 400,
Forest Hills, NY 11375, USA

Visit us on the World Wide Web at:
www.callistoreference.com

ISBN: 978-1-63239-965-6 (Hardback)

Cataloging-in-Publication Data

Agricultural science and management / edited by Jude Boucher.
p. cm.
Includes bibliographical references and index.
ISBN 978-1-63239-965-6
1. Agriculture. 2. Farm management. 3. Management. I. Boucher, Jude.
S493 .A37 2018
630--dc23

Table of Contents

Preface

Agricultural science is a branch of biology which studies the techniques practiced in the field of agriculture. It focuses on developing innovative methods to maximize crop yield while considering ecological as well as economical factors. Some of the commonly studied aspects of agricultural science include agronomy, soil degradation, agricultural productivity, agroecology, etc. This book attempts to understand the multiple branches that fall under the discipline of agricultural science and how such concepts have practical applications. With state-of-the-art inputs by acclaimed experts of this field, this book targets students and professionals.

The researches compiled throughout the book are authentic and of high quality, combining several disciplines and from very diverse regions from around the world. Drawing on the contributions of many researchers from diverse countries, the book's objective is to provide the readers with the latest achievements in the area of research. This book will surely be a source of knowledge to all interested and researching the field.

In the end, I would like to express my deep sense of gratitude to all the authors for meeting the set deadlines in completing and submitting their research chapters. I would also like to thank the publisher for the support offered to us throughout the course of the book. Finally, I extend my sincere thanks to my family for being a constant source of inspiration and encouragement.

Editor

1

Phylogenetic Relationship of Indonesian Water Yam (*Dioscorea alata* L.) Cultivars Based on DNA Marker Using ITS-rDNA Analysis

Purnomo[1], Budi Setiadi Daryono[2] & Hironobu Shiwachi[3]

[1] Laboratory of Plant Systematic, Faculty of Biology, Universitas Gadjah Mada, Indonesia

[2] Laboratory of Genetic, Faculty of Biology, Universitas Gadjah Mada, Indonesia

[3] Graduate School of Agriculture, Tokyo University of Agriculture, Tokyo, Japan

Correspondence: Purnomo, Laboratory of Plant Systematic, Faculty of Biology, Universitas Gadjah Mada, Indonesia. E-mail: nomo@ugm.ac.id

Abstract

Research on genetic diversity and intra-species classification of Indonesian *Dioscorea alata* L. based on morphological characters has been done, and the result shows that there are 4 sub-groups of green cultivars group, and 5 sub-groups of purplish-red cultivars group. The objectives of this study are to determine the phylogenetic relationship cultivars of *D. alata* Indonesia compared to *D. bulbifera* as nearest species as well as 3 cultivars from GenBank. The young leaves of 18 water yam cultivar accessions were collected from Java, Madura, South Sumatera, South Kalimantan, Centre Celebes (Sulawesi), Ternate, West Papua, and Nusa Tenggara islands of Indonesia. DNA Isolation was conducted using Phytopure reagent. DNA amplification was conducted using thermocycler, predenaturation at 95 °C 5 minutes, denaturation at 95 °C 1 minutes, annealing at 60 °C 3 minutes, and elongation at 72 °C 2 minutes along with 30 cycles. The PCR products were electrophoresed in 1.5% agarose and visualized under UV transiluminator. Fifty-five µl of PCR products with positive targeted band between 700-800 bp were sent to 1st Base Singapore for purification and sequencing of 18S, ITS-1, 5.8S, ITS-2, and 28S rDNA. The DNA sequences were compared and aligned by BioEdit program (version 7.0.5.2) and MEGA programs. Comparison of entire sequences of the tested samples were aligned by software ClustalW (version 1.83). Phylogenetic trees were based on hierarchical clustering of the alignments of 18S, ITS-1, 5.8S, ITS-2, and 28S rDNA and produced by Neigbor-Joining using MEGA 5 software of the bootstrap values (1000 replicates). The study result shows that *D. alata* cultivars have high genetic variability on ITS-1, 5.8S, and ITS-2 rDNA region. Groups of green and purplish red cultivars formed based on morphological characters are not formed based on ITS-rDNA markers. Sub-groups were formed based on ITS-rDNA molecular markers derived from both the green and purplish-red cultivar groups. This result revealed that two cultivar groups are not similar with RAPD and morphological characters.

Keyword: *Dioscorea alata* L., cultivars, relationship, molecular, ITS-1 & ITS-2 rDNA

1. Introduction

There are many cultivars of Indonesian water yam (*D. alata*), all of the cultivars have edible tuber with many kinds of shapes and colors. Based on morphological and molecular characters by RAPD analysis, Purnomo et al. (2012) classified Indonesian water yam cultivars into *D. alata* cultivar group 'green' and 'purplish-red'. *D. alata* cultivar group 'green' has 4 sub-groups, *i.e.* (1) white tuber flesh rounded to oblong tuber shape (uwi beras, uwi elus, uwi alas, uwi putih, uwi kemayung cultivars), (2) white and sweetish tuber flesh (uwi legi), (3) white tuber flesh ovate to oblong tuber shape (uwi putih cultivars from Ternate and Lombok), and (4) yellow ovate to oblong tuber shape (uwi butun, uwi kuning). *D. alata* cultivar group 'Purplish red' has 5 sub-group cultivars, i.e. sub-group of (1) white to yellow long cylindrical tuber shape (uwi luyung putih, uwi luyung kuning, uwi ulo, uwi ular, uwi kuning from Merauke), (2) purplish-red long cylindrical tuber shape (uwi luyung senggani), (3) white with purple ring tuber flesh (uwi bangkulit from many place in Indonesia), (4) Purple irregular tuber shape (uwi senggani, uwi ungu, uwi merah from many place in Indonesia), (5) yellow tuber flesh with purple tuber skin (uwi kuning from Pelaihari, Kalimantan).

Morphological characters and molecular by RAPD analysis can be conducted to the similarity relationship only. Therefore, the region of ITS-rDNA is helpful to phylogenetic analysis that can determine the ancestral, origin, and phylogenetic relationship analysis among cultivars. Currently, ITS-rDNA is widely used in taxonomy and molecular phylogenetics (Lin et al., 2007). ITS-1 and ITS-2 regions are variable and useful as a source of polymorphisms for distinguishing genetic variation among species within the genus or among populations.

The region of ribosomal DNA (rDNA) is the sources of valuable characters as *Angiospermae* phylogeny on *Internal Transcribed Spacer* (ITS) is located between 18S-28S nuclear rDNA genes. ITS-1 is located between 18S and 5.8S gen, while ITS-2 is located between 5.8S and 28S gen. The regions of 18S, 5.8S, and 28S rDNA are the conserved area, and the regions of ITS-1 and ITS-2 are the variable area. The variables area is used for genetic variability interspecies on the same genus or among the population because it has a high polymorphism. The regions of ITS 1 and ITS 2 are commonly used as character sources to determine the phylogenetic relationship. Amplification of target band on PCR product the region of ITS-1 and ITS-2 rDNA on more than 500 base pair (Baldwin et al., 1995; Baldwin & Markos, 1998; Lin et al., 2007, 2011).

ITS-1 and ITS-2 regions of rDNA analysis are used to identify molecular variability and phylogeny of *Colletotrichum* species from *Almond* and the other fruits (Freeman et al., 2000). ITS-1 and ITS-2 regions of rDNA analysis are used to determine the phylogeny of *Cercospora* and *Mycosphaerella* (Goodwin et al., 2001). The geographical distribution of *Waitea circinata* var. *circinata* on *annual bluegrass* in America can be detected by the sequence of ITS-rDNA region (Chen et al., 2009). Determination of short and long term of DNA barcodes is required for phylogeny analysis of terrestrial plants (Chase et al., 2005).

In the present study, ITS-1- and ITS-2-based analyses were used to ascertain the genetic relationship among 18 cultivar accessions of *D. alata* (water yam), 1 accession of *Dioscorea* sp., 3 water yam cultivars from GenBank, and 1 species of *D. bulbifera* as out group. The objectives of this study were to determine the phylogenetic relationship cultivars of *D. alata* Indonesia compared to *D. bulbifera* as nearest species and cultivars from GenBank.

2. Materials and Methods

2.1 Plant Materials

The young leaves of 18 water yam cultivar accessions from many places in Indonesia were collected from Java, Madura, South Sumatera, South Kalimantan, Centre Celebes (Sulawesi), Ternate, West Papua, and Nusa Tenggara islands (Table 1). Nucleotide series of 3 water yam cultivars from GenBank *Dioscorea alata* L. cultivar *Guangxi Yulin shanyao clone* 2 (FJ860067.1), *D. alata* L. bio-material *Hainan shanyao* (FJ860070.1), dan *D. alata* L. cultivar Guangxi *Shatianzhen shanyao* (FJ860065.1) were collected from China.

Table 1. Accession numbers, cultivar names, accession origin, and morphological characters of *Dioscorea alata*

Accession number	Cultivar name (Indonesia)	Accession origin	Morphological characters (Stem nodes, wing stem, leave nerves color), (tuber shape), and (tuber flesh color)
06	*Uwi beras*	Rembang, Central Java	Green, rounded, white
10	*Uwi putih*	Purwodadi, Central Java	Green, rounded, white
17	*Ubi putih*	Buon, Luwuk, Central Sulawesi	Green, rounded-cylindrical, white
21	*Uwi legi*	Sewon, Bantul, Yogyakarta	Green, oblong, white
25	*Uwi butun*	Gunung Kidul, Yogyakarta	Green, ovate, yellowish white
29	*Uwi Luyung putih*	Sewon, Bantul, Yogyakarta	Green, length cylindrical, white
35	*Uwi Luyung kuning*	Kulon Progo, Yogyakarta	Green, length cylindrical, yellow
37	*Uwi ulo*	Karebet, Sendangsari, Bantul, Yogyakarta	Green, length, white-yellow
42	*Uwi kuning*	Sleman, Yogyakarta	Green, ovate-oblong, yellow
48	*Uwi bangkulit*	Serongga, Batulicin, South Kalimantan	Purplish red, ovate, white with purple ring
50	*Uwi Luyung senggani*	Karebet, Sendangsari, Bantul, Yogyakarta	Purplish red, length cylindrical, purple
52	*Uwi senggani*	Karebet, Sendangsari, Bantul, Yogyakarta	Purplish red, irregular, purple
55	*Uwi ungu*	Pelaihari, South Kalimantan	Purplish red, irregular, purple
62	*Obi item*	Pamekasan, Madura, East Java	Purplish red, oblong, purple with blackish dot
64	*Obi violet*	Bangkalan, Madura, East Java	Purplish red, oblong, violet
70	*Ubi ungu*	Banggai, Central Sulawesi	Purplish red, oblong-cylindrical, violet with white ring
78	*Gembolo* (*D. bulbifera*)	Wonosadi, Gunung Kidul, Yogyakarta	Dark green, rounded with 3-6 branches, grayish white
107	*Ubi hutan (Dioscorea sp.)*	Kinton-Toili, Luwuk, Central Sulawesi	Light green, length cylindrical, white-white ash
138	*Ubi ungu*	Lombok, Nusa Tenggara	Purplish red, ovate-rounded, purple
142	*Ubi ungu*	Merauke, West Papua	Purplish red, oblong-cylindrical, purple

2.2 DNA Extraction

DNA Isolation is conducted by *Phytopure* (Daryono & Natsuaki, 2002). The fresh young leaf is sliced and then grounded into powder with liquid nitrogen. Powder is brought in 1.5 μl tube, 400-500 μl of phytopure I reagen is added and shaken by hand, and then 75-100 μl phytopure II reagen is added and shaken by hand, and then incubated at 65 °C for 10 minutes on waterbath, and then on ice for 20 minutes, and then 400-500 μl cold chloroform is brought in, and then 50-70 μl phytopure resin was added in the centre carefully, and then centrifuged at 3000 rpm for 10 minutes. The supernatans in the new tube 1.5 ml is removed, cold isopropanol with the same volume with supernatans is added and shaken well by hand, centrifuged at 10.000 rpm for 10 minutes, white DNA pellet on the bottom. Supernatans was thrown and DNA pellet washed with 100 μl ethanol 70% added and centrifuged at 10.000 rpm for 5 minutes. Ethanol and dried DNA pellet were eliminated with the wind, and then 1xTE buffer was added and it was kept in the freezer at -12 °C.

The purity of DNA was determined with *GeneQuant* (Life Science, Ltd., UK) with ratio on spectrophotometer at 260 nm (absorbance optimum detection for DNA) and 280 nm from DNA samples, DNA purity refers to standard DNA purity 1.8-2.2 on 260/280 nm (Sambrook et al., 1989; Googwin et al., 2001).

2.3 Polymerase Chain Reaction (PCR) Amplification

Each PCR reaction (25 μl) on tube 200 μl consisted of Mega Mix Blue 22 μl, forward primer 5-GATCGCGGCGGCGACTTGGGCGGTTC-3 1 μl, reverse primer 5-GGTAGTCCCG CCTGACCTGGG-3 1 μl, and DNA template 1 μl. The use of primer was according to Li et al. (2011) and Muellner et al. (2008). PCR reaction in the tube was centrifuged at 8000 rpm for 30 seconds. Amplification was conducted by thermocycler, predenaturation at 95 °C for 5 minutes, denaturation at 95 °C for 1 minute, annealing at 60 °C for 3 minutes, elongation at 72 °C for 2 minutes. PCR reaction was conducted on 30 cycles. The PCR products were electrophoresed in 1.5% agarose with good view (modification of ethidium bromide) staining and visualized under UV transilluminator compared to DNA ladder (marker). Photography is done with digital cameras.

2.4 Nucleotide Sequencing

Fifty-five (55) μl PCR products with targeted band positive between 700 and 800 bp were sent to 1st Base Singapore for purified and sequencing of 18S, ITS-1, 5.8S, ITS-2, and 28S rDNA.

2.5 Sequence Alignment and Phylogenetic Tree

The DNA sequences were compared and aligned by BioEdit program (version 7.0.5.2) and MEGA programs and further verified by comparison with sequences of other cultivars and *D. bulbifera* by BLAST (version 2.2.2.4) search on the website of the National Centre for Biotechnology Information (NCBI). A comparison of the entire sequences of the tested samples was made by software ClustalW (version 1.83). Phylogenetic trees were based on hierarchical clustering of the alignments of 18S, ITS-1, 5.8S, ITS-2, and 28S rDNA and produced by Neigbour-Joining using MEGA 5 software of the bootstrap values (1000 replicates). For outgroup sequences *D. bulbifera* was chosen because morphologically this species has a high morphological similarity (closed relationship).

3. Result and Discussion

3.1 ITS rDNA Band Target

Visuals of electrophoresis of PCR product under ultraviolet (UV) are shown in Figure 1. Figure 1 shows the size of ITS target band of 19 water yam (*D. alata*) cultivars and *D. bulbifera* species. Accession numbers 01, 05, 23, and 40 in Figure 1 are without a real band and also they have no representative sequence because of the number of nucleotides less than 400. On a number of 20 accessions composed of 18 cultivars, *D. alata, Dioscorea* sp., and *D. bulbifera,* in the ITS-rDNA region by electrophoresis the target bands are detected on 700-800 bp size. Those results are suitable to the last ITS-rDNA research, especially on *Angiospermae* (Baldwin & Markos, 1998; Balwin et al., 1995; Hidayat & Pancoro, 2001; Chase et al., 2005). The tick band indicates that those regions are the optimal amplification of ITS-rDNA.

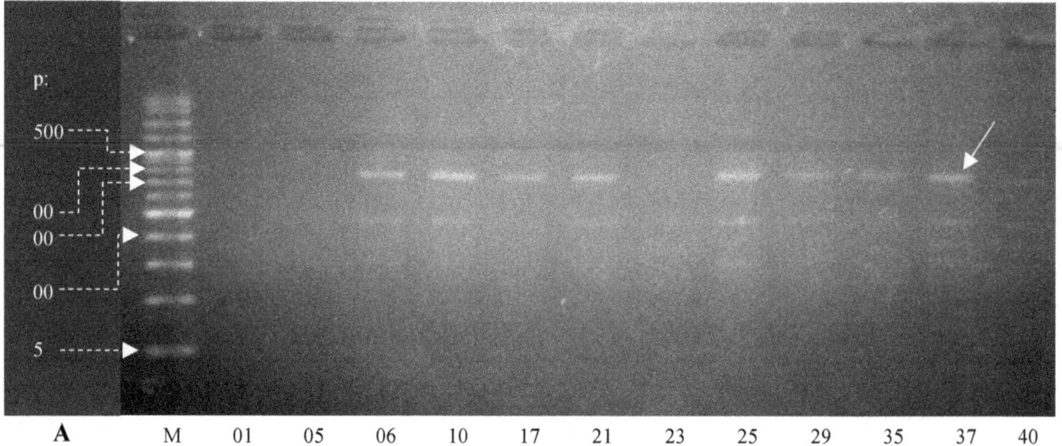

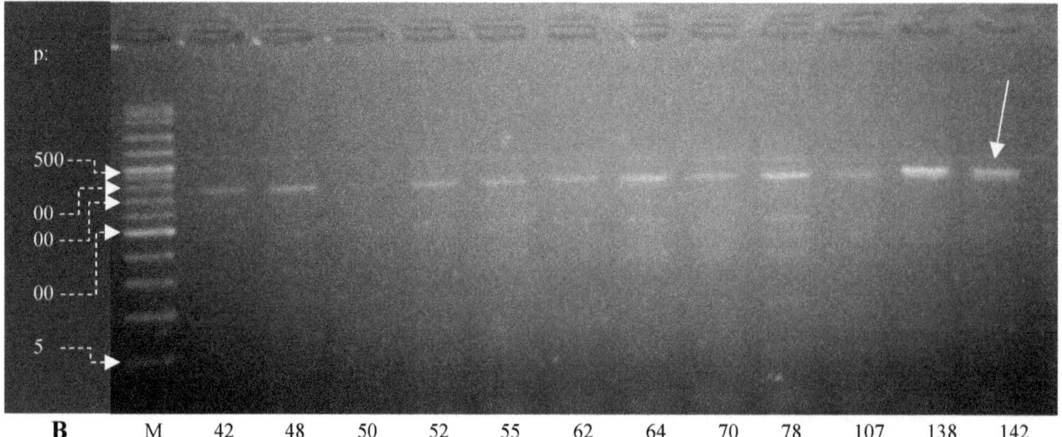

Figure 1. Electrophoresis product PCR for ITS target band on between 700 and 800 bp (single arrow shows the ITS bend). Legend: A = accession number 01-40, B = accession number 42-142. M = DNA marker 1 Kb (Fermentas), series number of 01, 05 until 142 = accession number of Indonesia *D. alata* cultivars

3.2 rDNA Sequence Analysis

ITS-rDNA forward primer 5-GATCGCGGCGGCGACTTGGGCGGTTC-3 and reverse primer 5-GGTAGTCCC GCCTGACCTGGG-3 producing nucleotide series of 19 water yam cultivars and *D. bulbifera* species. 18S, ITS-1, 5.8S, ITS-2, and 28S rDNA sequences of 19 *D. alata* cultivars, *D. blbifera*, and 3 *D. alata* cultivars from GenBank have 457-831 nucleotides. From the longest 831 nucleotides conserve region are 6.38% and variable region are 93.62%. Almost all nucleotides in the variable region experience the kind mutation, *i.e.* deletion, insertion, transition, and transverse. All cultivars have 100% homology compared with water yam from GenBank and 99.98% with *D. bulbifera*.

3.3 Sequence Alignment/Nucleotide Variation

Commonly in the species of *Angiospermae*, the mutation occurred on in ITS-1 and ITS-2 only, but in all Indonesian *D. alata* cultivars mutation occurred on ITS-1, ITS-2, also in 5.8S, 18S, and 28S genes. Almost all nucleotides in the variable region experience the kind of mutation, *i.e.* deletion, insertion, transition, and transverse mutation. Intra-species classification into 2 groups of monophyletic is more supported by a mutation in the insertion and deletion type, and into sub-groups is more supported by transition and transverse.

3.4 Sequence Coefficients of Identity

The analysis on the sequence identity matrix using ITS-1 and ITS-2 sequences showed that the identity percentages (genetic similarity) among 20 accession of Indonesia *D. alata* and 3 cultivars from GenBank had a range of 72.92 to 99.01 (Table 2). The genetic distance in ITS among 20 accession of Indonesia *D. alata* and 3 cultivars from GenBank was 0.00 to 0.18. According to the data, a cultivar of *uwi luyung putih*, *uwi luyung kuning*, and *uwi ulo* (genetic similarity = 0.00; genetic distance = 98.40-98.80) has the highest relatedness to all cultivars, and on both the 3 cultivars morphologically has a long cylindrical tuber shape. There is also the cultivar of *uwi senggani* and *uwi violet* (Madura) (99.01; 0.00) and also the cultivar of *uwi legi* and *uwi violet* (98.89; 0.00) (Table 2).

3.5 Phylogenetic Trees

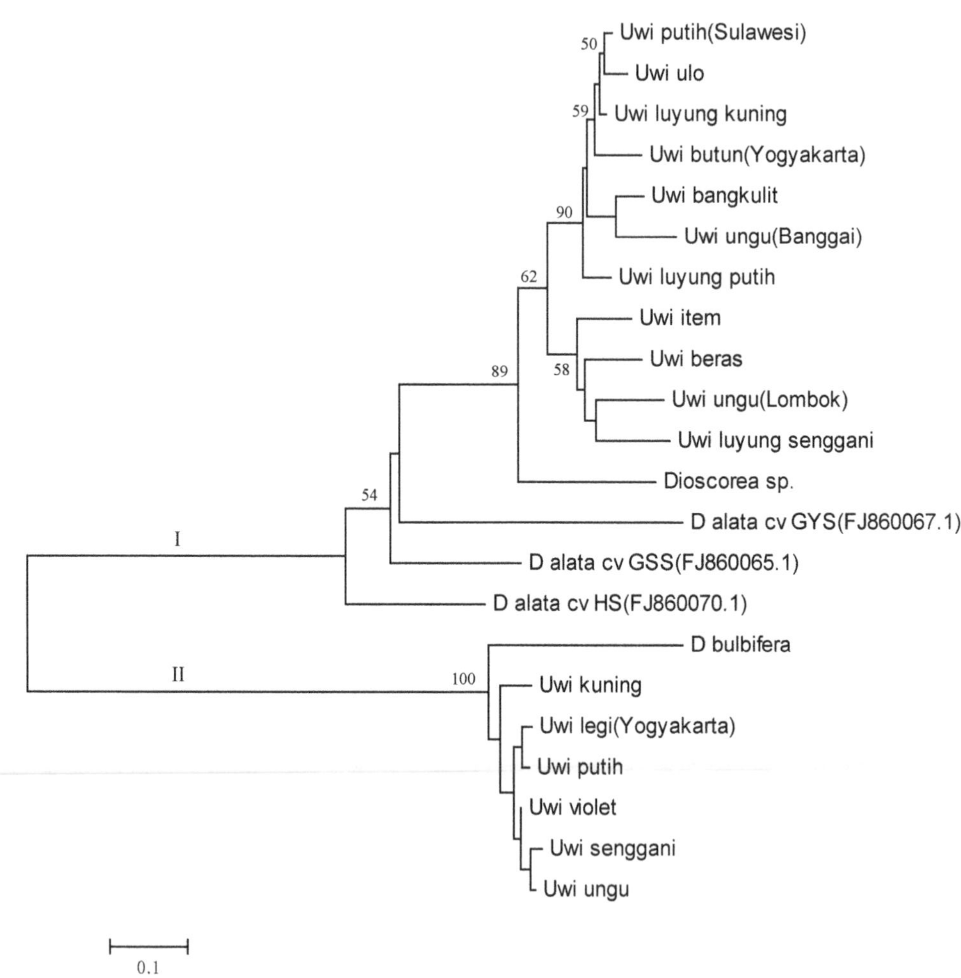

Figure 2. Phylogenetic tree of Indonesia *D. alata* cultivars compared with *D. bulbifera* based on ITS-rDNA marker generated by Neighbor-Joining

Phylogenetic tree based on ITS-1 and ITS-2 (Figure 2) was generated by Neighbor-Joining (NJ). Bootstrap values are determined by Kimura 2 method, bootstrap value > 50 indicate the reliable grouping so that the relationship between accession and their evolution history can be created. Relationships are indicated by the length clade of the phylogenetic tree that is illustrated genetic distance and similarity between accessions (Tamura et al., 2011).

Figure 2 shows that *D. bulbifera* have a close relationship with 23 *D. alata* cultivars, including *Dioscorea* sp., and *D. alata* cultivars have high genetic variability on 18S, ITS-1, 5.8S, ITS-2 and 5.8S rDNA regions. There are 2 monophyletic clades that occurred on the phylogenetic tree; clade I was a group composed of 11 cultivars; *D. alata* Indonesia has a close relationship with 3 *D. alata* cultivars from GenBank and *Diosocrea* sp., and clade II, composed of 7 cultivars of *D. alata* Indonesia, had a closer relationship with *D. bulbifera.* Clade I has 92.28-93.33% genetic similarity, and 0.06-0.07 genetic distances. Clade II has 74.63-77.72 genetic similarity and 0.16-0.18 genetic distances. The phylogenetic relationship on clade I is to prove that *Dioscorea* sp. includes the species of *D. alata.*

The genetic variability in *D. alata* is dominantly caused by natural gene flow rather than vegetative reproduction and by a natural hybrid between species of *Dioscorea.* Water yam may be domesticated in the Indochina region with *D. hamiltonii* (J. D. Hook) and *D. persimilis* Prain & Burk. The cultivation of water yams was started using wild cultigens (Lebot et al., 1998) and then continued using vegetative reproduction or clone (soma clonal) from tubers. The genetic distance between *D. alata* cultivars is very closed (nearest); it was known that each cultivar is the result of soma clonal reproduction from a primitive form. It was in accordance with the research of van den Brouche et al. (2015) and Chair et al. (2016), which states that *D. alata* cultivars in Vanuatu is soma clonal engineered and selected as cultivars.

Phylogeny tree branches (clades; lineages) that form the two branches I and II have a far relationship due to mutational events in each cultivar. All of the mutation types such as insertion, deletion, transition, and transvertion occurred, and the insertion and deletion dominantly. The difference of nucleotide sequence of ITS-rDNA region is expressed to the basic tuber shape of *D. alata* cultivars; clade I has a short to long cylindrical tuber shape, and clade II has ovate to oblong or irregular tuber shape.

The phylogeny tree branches and sub-branches showed no grouping based on geographic region, such as the clusters that occur based on molecular characters with RAPD analysis. Establishment of branches I and II based on molecular characters with ITS-rDNA analysis does not have the pattern of group and sub-group cultivars, such as the grouping based on morphological characters (Purnomo et al., 2012) and RAPD (Purnomo et al., 2016). Distinguishing between branches I and II based on analysis of rDNA ITS regions is predominantly due to the insertion and deletion mutation types. The results show compliance with the research of identification of *Durio* spp. in Kalimantan with the same marker (Mursyidin & Qurrohman, 2012), and the difference of sub-branches on each branch are dominantly due to the insertion, deletion, transition, and transvertion mutation types. These results are consistent with the research on the identification of 10 strains of algae *Porphyra haitanensis* experiencing the same type of mutation (Chen et al., 2010).

Sub-groups are formed based on ITS-rDNA molecular markers derived from both the green and purplish-red cultivar groups based on the results of clustering by morphological characters and RAPD analysis so that the samples used in ITS-rDNA analysis represent each cluster that was formed. Hence, the two branches are polyphyletic.

Acknowledgements

This research was supported by I-MHERE Project of the Faculty of Biology Universitas Gadjah Mada Indonesia 2010, through research grant number UGM/BI/1157/I/ 05/04 and the Directorate of Indonesian Higher Education through Sandwich-like Program 2011 for molecular analysis in Tokyo University of Agriculture, Tokyo, Japan.

References

Baldwin, B. G., & Markos, S. (1998). Phylogenetic Utility of the External Transcribed Spacer (ETS) of 18S-26S rDNA: Congruence of ETS and ITS Trees of *Calycadenia* (Compositae). *Molecular Phylogenetics and Evolution, 3*(10), 449-463. https://doi.org/10.1006/mpev.1998.0545

Baldwin, B. G., Sanderson, M. J., Porter, J. M., Wojeicchowski, M. F., Campbell, C. S., & Donoghue, M. J. (1995). The Its Region of Nuclear Ribosomal DNA: A Valuable Source of Evidence on Angiosperm Phyllogeny. *Annals of Missouri Botanical Garden, 82*(2), 247-277. https://doi.org/10.2307/2399880

Chair, H., Sardos, J., Supply, A., Mournet, P., Malapa, R., & Lebot, V. (2016). Plastid Phylogenetics of Oceania Yams (*Dioscorea* spp., *Dioscoreaceae*) Reveals Natural Interspecific Hybridization of the Greater Yam (*D. alata*). *Botanical Journal of Linnaean Society, 180*(3), 319-333. https://doi.org/10.1111/boj.12374

Chase, M. W., Salamin, N., Wilkinson, M., Dunwell, J. M., Kesanakurthi, R. P., Haidar, N., & Savolainen, V. (2005). Land Plants and DNA Barcodes: Short-term and Long-term Goals. *Phil. Trans. R. Soc., B*, 1-7. https://doi.org/10.1098/rstb.2005.1720

Chen, C. M., de la Cerda, K. A., Kaminski, J. E., Douhan, G. W., & Wong, F. P. (2009). Geographic Distribution and rDNA-ITS Region Sequence Diversity of *Waitea circinata* var. *circinata* Isolated from Annual Bluegrass in the United States. *Plant Disease, 93*(9), 906-911. https://doi.org/10.1094/PDIS-93-9-0906

Chen, C. S., Xie, C., Ji, D., Liang, Y., & Zhao, L. M. (2010). Molecular divergence and application of the ITS-5.8S rDNA and RUBISCO spacer in *Porphyra haitanensis* Chang et Zheng (*Bangiales, Rhodophyta*). *Aquacult Int., 18*, 1045-1060. https://doi.org/10.1007/s10499-010-9322-y

Daryono, B. S., &Natsuaki, K. T. (2002). Application of Random Amplified Polymorphic DNA Markers for Detection of Resistant Cultivars of Melon (*Cucumis melo* L.) Againts Curcubit Viruses. *Acta Horticulturae, 588*, 321-329. https://doi.org/10.17660/ActaHortic.2002.588.52

Freeman, S., Minz, D., Jurkevitch, E., Maymon, M., & Shabi, E. (2000). Molecular Analyses of *Colletotrichum* Species from Almond and Other Fruits. *Phytopathology, 90*(6), 608-618. https://doi.org/10.1094/PHYTO.2000.90.6.608

Goodwin, S. B., Dunkle, L. D., & Zismann, V. L. (2001). Phylogenetic Analysis of *Cercospora* and *Mycosphaerella* Based on the Internal Transcribed Spacer Region of Ribosomal DNA. *Ecology & Population Biology, 91*(7), 648-658.

Hidayat, T., & Pancoro, A. (2001). Studi Filogenetika Molekuler Anacardiaceae Berdasarkan pada Variasi Urutan Daerah *Internal Transcribed Spacer*. *Hayati, 8*(4), 98-101.

Lin, T. C., Hsieh, C. C., Agrawal, D. C., Kuo, C. L., Chueh, F. S., & Tsay, H. S. (2007). ITS Sequence Based Phylogenetic Relationship of Dangshen Radix. *Journal of Food and Drug Analysis, 15*(4), 428-432.

Lin, T. C., Yeh, M. S., Cheng, Y. M., Lin, L. C., & Sung, J. M. (2011). Using ITS2 PCR-RFLP to Generate Molecular Markers for Authentication of *Sophora flavescens* Ait. *J. Sci. Food Agric.* Research Article Society of Chemical Industry. Wiley Online Library.

Malapa, R., Arnau, G., Noyer, J. L., & Lebot, V. (2005). Genetic Diversity of the Greater Yam (*Dioscorea alata* L.) and Relatedness to *D. nummularia* Lam. and *D. transversa* Br. as Revealed with AFLP Markers. *Genetic Resources and Crop Evolution, 52*(7), 919-929. https://doi.org/10.1007/s10722-003-6122-5

Muellner, A. N., Samuel, R., Chase, M. W., Coleman, A., & Stuessy, T. F. (2008). An evaluation of tribes and generic relationships in Melioideae (Meliaceae) based on nuclear ITS ribosomal DNA. *Taxon, 57*(1), 98-108.

Mursyidin, D. H., & Qurrohman, M. T. (2012). Kekerabatan Filogenetik 15 Jenis Durian (*Durio* spp.) Berdasarkan Analisis Bioinformatik Gen 5.8S rDNA dan ITS Region. *Bioscientiae, 9*(1), 45-54.

Purnomo, B. S., Rugayah, D., Sumardi, I., & Shiwachi, H. (2012). Phenetic Analysis and Intraspesific Classification of Indonesia Water Yam (*Dioscorea alata* L.) Based on Morphological Characters. *SABRAO Journal of Breeding and Genetics, 44*(2), 277-291.

Sambrook, J., Fritsch, E. F., & Maniatis, T. (1989). *Molecular Cloning: A Laboratory Manual* (2nd ed.). N. Y., Cold Spring Harbor Laboratory, Cold Spring Harbor Laboratory Press.

Tamura, K., Peterson, D., Peterson, N., Stecher, G., Nei, M., & Kumar, S. (2011). MEGA5: Molecular Evolutionary Genetics Analysis Using Maximum Likelihood, Evolutionary Distance, and Maximum Parsimony Methods. *Molecular Biology and Evolution, 28*, 2731-2739. https://doi.org/10.1093/molbev/msr121

Van den Broucke, H., Mournet, P., Vignes, H., Chair, H., Malapa, R., & Duval, L. V. (2015). Somaclonal variants of taro (*Colocasia esculenta* Schott) and yam (*Dioscorea alata* L.) are in corporated into farmers' varietal portofolios in Vanuatu. *Genetic Resources and Crop Evolution, 63*(3), 495-511. https://doi.org/10.1007/s10722-015-0267-x

2

The Evaluation of Asymmetry in Price Transmission and Market Power in Iran Sugar Production Industry

Mohammad Omrani[1], Mohammad Nabi Shahiki Tash[1] & Ahmad Akbari[1]

[1] School of Management & Economics, University of Sistan and Baluchestan, Zahedan, Iran

Correspondence: Mohammad Omrani, School of Management & Economics, University of Sistan and Baluchestan, Zahedan, Iran. E-mail: m_omrani82@yahoo.com

Abstract

Asymmetric price transmission in production institutes can be a reason for the existence of market power. In this regard, by the help of an integrative sample, this study simultaneously analyzes the price transmission between production markets and retails markets of sugar in the presence of the parameter of market power. For this purpose, the seasonal data of the required variables during the years 1995-2013 was used. In order to meet the study objectives, the behavior of the retails price of sugar was evaluated in the framework of two regimes of different changes which were compatible with nature of presenting agricultural products that are widely supplied in the harvest seasons. Also, the inverse elasticity of product supply was used as the parameter of market power. Considering the more probable regime (second regime), it was realized that the marketing elements at the wholesale level intend to price transmission increase more intensively than price decrease to the retail trades level. In this regime, asymmetric price transmission and existence of market power were approved. In the first regime, market power was much more and considering the fact that it has the 30% probability of happening, it has more compatibility with the seasons when supply is abundant. In these seasons, sugar production and refining units increase the market power and create a monopsony market in buying the inputs.

Keywords: price transmission, market power, sugar production

1. Introduction

In the agricultural products, sugar as a strategic product is always considered by governments. Sugar for consumption is supplied through domestic production and import. By counting the annual consumption per capita as 30 kg sugar in the country, the level of annual sugar requirement is measured as 2.3 million tons. Approximately half of this quantity is produced domestically and the other half is supplied by import (Customs Department of Islamic Republic of Iran, 2014). Therefore, it is required that more efforts are made for more identification of domestic markets so that appropriate policies can be taken. For this purpose, it is required that by doing more studies in this field, more data be collected and by the use of appropriate criteria, the market structure and market power in the transformational industries of agriculture sector be studied more. In addition to studying the market structure of these industries, the way of price transmission process in the chain of supplying goods has to be studied as well. This is because two of the most effective factors on the welfare of producers are marketing of an item and its consumers. The methodology of price transmission of a product is affected by nature and structure of the market, so much so that storability of the products, the existence of non-competitive structures, and using market power influence the way of price transmission. Simultaneous use of market structure concepts and the pattern of price transmission are explained in studies of Liang (1989), and Canaan and Cotterill (2006).

In the empirical surveys, the utmost important reason for asymmetric price transmission is mentioned to be the market power of productive institutes. Production industries of agricultural products do not usually provide all conditions of total competition in the market like an affluent number of producers and non-concentration on industry (Meyer & Van Cramon-Taubadel, 2004). Also, the empirical studies have shown that in practice, the market of agricultural products and food industry are non-competitive (Brown & Yücel, 2000; Bornstein et al., 1997; Bailey & Brorsen, 1989).

If processing and marketing institutes of a product use their market powers, they can prevent full transmission of changes in the initial product price and inputs of marketing to the final product. In other words, it is possible that the effect of increasing or decreasing of producers' price would not be symmetric to the price of the users. Asymmetry price transmission, by influencing the market margin, sometimes brings huge profits for the marketing brokers and by decreasing the welfare of the producers it decreases the efficiency of the marketing system.

Food industries and transformational industries of agriculture sector include 18.3% of the whole industries of the country and in addition to that, 15.1% of employment, 8.93% of investment and 9.64% of added value of industry belong to this sector (Iranian Statistics Center, 2014). Considering the few number of processing industries in the market of transformational industries and food industry in Iran, existence of non-competitive markets and asymmetric price transmission in these industries is probable and subject to questions. In case there is asymmetry in price transmission, price fluctuations lead to an increase of market margins. High margins of the markets of food and agricultural products which are not typically compatible with the operated marketing have always been subjected to considerations. It is so much that according to carried out studies, approximately 50% of the paid price of the consumers for the product is for the marketing expenses (Dehdashti & Seidzadeh, 2006). Therefore, not only is it required that the quantity of competition in the transformational part of agriculture sector receive more attention, considering the concept of asymmetrical transmission too is significantly important.

Pultzman (2000) in a comprehensive study of 282 different products including 120 agriculture products stated that asymmetric price transmission is more a rule than an exception. Therefore, the presented standard theory for the markets is not correct, because this theory is unable to predict and explain an asymmetrical modification of price. On the other hand, Gauthier and Zapata (2001), and Van Cramon-Taubadel and Meyer (2000) suggested that in dealing with the issue of asymmetry, because of methodological problems related to empirical tests, it is necessary to stay on the safe side. Two issues have made asymmetric price transmission highly important. The first one is doubt about the correctness of economic theories and the second one is the necessity of change in the previous welfare deductions (Meyer & Cramon-Taubadel, 2004). This issue itself, in its own turn, critically challenges the selected policies. Asymmetric price transmission can appear in the speed or size or in both of them.

Most carried out studies on price transmission have mentioned the structure of non-competitive market as the reason for asymmetry (Meyer & Cramon-Taubadel, 2004). On the other hand, some researchers like Ward (1982) believe in condition of market power and monopoly, worry about the decrease of the market's share after an increase of price will lead to faster transmission of a decrease in price compared to its increase. So far, few empirical studies have been conducted on testing the relationship between the market power and asymmetric price transmission. The existence of modification costs and some executive costs arising from modification of the activity volume are mentioned as other main sources of asymmetrical transmission (Meyer & Cramon-Taubadel, 2004). Of course, highness of execution costs and modification can be the reason of asymmetric price transmission, because it has been identified that averagely 27% to 35% of the net profit margin consist of these costs (Dutta et al., 1999; Luis et al., 1997).

In most studies, the asymmetric price transmission is observed as the faster and more perfect price transmission increase compared to price decrease (positive price transmission). However in the beef market of the US, Bailey and Brorsen (1998) showed that in short-run, it is possible that the margins in the packing units could reduce while struggling for maintaining the activity of unit in the surface (or close to) full capacity. In Iran as well, it was clarified that in the chicken market (Husseini & Nikokar, 2006; Husseini et al., 2008; Ghadami Kouhestani et al., 2010) and beef market (Husseini & Ghahreman Zadeh, 2006; Nikokar et al., 2010) there is asymmetric price transmission between the farm price and the retails price, and price increase is transmitted faster than its decrease from the farm to the retails market. For sugar as well, Kilima (2006) studied asymmetry of price transmission from the global market to the domestic market in Tanzania. His findings showed that price transmission between the global market and the domestic market of this product is asymmetry. Of course about the agriculture products as the comprehensive study of Pultzman (2000) showed as well, asymmetric price transmission is generally more common than symmetric price transmission. In this regard, Aguiar and Santana (2002) stated in a general review of the studies that about the agricultural products, it is generally expected that the price transmission process is observed as asymmetry. Of course, there are some cases of symmetry as well. For example, Bakucs and Ferto (2006) evaluated price transmission between the farm price and retails price of pork meat in Hungary in the short-run and in the long-run as symmetry. Jazghani et al. (2011), in the marketing chain of rice in Iran, studied the vertical price transmission. Their studied scales showed that the price changes of producers immediately are transmitted to the prices of wholesale and retails. In the used models, transmission from the producer to retails price as well as from wholesale to retails price was asymmetry, but price transmission from producer to wholesaler is symmetry.

Efforts of different studies were usually on the sources of creating asymmetric price transmission. Some examples are Ward (1982), Bailey and Brorsen (1989), and Damania and Yang (1998) which count the market power as significant. Also, Bulk et al. (1998), as well as Brown and Yücel (2000), introduced collusion for more profit as a factor of asymmetrical transmission. Bor et al. (2014), also by the use of the error correction model, studied asymmetry in price transmission from the farm to retails in the milk market in Turkey. The results showed positive asymmetry in price transmission from the farm to the retails price. It meant that increase of prices on the farm transmitted to the retail prices faster than their decrease, hence the welfare of the consumers decreased. In addition to that, the results showed that the market power in the marketing chain of milk in this country existed which led to asymmetry in price transmission. Also, the results of the studies of Mc Loren (2013), Digal and Ahmadi Esfahani (2002) and Wang et al. (2006) showed that the market power and defected competition lead to asymmetry in price transmission in the processing sector of foodstuff.

For clarifying the way of price transmission, by the use of an integrated pattern in this study, asymmetric price transmission was simultaneously used with the market structure in the sugar production industry. It has to be mentioned about sugar that considering the fact that it is significantly important among the consumed products of families, it has a lot of use as well in the agriculture transformational industries and production of other food industry. Although the global average of sugar consumption is annually 23 kg per capita, this index has gone beyond the world standard in Iran and has reached 30 kg (FAO, 2014).

2. Theoretical Principles

In this study, consider the sugar marketing industry as a firm, which produces a homogenous product (q) using agricultural input (x) along with other inputs (m) and sells its product in a competitive market at a price p. The market for nonagricultural inputs such as labor, electricity and etc is likely to be competitive because the share of the industry is much smaller than the overall size of the market. However, an individual firm may enjoy market power in a regional-agricultural input market or in national output market.

Following Schroeter and Azam (1991), we assume that the marketing cost function is separable in agricultural input and marketing inputs. We also assume that the relationship between the alone input and output is one of the fixed proportions (*i.e.* $q = \lambda x$, where, $\lambda = 1$). Then the profit function (π) of the i-th marketer/retailer in the j-th region can be expressed as:

$$\pi_{ij} = pq_{ij} - w_i(Q_j, z)q_{ij} - c_{ij}(q_{ij}, v) \tag{1}$$

Where, q_{ij} is the firm input as well as output quantity, $w_j(Q_j, z)$ is the price of agricultural inputs in the region j, z is the vector of supply-shifting exogenous variables, v is the vector prices for non-agricultural inputs and $c_{ij}(q_{ij}, v)$ is the processing cost function of the i-th firm in the j-th region. The first-order condition for firms profit maximization is:

$$\frac{\partial \pi_{ij}}{\partial q_{ij}} = (p - w_j) - \frac{\partial w_j \partial Q_j}{\partial Q_j \partial q_{ij}} q_{ij} - \frac{\partial c_{ij}}{\partial q_{ij}} = 0 \tag{2}$$

Converting Equation (2) to elasticities, the retail can be expressed as:

$$p = w_j + \theta_{ij}(\varepsilon_j^{-1} Q) + mc_{ij} \tag{3}$$

Where,

$\theta_{ij} = (\partial Q_j / \partial q_{ij})(q_{ij}/Q_j)$ is the regional input market conjectural elasticity of firm i in the region j;

$\varepsilon_j = (\partial Q_j / \partial w_j)(Q/Q_j)$ is the slope of input supply function in the region j times the inverse of region j national market share;

$Q = \sum Q_j$ is the total national input/output quantity; and,

mc_{ij} is the marginal processing cost of the firm i in the region j.

Following Schroeter and Azam (1991), we assume that ε_j are the same and equal to the national value ε.

Equation (3) is multiplied by q and the result is summed over all firms within the region and over all regions, then divided by Q, we have:

$$\sum_i \sum_j p \frac{q_{ij}}{Q} = \sum_i \sum_j w_j \frac{q_{ij}}{Q} + \sum_i \sum_j \theta_{ij}(\varepsilon_j^{-1} Q) \frac{q_{ij}}{Q} + \sum_i \sum_j mc_{ij} \frac{q_{ij}}{Q} \tag{4}$$

Let $\varphi_{ij} = q_{ij}/Q$, then with the assumptions made above, Equation (3) can be written as below:

$$\sum_i\sum_j p\varphi_{ij} = \sum_i\sum_j w_j\varphi_{ij} + \varepsilon_j^{-1}Q\sum_i\sum_j \theta_{ij}\varphi_{ij} + \sum_i\sum_j mc_{ij}\varphi_{ij} \quad (5)$$

if P, θ, W and MC show the average weight values, the mean value of the price shown in the relation (5) could be written in this form:

$$P = W + MC + \theta(\varepsilon^{-1}Q) \quad (6)$$

Equation (6) shows the optimal behavior of firm with monopsony power in the agricultural input market but selling in a competitive output market and purchasing non-agricultural inputs in competitive markets. The conjectural elasticity parameter θ measures the extended monopsony power exercised by the firm. However, when the farm input market is perfectly competitive, the conjectural elasticity parameter reduced to zero and the farm retail price relationship becomes $P = W + MC$.

An alternative Equation (6) that facilitates testing for imperfect competition is:

$$P = MC + W\left(\frac{\tilde{\varepsilon} + \theta}{\tilde{\varepsilon}}\right) \quad (6a)$$

Where, $\tilde{\varepsilon} = (\partial Q/\partial W)(W/Q)$ is the price elasticity of the vector of aggregate farm supply curve. From Equation 6a, it is immediately clear that when $\theta \succ 0$ (i.e. middlemen exercise oligopsony power), a one unit increase in the farm price causes the retail price to increase by more than one unit (*i.e.* $\frac{\partial P}{\partial W} \succ 1$).

Given this theoretical prediction a simple T-test will be constructed to evaluate the overall (long-run) impact of imperfect competition on farm-retail price relationship implied in Equation (6a).

3. Methodology Research

Most agricultural products are perishable, exhibit a seasonal production pattern, and the supply function is relatively price inelastic in the short-run owing to production acreage being fixed by decisions made in the past. For these reasons, farmer and buyers bargain over a fixed quantity in each trading period. These features are posited to give rise to multiple pricing regimes, which we test using the following specification:

$$\text{Regime1}: P = \beta_{11}W + \beta_{12}MC + \theta_1(\varepsilon^{-1}Q)$$

$$\text{Regime 2}: P = \beta_{12}W + \beta_{22}MC + \theta_2(\varepsilon^{-1}Q) \quad (7)$$

This characteristic of agricultural crops market is consistent with the pricing behavior for fresh lettuce observed by Sexton and Zhang (1996). In that study two pricing regimes are identified: a harvest- or peak season regime where price is equal to harvesting cost, and an off-peak regime where price exceeds harvesting costs and is determined as a result of bargaining between buyers and producers.

Our model generalizes Sexton and Zhang's in that a finite mixture estimation procedure is used that permits both the producer–retail price relationship and the market power parameter to vary between regimes. An advantage of the mixture procedure is that the pricing regimes do not have to be imposed a priori, but rather can be identified through the properties of the data. If the data indicate more than one regime, auxiliary regressions can be used to identify factors that may explain regime membership.

In general, a finite mixture distribution of a price is expressed as:

$$f_i(P_i) = \tau_1 f_{i1}(P_i) + \tau_2 f_{i2}(P_i) + \ldots + \tau_k f_{ik}(P_k) \quad (8)$$

Where, $\tau_j \succ 0$, $\sum \tau_j = 1$ and $\int f_j(m)dm = 1$ for all j. Thus, the mixture density function is a probabilistically weighted average of component densities, f_j. Assuming that agricultural product prices are normally distributed, a two regime pricing model can be expressed as:

$$f_i(P_i|\theta) = \tau\varphi_1(P_i|\mu_1, \sigma_1) + (1-\tau)\varphi_2(P_i|\mu_2, \sigma_2) \quad (9)$$

Where, φ_i are normal density functions and $\mu_j = X_i\beta_j$ are vectors of explanatory variables and associated parameters. Using this framework, the two-regime agricultural product pricing model defined in Specification 7 can be expressed as,

$$P = \beta_{11}W + \beta_{12}MC + \theta_1(\varepsilon^{-1}Q_1) + e_1 \quad \text{with a probability of } \tau$$

Or,

$$P = \beta_{21}W + \beta_{22}MC + \theta_2(\varepsilon^{-1}Q_2) + e_2 \quad \text{with a probability of } 1-\tau \quad (10)$$

Where, e_j are independent and identically distributed (i.i.d.) error terms. In Equation (10), the marginal cost function, MC, is obtained from the cost function, C, which is defined as a translog cost function. We use this function form because it has appropriate features including equality in prices and convexity in the product. In this function form, characteristics of concavity in prices, symmetry and monotony can be imposed and tested (Richards et al., 2001).

Based on these empirical tests and the theoretical relationship derived in Equation (10), the producer-to-retail price transmission process for the sugar marketing industry was specified as:

$$P_t = \begin{cases} \sum_{i=0}^{n} \beta_{11i}^{+} \Delta w_{t-i}^{+} + \sum_{i=0}^{n} \beta_{11i}^{-} \Delta w_{t-i}^{-} + \beta_{12} MC + \theta_1(\varepsilon^{-1}Q) + e_1 \\ \sum_{i=0}^{n} \beta_{21i}^{+} \Delta w_{t-i}^{+} + \sum_{i=0}^{n} \beta_{21i}^{-} \Delta w_{t-i}^{-} + \beta_{22} MC + \theta_2(\varepsilon^{-1}Q) + e_2 \end{cases} \tag{11}$$

$$\begin{cases} \Delta w_t^{+} = \sum_{t=1}^{T} \max(W_t - W_{t-1}, 0) \\ \Delta w_t^{-} = \sum_{t=1}^{T} \min(W_t - W_{t-1}, 0) \end{cases}$$

Where, superscripts + and - denote cumulative values of rising and falling producer prices which are computed using the Wolffram (1971) methodology as modified by Houck (1977).

A formal test that the price transmission mechanism is symmetric is given by:

$$\begin{aligned} H_N &: \sum_{i=0}^{n} \beta_{ji}^{+} = \sum_{i=0}^{n} \beta_{ji}^{-} \\ H_A &: \sum_{i=0}^{n} \beta_{ji}^{+} \neq \sum_{i=0}^{n} \beta_{ji}^{-} \end{aligned} \tag{12}$$

The null hypothesis in Equation (12) is a test of linear restrictions and a t-test is appropriate. To test whether a single- or a two-regime model fits the data better, a corrected log likelihood ratio test as suggested by Wolfe (1971) was used. Although it is not a formal specification test, the simple t-test on mixing-weight parameter (τ) can also be used to test the significance of two-regime model.

To match the theoretical model with the empirical observations, various statistical procedures are followed. First, a unit-root test was conducted to determine whether seasonal producer-retail prices are stationary at seasonal frequencies. Second, Akaike's Information Criterion (AIC) was used to determine the optimum lag length (Akaike, 1974). Third, the model was tested for causality between producer prices and retail prices as proposed by Geweke et al. (1975).

4. Data

The data used in this study was taken from the Iranian Statistics Center database and from the Central Bank covering period of 1995-2013. This data includes the seasonal amounts of the time series of variables of wages, the value of sugar beet and sugarcane, price of sugar beet and sugarcane, value of energy products, capital stock, output quantity, producer price index and consumer price index. The final output is sugar that is considered in monetary values. Sugar beet and sugarcane, measured by their monetary values are as intermediate inputs in production of output. Other intermediate inputs are wages, capital stock and energy products. Regarding the different type of energy including electricity, gas and oil products are aggregated using their cost. Capital stock captures the monetary value of facilities and capital goods used by sugar producing firms. Producer price index (PPI) includes the weighted average changes in price of manufacturing. Consumer price index is also known as CPI.

5. Results and Discussion

Considering the fact that the used data was in a time series, first, their statistical behavior in variables of stationary was evaluated by the use of unit root test. The stationary test of the variables was conducted under two assumptions of the existence of intercept, and the existence of intercept and trend. The achieved results from this test showed that the applied variables have stationary behavior. Also in another part of the study, was considered the causality relationship between retails price and the producer price and it was recognized that the direction of price transmission was from producer to the retail price. Because of this, in the achieved specifications, the variable of consumer price index is considered as a function of the producer price index. It has to be mentioned that in analyzing the results, price index means sugar price index which is shortly referred to as price index or price.

In the Table 1, the results achieved from a two-regime specification have been presented. It has to be mentioned that the variables have been used in logarithmic form; because of this, they can be used as elasticity. The presented results show that the quantity of the statistic logarithm of the likelihood for the two-regime pattern is higher than the single regime pattern. The test of the difference between these two patterns according to the mentioned statistic also confirms the superiority of two-regime pattern. Of course, comparing the findings of the two patterns also clearly shows the superiority of the two-regime pattern. The quantity of the statistic τ showed that the probability of appearance of the first regime is 30% and the second regime with a probability of 70% will appear. Therefore, more concentration could be made in the second regime.

In the first regime, except for the decreasing amounts or series of producer that have negative effects on the retail prices, of course, this is also compatible with the principles and theory of price transmission, other variables have positive effects on retails price of sugar. Also, the series of price increases in the producer level get transmitted to retails level with a higher coefficient and higher statistical significance compared to series of price decreases in the producer level. On the other hand, the lags of series of increases and decreases of producer price affect the price in the retails level with considerably less absolute coefficient compared to the current producer price. This is while except for the first lag in the price increases of the producer, other cases have not even statistically become significant. It means that in the first regime, price in the retails level is affected more than the prices of the current period.

About the affecting of the producer price index, it has to be mentioned that considering the existence of positive intercept which is statistically significant, there is always a difference between these two prices and, depending on the producer price index, this difference is inclined to increase. This inclination to increase can be a prerequisite of imposing the market power too. Of course for market power, clearly, the variable coefficient of inverse elasticity of product supply has been used. The coefficient of this variable shows the distance between the marginal cost and price. In the first regime, the coefficient amount of this variable is 66% and positive. Therefore, according to this finding, it can be said that in the first regime, the sugar producing units have market power and are able to increase the price of supplied product in the retails market beyond the increase in the marginal cost. The increase in the marginal cost of production leads to increase of price in the retail level. As it is observed, it is expected that in return for 1% increase in the marginal cost of production at the producer level, the retails price index increases 3.6%.

In the second regime, like the first regime, market power has a high statistical significance, but the variable coefficient of inverse elasticity of supply is less than that of the first regime and has a significant difference with its amount in the previous regime. In the second regime, the effect of marginal cost is positive like in the first regime, it has high statistical importance, and the absolute value of its coefficient is high, therefore, increase the marginal cost will have a positive and considerable effect on the price in the retail level. It was also clarified that price increase in producer leads to price increase in retails and this effect happens in the current period and one delay period. On the other hand, the effect of amounts or the decreasing series of producer price index also is transmitted during the current period and one delay period. In addition to that, it has a reverse effect, and in terms of coefficient absolute value and statistical significance, it has less effect compared to amounts or increasing series of the producer price. In the study of Acharya et al. (2010) also similar findings are observed.

Table 1. The achieved results from measuring the two-regime pattern of price transmission from the producer to the retails market in the sugar production industry

Variable	Regime 1		
	Coefficient	Standard deviation	Statistic Z
Intercept	9.8295***	1.7729	5.54
The increasing series of producer price	5.8962***	0.9517	6.19
The decreasing value of producer price	-2.5719*	1.5731	-1.63
Inverse elasticity of product supply	0.6607***	0.0811	8.14
Marginal cost	3.5564***	0.0906	39.23
The first lag of increasing series of producer price	1.4954**	0.6641	2.25
The first lag of the decreasing series of producer price	-0.9775	1.5751	-0.62
The second lag of increasing the series of producer price	0.4542	0.4626	o.98
The second lag of decreasing series of producer price	-1.5368	1.6502	-0.93

Table 1. Continued

Variable	Regime 2		
	Coefficient	Standard deviation	Statistic Z
Intercept	1.6394	1.0398	1.57
The increasing series of producer price	2.8601***	1.0632	2.68
The decreasing value of producer price	-1.8854*	1.0644	-1.77
Inverse elasticity of product supply	0.1033***	0.0342	3.01
Marginal cost	1.6681***	0.1118	14.91
The first lag of increasing series of producer price	5.4662***	1.4523	3.76
The first lag of the decreasing series of producer price	-1.7670*	1.0673	-1.65
The second lag of increasing the series of producer price	-2.1687	2.5754	-0.84
The second lag of decreasing series of producer price	-1.3857	1.1996	-1.15
Statistic	Log Likelihood	Q (1)	Q(2)
	143.54	12.71(0.001)	13.34(0.003)

Note. *, ** and *** are respectively significance in the levels 10%, 5% and 1%.

Source: Research findings.

In the Table 2 also the results of the integrative regime have been presented. According to the statistics of diagnosis, this specification compared to the previous specification has lower descriptiveness. In this way, the statistic of likelihood logarithm for this specification is meaningfully lower than the two-regime specification. In addition to that, it is observed that except for the variable of the marginal cost, other variables do not have a significant effect on retails price index. However, the considerable point is the existence of significance coefficient for marginal cost variable. It means that even by keeping in mind the one-regime pattern, the effect of the marginal cost is observable in the retail price level. The achieved coefficient for this variable in the one-regime pattern is approximately its average value in the two-regime pattern.

Table 2. The achieved results from measuring the one-regime pattern of price transmission from the producer market to the retails market in the sugar production industry

Variable		Coefficient	Standard deviation	Statistic t
Intercept		-1.1309	2.5631	-0.44
The increasing series of producer price		1.4704	2.3546	0.62
The decreasing value of producer price		1-.0144	3.0643	-0.33
Inverse elasticity of product supply		0.1027	0.0933	1.09
Marginal cost		2.6124***	0.2329	11.21
The first lag of increasing series of producer price		2.3785	2.6965	0.88
The first lag of the decreasing series of the producer price		-1.4675	3.0531	-0.48
The second lag of increasing the series of the producer price		4.9334	2.5141	1.96
The second lag of decreasing series of the producer price		-3.1471	3.3561	-0.93
Statistic	R^2	Log Likelihood	Q(1)	Q(2)
		-52.88	63.65(0.001)	114.73(0.001)

Note. *, ** and *** are respectively significance in the levels 10%, 5% and 1%.

Source: Research findings.

In Table 3, the amounts of elasticity in the short-run and long-run and also their difference for every one of the decreasing and increasing series are presented. As it is observed, the integrative one-regime pattern in both long-run and short-run shows that the producer price does not have a significant effect on retails price; in other words, price changes are not transmitted from producer to retail. Of course, this does not mean that there is no relationship between these two series; because in the two-regime pattern this relationship is obvious.

The short-run coefficients compared to the long-run coefficients have higher absolute values and also long-run coefficients have positive effects. It means that by an increase of price in the long-run, in the producer level, the price in the retails level increases as well, and vice versa. About both regimes, in addition to the difference in the direction of affecting of the two increasing and decreasing series, according to the absolute value of the coefficient, the effect of increasing series is approximately two times more than the effect of decreasing series. Therefore, it is expected that the decreasing series have the reverse effect on the producer price index and after a decrease of the producer price, the price in retails level increases. This is while the increase in the producer price also shows a positive effect on the consumer (or retail) price index. In other words, in the short-run, in the second regime, the market elements have tendencies to maintain retails prices in high levels. Of course, in every two regimes, the coefficient difference of two increasing and decreasing series is statistically significant.

The amounts of short-run coefficients in both regimes compared to the equivalent long-run amounts show bigger values. It is so much that about the series of increasing the producer price, the amounts of the coefficients have increased more than 5 times, but about the decreasing series, in addition to the fact that its coefficient sign is positive, the coefficient amount also in the regime is more than triple, but in the second regime, the absolute value of long-run coefficient is approximately double more than its amount for the short-run coefficient. Also unlike in the short-run, in the long-run, in both regimes, the direction changes of the two price indexes of producer and retails are convergent , of course, the coefficient value of increasing series in the second regime has a negative sign but it has not become significant. The amounts of the statistic t also show that in both regimes, the efficiency of increasing series is significantly superior to the efficiency of the decreasing series and difference of coefficients has high statistical significance in the long-run.

About the low-value coefficient, it has to be explained that these values, in fact, show the amount of change in the slope of the related variable; whereas between price in the producer and retails, there is a significant distance as margin, and existence of significant coefficients that mostly have positive signs means that the market elements at the producer level are able to increase the distance or margin between producer price values and those of the retail by imposing the market power.

Table 3. Price transmission elasticity between producer and retail markets

		Integrative regime	First regime	Second regime
Short-run	Increasing series of the producer price	1.470	5.896***	2.860***
	Decreasing series of the producer price	-1.014	-2.571*	-1.885*
	Difference	2.484	8.460	4.745
	Statistic t	0.608	4.40	2.98
Long-run	Increasing series of the producer price	0.126	1.028***	-0.412
	Decreasing series of the producer price	1.168	0.885***	3.620***
	Difference	1.0418	0.1422	-4.033
	Statistic t	-1.446	0.288	-2.603

Note. *, ** and *** are respectively significance in the levels 10%, 5% and 1%.

Source: Research findings.

6. Conclusion and Suggestions

This study was carried out with the objective of analyzing price transmission between production market and retails market of sugar. In this analysis, clearly, some deductions were made for market power. Most studies, especially about sugar, have evaluated price transmission between domestic market and the global markets. One of these studies is Cilima (2006) according to whose findings, price change in the global markets has affected the sugar price in the domestic markets of Tanzania, and this price transmission between global and domestic markets was evaluated as asymmetry. Unlike the above-mentioned studies, in this study, the price transmission pattern was evaluated in two levels of the market, including the wholesale or producer level and the retail level. However, the most important distinct aspect of this study is preparing more tools for analyzing market power; because as Pultzman (2000) carried out a comprehensive study over 282 different products including 120 agricultural crops, he mentioned that monopoly and market power can be elements of asymmetric price transmission. In other words, if the asymmetric price transmission is approved, it can be regarded as a probability on the existence of market power. In addition to the tools of deducing from market power, in terms of the applied specification too, this study helps the existing literature. It is in this way that in this study, sugar's retails price behavior has been evaluated in the framework of two different regimes of changes. Different behavior is specifically compatible with the nature of supplying agricultural products that have wide supplies in the harvest seasons. However in terms of tools, in the applied pattern in this study, in addition to the fact that according to the price transmission pattern about market price it was deducted, another tool was using the inverse of elasticity supply. By the use of this tool, it was identified that significantly there is a distance between price and marginal cost and therefore it can be a reason for the existence of market power. Especially by relying on the more probable regime (second regime), it was clarified that the marketing elements at the producer level are inclined to transmission price increase to the retails level with higher intensity compared to the price decrease. Asymmetric price transmission and existence of market power were approved in conditions where it was clarified that determining the behavior of retails price of sugar is not the same in the whole period; and considering the inelastic supply of product in the short-run, it is required that change in the supply behavior should be considered as well. It seems that the first regime altogether according to both achieved deductions for market power-inverse elasticity coefficient of supply and the price transmission pattern- shows much more market power compared to the second regime. It also seems that the first regime has 30% more probability of happening which is most accordant with the seasons of affluent supply when considering the limitedness of production institutes, production, and processing of sugar which has high activity scale as well, imposes market power and creates a monopsony market. However, by getting away from the harvest season and by a decrease of product supply, their market power decreases. On the basis of this analysis, it is advised that by preparing the beds for the operation of small institutes in order to increase competition in the market, production, and processing of sugar is to be carried out.

References

Acharya, R. N., Henry, W. K., & Steven, B. C. (2011). Asymmetric farm – Retail price transmission and market power: A new test. *Applied Economics, 43*, 4759-4768. http://dx.doi.org/10.1080/00036846.2010.498355

Aguiar, D. R. D., & Santana, J. A. (2002). Asymmetry in farm to retail price transmission: Evidence from Brazil. *Agribusiness, 18*(1), 37-48. http://dx.doi.org/10.1002/agr.10001

Akaike, H. (1974). A new look at the statistical identification model. *IEEE Transactions Automatic Control, 19*, 716-23. http://dx.doi.org/10.1109/TAC.1974.1100705

Bailey, D., & Brorsen, B. W. (1989). Price Asymmetry in spatial fed cattle markets. *Western Journal of Agricultural Economics, 14*(2), 246-252.

Bakucs, L. Z., & Ferto, I. (2006). Marketing margins and price transmission on the Hungarian beef market. *Food Economics, 3*, 151-160. http://dx.doi.org/10.1080/16507540601176075

Bor, Ö., Ismihan, M., & Bayaner, A. (2014). Asymmetry in farm-retail price transmission in the Turkish fluid milk market. *International Journal of new Media, 2*.

Bornstein, S., Cameron, A. C., & Gilbert, R. (1997). Do Gasoline Prices respond asymmetrically to Crude Oil Price Changes? *Quarterly Journal of Economics, 112*, 305-339. http://dx.doi.org/10.1162/003355397555118

Brown, S. P. A., & Yücel, M. K. (2000). Gasoline and Crude Oil Prices: Why the Asymmetry? *Federal Reserve Bank of Dallas, Economic and Financial Review* (3rd Quarter, pp. 23-29).

Canaan, B., & Cotterill, R. W. (2006). Strategic pricing a differentiated product oligopoly model: Fluid milk in Boston. *Agricultural Economics, 35*, 27-33. http://dx.doi.org/10.1111/j.1574-0862.2006.00136.x

Damania, R., & Yang, B. Z. (1998). Price Rigidity and Asymmetric Price Adjustment in a repeated Oligopoly. *Journal of Institutional and Theoretical Economics, 154*, 59-679.

Dehdashti, Sh., & Seyed Zadeh, H. (2006). The relationship between applying elements of integrating marketing and merchantability of farmed fish from the perspective of the consumers, case study: The city of Ilam. *Seasonal of Agricultural Economy and Development, 4*(53).

Digal, L. N., & Fredoun, Z. A.-E. (2002). Market power analysis in the retail food industry: A survey of methods. *The Australian Journal of Agricultural and Resource Economics, 46*(4), 559-584. http://dx.doi.org/10.1111/1467-8489.00193

Dutta, S., Bergen, M., Levy, D., & Venable, R. (1999). Menu costs posted prices, and multiproduct retailers. *Journal of Money, Credit, and Banking, 31*(4), 683-703. http://dx.doi.org/10.2307/2601217

FAO. (n.d.). *Statistical Database*. Retrieved from http://www.fao.org

Gauthier, W. M., & Zapata, H. (2001). *Testing symmetry in price transmission models*. Louisiana State University, Department of Agricultural Economics &Agribusiness, Working Paper.

Ghadami Kouhestani, M., Nikokar, A., & Dorandish, A. (2010). The threshold pattern of price transmission in the Iranian chicken market. *Seasonal of Agricultural Economy and Development, Agriculture Sciences and Industries, 24*(3), 384-392.

Houck, J. P. (1977). An Approach to specifying and estimating nonreversible Functions. *American Journal of Agricultural Economics, 59*, 570-572. Retrieved from http://links.jstor.org/sici?sici=0002-9092%28197708%2959%3A3%3C570%3AAATSAE%3E2.0.CO%3B2-%23

Husseini, S. S., & Dorandish, A. (2006). Price transmission pattern of Iranian pistachio in the global market. *Iranian Journal of Agricultural Sciences, 2-37*(1), 153-145.

Husseini, S. S., & Ghahreman Zadeh, M. (2006). Asymmetrical modification and price transmission in the Iranian red meat industry. *Seasonal of Agricultural Economy and Development, 53*, 1-22.

Husseini, S. S., & Nikokar, A. (2006). Asymmetrical price transmission and its effect on the market margin in the Iranian chicken industry. *Iranian Journal of Agricultural Sciences, 2-37*(1), 1-9.

Husseini, S. S., Salami, H., & Nikokar, A. (2008). The pattern of price transmission in the Iranian chicken market. *Journal of Iranian Agricultural Sciences, 2*(1), 1-21.

Jezghani, F., Moghaddasi, R., Yazdani, S., & Mohamadinejad, A. (2011). Price Transmission Mechanism in the Iranian Rice Market. *International Journal of Agricultural Science and Research, 2*(4).

Kilima, F. T. M. (2006). *Are Price Changes in the World Market Transmitted to Markets in Less Developed Countries? A Case Study of Sugar, Cotton, Wheat and Rice in Tanzania*. Department of Agricultural Economics and Agribusiness, Sokoine University of Agriculture, Morogoro, Tanzania, No. 160. http://dx.doi.org/10.2139/ssrn.925683

Levy, D., Bergen, M., Dutta, S., & Venable, R. (1997). The magnitude of menu costs: Direct evidence from large U.S. supermarket chains. *Quarterly Journal of Economics, 112*(3), 791-825. http://dx.doi.org/10.1162/003355397555352

Liang, J. N. (1989). Price reaction functions and conjectural variations: An application to the breakfast cereal industry. *Review of international Organization, 4*(2), 31-58. http://dx.doi.org/10.1007/BF02284668

McLaren, A. (2013). *Asymmetry in price transmission in agricultural markets.* Working Paper Series, University De Geneve.

Meyer, J., & Cramon-Taubadel, S. V. (2004). Asymmetric Price Transmission: A Survey. *Journal of Agricultural Economics, 55*(3), 581-611. http://dx.doi.org/10.1111/j.1477-9552.2004.tb00116.x

Nikokar, A., Husseini, S. S., & Dorandish, A. (2010). The pattern of price in the Iranian beef industry. *Seasonal of Agricultural Economy and Development, Agriculture Sciences and Industries, 24*(1), 23-32.

Peltzman, S. (2000). Prices rise faster than they fall. *Journal of Political Economy, 108*(3), 466-502. http://dx.doi.org/10.1086/262126

Richards, T. J., Patterson, P. M., & Acharya, R. N. (2001). Price behavior in a dynamic oligopsony: Washington processing potatoes. *American Journal of Agricultural Economics, 83*, 259-71. http://dx.doi.org/10.1111/0002-9092.00154

Schroeter, J. R., & Azzam, A. (1991). Marketing margins, market power, and price uncertainty. *American Journal of Agricultural Economics, 73*, 990-9. http://dx.doi.org/10.2307/1242426

Sexton, R. J., & Zhang, M. (1996). A model of price determination for fresh produce with application to California iceberg lettuce. *American Journal of Agricultural Economics, 78*, 924-34. http://dx.doi.org/10.2307/1243849

Von Cramon-Taubadel, S., & Meyer, J. (2000). *Asymmetric Price Transmission: Fact or Artifact?* University Göttingen, Institute for Agricultural Economy, Working Paper.

Wang, X., Habtu, T. W., & Tony, R. (2006). *Price Transmission, Market Power and Returns to Scale.* Land economy working paper, SAC Edinburgh.

Ward, R. W. (1982). Asymmetry in retail, wholesale and shipping point pricing for fresh vegetables. *American Journal of Agricultural Economics, 62*, 205-212. http://dx.doi.org/10.2307/1241124

Website of the Customs Department of the Islamic Republic of Iran. (n.d.). Retrieved from http://www.irica.gov.ir

Wolfe, J. H. (1971). A Monte Carlo study of the sampling distribution of the likelihood ratio for mixtures of normal distribution. *Tech. Bull. STB 72*. US Naval Personnel and Training Research Laboratory, San Diego.

Wolffram, R. (1971). Positivistic measures of aggregate supply elasticities: Some new approaches some critical notes. *American Journal of Agricultural Economics, 53*, 356-359. http://dx.doi.org/10.2307/1237462

3

E2, an Aquatic Hazard Worldwide

Nagwa Elnwishy[1,2] & Nada Sedky[1]

[1] Zewail City of Science and Technology, Egypt

[2] Biotechnology Research Center, Suez Canal University, Egypt

Correspondence: Nagwa Elnwishy, Zewail City of Science and Technology, Egypt. E-mail: nelnwishy@zewailcity.edu.eg

Abstract

Endocrine disruptors are defined as exogenous agents that alter the function of endocrine system, which in turn, causes adverse health effects in an intact organism or its progeny. Of these compounds, 17β-estradiol is of primary importance, since it is physiologically present in both men and women, as well as, being produced synthetically as a component in some pharmaceutical products. Once it reaches the aquatic environment through domestic sewage, ground water and streams, it makes a serious threat to the aquatic life. The review tackles the biological significance of these compounds as well as the danger that they present to the surrounding environment, areas at which these compounds have been detected worldwide, the methods used in detection and fundamentally significant solutions to get rid of this hazard using different methods such as; the bioremediation process.

Keywords: 17β-estradiol (E2), pollution, aquatic organisms, bioremediation

1. Introduction

The presence and effect of pharmaceutical compounds in the aquatic ecosystems has been one of the emerging concerns to the environment. One of these EDCs is E2, which has been a matter of delving since the 30s (Cook et al., 1934; Tawfic, 2006). The major sources of aquatic contamination by E2 are excretion from female human bodies and live stocks (Narender & Cindy, 2009), and synthetic estrogenic chemicals. Reaching aquatic systems, it may lead to severe damaging effects to the aquatic organisms' reproductive system, and reduction of the total aquatic reproduction (Versteeg et al., 2005).

1.1 Estrogenic Pharmaceutical Residues

Pharmaceuticals are in fact a group of a large list of rising contaminants that have been detected around the globe and used largely in up to tons per year (Boxall, 2004). They have been found in waste water, surface and groundwater, and drinking water. The most familiar source by which these compounds bust in the environment is via treated and untreated wastewater, or via urban or agricultural runoff (Shane et al., 2011).

The effect of these pharmaceuticals on the environment can be clearly demonstrated through the following examples; very tiny concentrations of 17α-ethynylestradiol and fluoxetine were able to cause a remarkable decline in the growth rates for *Physa pomilia* snails (Luna et al., 2013), traces of Clotrimazole that are similar to those present in nature rendered the algal 14α-demethylase unfunctional in laboratory trials (OSPAR-Commission, 2013) and diclofenac in low amounts (1 ug/L) caused deleterious effects on the kidney and intestine of the rainbow trout (Mehinto et al., 2010).

Most organisms that live in the sea have an innate phenomenon called "Smellscape"; where these organisms use the natural chemicals present in the sea for signalling and other functions. As a matter of fact, many pharmaceuticals were found to have a disruptive effect on this process, owing to their structural likeliness to the original compounds (Klaschka, 2008).

Generally, pharmaceuticals and compounds derived from personal care products (PCPs), such as antibiotics, caffeine, contraceptives, chemotherapeutics, narcotics and painkillers, are all found in urban streams. In fact, once these compounds are released into the aquatic environment, they are diluted, to solids, degraded biologically or by photolysis. Some compounds can persist and can be available in drinking water even after treatment.

For instance, different groups of pharmaceuticals were detected in urban waste water in Spain (Gracia-Lor et al., 2012). It was also found that even when wastewater is being treated by waste water treatment plants (WWTP), there is a high possibility of potential transport of the compounds, as well as other organic wastewater compounds, to the groundwater and streams.

Different EDCs were added by the Environmental Protection Agency (EPA) to limit the levels of pharmaceutical residues in water, but only four of the compounds in the EPA list were pharmaceuticals, three of them belong to birth contraceptives and one is an antibiotic (EPA, 2009).

Natural estrogens exist naturally in human and animal bodies in certain amounts. Synthesis of natural estrogens occurs predominantly in the ovary (Yaghjyan & Colditz, 2011) in premenopausal women and in peripheral tissues in postmenopausal women (Halm et al., 2004). Part of active estrogens is also manufactured from circulating estrone sulfate or 17β-estradiol sulfate as a result of de-conjugation by sulfatase (Chetrite et al., 2000; Ogunleye & Holmes, 2009; Zhou et al., 2011). Local release of biologically active estrogens from conjugates and their further metabolism extend peripheral tissues' response to estrogen (Sawssan et al., 2015).

Unlike synthesized estrogens, natural estrogens remain in the blood for a short time; maximum few hours, then they are broken down in the liver by enzymes and are either extracted or used to build up molecules thereafter (Katie, 2008). On the other hand, synthesized estrogens are more stable in the bodies and take longer time to breakdown. So, they are emanated daily in urine and feces by fish, homo sapiens, and wild animals and then accumulated in domestic sewages (Kolodziej et al., 2003).

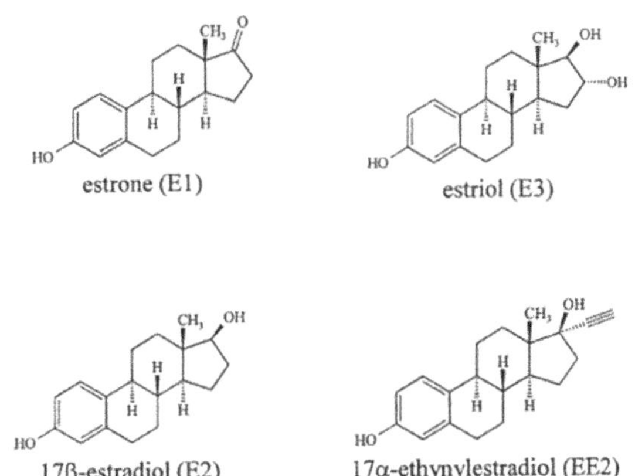

Figure 1. Structure of estrogens (Kuch & Ballschmiter, 2000)

There are 3 types of these estrogens which are commonly used in application for medicinal use, also known as bio-identical hormone replacement therapy (Holtorf, 2009). These types are estrone (E1), estradiol (E2) and estriol (E3) as shown in Figure 1. Estrone and estradiol are the two main types which cause most of the health problems associated with estrogen use.

It is worth noting that pharmaceutical residues are one of the dangerous residues which have not received proper attention in developing countries like Egypt for instance. Infact, Egypt heavily commercialized these medical compounds in early eighties, mainly due to the massive increase of population and the limited natural resources which is currently over 91 million citizens (CAPMAS, 2016) Unlike the case in other countries, only very few people in Egypt didn't use any contraceptive method in spite of sexual exposure and an expressed desire to avoid pregnancy (Sultan et al., 2010).

1.2 Role of E2

E2 is one of the powerful natural estrogens. It is found in men and women. It is believed to have a role in physiological activities and the reproductive process. For instance, E2 as well as other estrogens are responsible for the stimulation of the sprouting of sex organs and the development of secondary sexual characteristics, and they also influence the gonadotropin production (Chimento et al., 2014). In addition, estradiol concentration can be affected by equine chorionic gonadotropin (eCG) (Fu et al., 2014).

In female organisms, E2 is produced basically by the developing follicle of the ovary. But in male organisms, the role of E2 is not fully clear though it appears to be indulged in the control of gonadotropin secretion (Haring et al., 2012). E2 was believed to have an inhibitory effect on the pituitary gonadotropin hormone either by direct inhibition or by indirect inhibition of the Gonadotropin-releasing-hormone (GnRH) (Sandeep et al., 2011; Ten Kulve et al., 2011).

It was found that the antioxidant activity of E2 and other steroid hormones play a role in the neuronal protection in neuronal cells; hence, E2 may have implications for the prevention and treatment of Alzheimer (Xu et al., 2016). The latter indicated that the neuronal protection afforded by E2, and which may help in the prevention of Alzheimer's disease, was estrogen receptor-independent.

Estradiol, as well as, the other estrogens has an important influence on big animals' growth, therefore, it is added to progesterone or testosterone in cattle feeding to boost the cattle's mass.

In plants, E2 and esterone were found to be able to improve chickpea plant growth and play a key role in controlling its biochemical parameters to survive under harsh conditions (Erdal & Dumlupinar, 2011).

2. Extraction and Detection of E2

There are different methods to determine E2 and other similar compounds; chemical method used for identification and quantification, and bio analytical method for evaluation of the compounds' activity (Birkett & Lester, 2003) in Table 1. The most common bio monitoring method to identify the exposure of E2 in the aquatic systems is Vitellogenin (Vtg) (Seifert et al., 2003; Navas & Segner, 2006) which is a phospholipoprotien egg yolk protein found in juvenile and male fish (Denslow et al., 1999) and spiggin which is the androgen-induced glue protein in stickleback (Katsiadaki et al., 2002; Hahlbeck et al., 2004).

Analysis of estradiol and other estrogens is commonly achieved by HPLC or GC. However, LC-MS and GC-MS are more accurate techniques that have been used in recent years. Different extraction protocols, using Soxhlet extraction (Petrovic et al., 2001) sonication, supercritical fluid extraction (SFE), accelerated solvent extraction (ASE) (Petrovic et al., 2001) or microwave- assisted extraction (MAE), have been established for these compounds, before application of any of these analysis instruments as shown in Table 1. Generally, a complete chromatographic separation is achieved on a C18 column when applied on HPLC or LC-MS.

Table 1. Summary of common extraction and detection methods of 17β-estradiol

Source	Extraction and analysis technique	Reference
Water	ELISA	(Tanaka et al., 2004)
	Solid phase extraction/ELISA	(Matsumoto et al., 2005)
	molecularly imprinted polymer/HPLC	(Le Noir et al., 2007)
	Gas chromatography-mass spectrometry	(Kosjek et al., 2007)
	Pre concentrated on LiChrolut RP-18 cartridges/(SPE)/LC-MS	(Rodriguez-Mozaz et al., 2004)
	Pre concentration centrifugation/HPLC-UV	(Wang et al., 2006)
	Separation on Zorbax SB-CN/LC-UV	(Havlíková et al., 2006)
	SPE with Oasis/(LC-(ESI)MS-MS)	(Pedrouzo et al., 2009)
	EDS-1 cartridge using Aqua Trace Automatic SPE system/LC-MS-MS	(Isobe et al., 2003)
	SPE/LC-ESI-MS	(Rodriguez-Mozaz et al., 2004)
	SPE/ELISA	(Hintemann et al., 2006)
	HPLC fluorescence	(Panter et al., 2006)
	Chip separation utilizing micellarelectrokinetic chromatography (MEKC)	(Collier et al., 2005)
Soil	Soxhlet extraction in methanol/GC-MS	(Petrovic et al., 2001)
	Dionex, Thermo-stated Column Compartment TCC-100/HPLC	(Scherr et al., 2009)
Fish	SEP/non-radioactive HPLC	(Delvoux et al., 2007)
	Pre-column TMS/GC-MS	(Zou et al., 2007)

The advantage of using HPLC over GC is avoiding two steps occur in GC; the enzymatic hydrolysis step and the derivatization step. The enzymatic hydrolysis step is significant in GC for the immunoassay analysis of conjugated and unconjugated estrogens and progestogens, and the derivatization step that usually precedes a subsequent GC analysis is avoided (Petrovic et al., 2001).

However, the clean-up step for both GC and HPLC techniques is commonly dependent either on solid phase extraction (SPE) or on solid liquid adsorption chromatography in open columns using a miscellany of adsorbents. The most commonly used for column chromatography are modified Silica, Florisil, Alumina and different types of carbon are predominantly used for column chromatography whilest C18, NH_2 or CN modified are more widespread among SPE.

3. Global Detected Levels of E2 and Estrogenic Residues

E2 was detected in many locations worldwide; the investigation proved its presence in drinking water is very harmful to humans (Caldwell et al., 2009). Similar contamination was also reported in the west bank, within realm of possibility of pollution by sewage water, and along the Jordan River close to drainage (Barel-Cohen et al., 2006).

Different monitoring studies have detected different levels of a wide range of pharmaceuticals, including hormones, steroids, and antibiotics in soils, surface waters and groundwater (Hirsch et al., 1999; Kolpin et al., 2002). E2 and its derivatives were determined in marine environments, drinking waters and rivers in various streamlined countries even where tremendous safety measures of water and environmental safety are considered. For instance, in India, pharmaceutical residues were detected in the effluent in wastewater treatment plant (Larsson et al., 2007). Also in USA, 12 similar compounds and personal care products (PCPs) have been detected to discharge in the Mississippi river waters in New Orleans, Louisiana (Zhang et al., 2007).

Also, it was detected in the surface waters in Germany, Italy, Netherlands and the United Kingdom with levels ranged between 5.5 and 12 ng/L. The utmost of the reported concentrations so far range from the lowest quantity of a substance that can be detected to around 4 ng/L (Kolodziej et al., 2003). Yet, the metabolite of estradiol, estrone, was detected at concentrations up to 17 ng/L.

Although studies in the UK have did not detect estrogenic compounds in drinking water (Harries et al., 1996; Harries et al., 1997), they were detected in raw domestic sewage discharged into rivers (Desbrow et al., 1998; Rujiralai et al., 2011) and waste water in South Korea in ranges of 1.2-10.7 ng L^{-1} (Ra et al., 2011), China (Liu et al., 2011; Lu et al., 2011; Zhou et al., 2011), The Netherlands (Belfroid et al., 2006), Italy (Pojana et al., 2004; Pojana et al., 2007), Germany (Körner et al., 2001; Matsumoto et al., 2005; Hintemann et al., 2006) and was also detected in the drinking water in some parts of USA (Caldwell et al., 2009), as summurized in Table 2.

Table 2. Selected worldwide detected levels of estrogens

Type	Conc.	Source
Fish and Crustaceans	4.57 μg/kg	Japanese mackerel (Zou et al., 2007)
	96 μg/kg	Snails (Elnwishy et al., 2012)
	36 μg/kg	Catfish (Elnwishy et al., 2012)
	Nd	Tilapia (Elnwishy et al., 2012)
	4.63 μg/kg	Crucian (Zou et al., 2007)
	0.08 mg/g	Tilapia (Jiang et al., 2009)
	4.7 μg/kg	Greasy-back shrimp (Zou et al., 2007)
	0.0783 mg/g	Prawn (Jiang et al., 2009)
	26.4-77.1 ng/L	Surface water (Hintemann et al., 2006)
	4.1×10^3 ng/L	Sewage (Zhou et al., 2011)
	12 ng/L	Effluent from (STP) (Rujiralai et al., 2011)
	Nd	Waste water (Liu et al., 2011)
	1.2-10.7 ng/L	WWTP (Ra et al., 2011)
	75.2 ng/L	bottled mineral water, Germany (Wagner & Oehlmann, 2009)
	0.8-150 ng/L	Water, Netherlands (Vethaak et al., 2005)
	0.06-67 pM	River water, Japan (Matsumoto et al., 2005)
	1-191 ng/L	effluents from sewage treatment plants (Pojana et al., 2004)
Sediment	200 pg/g	Fresh water sediment (Petrovic et al., 2001)
	0.3 μg/kg	Lake Temsah (Elnwishy et al., 2012)
	0.9-2.6 ng/g	River sediment (Gong et al., 2011)
	3.1-289 μg/kg	Sediment (Pojana et al., 2004)
Big animals	4-28 ng/g	Cattle Manure (Andaluri et al., 2011)
	104-262 μg/kg	Dairy cattle feces (Wei et al., 2011)
	45-926 μg/kg	Beef cattle feces (Wei et al., 2011)
	40 pgm/L	Male bovine (Biddle et al., 2007)
	44 pgm/L	Females bovine (Biddle et al., 2007)

Also, it was detected in prawn and fish(Jiang et al., 2009), Japanese Spanish mackerel *Scomberomorus niphonius*, bivalves and snails (Zou et al., 2007), and in shellfish in France (Lagadic et al., 2007). Estrogens were detected in Llobregat catchment area in Spain in water samples at low levels between 2-5 ng L^{-1} (Brix et al., 2010). Furthermore, in Spain, detection was reported in sediments of rivers (Petrovic et al., 2001; Lopez et al., 2002; Gong et al., 2011). Lately, it was detected for the first time in Egypt (Elnwishy et al., 2012).

4. Impact of 17β-Estradiol and Other Estrogens on Aquatic Life

Despite the presence of these synthesized compounds in tiny amounts in the environment, the impact of their traces is very huge on marine animals and on homo-sapiens (Hansen et al., 1998). Synthesized estrogens have been suggested as one of the major groups of substances that cause endocrine disruption in wildlife (Lee & Liu, 2002), yet, there is not much information on the effect of these estrogens on the environment in spite of their ultimate impact on the ecosystem.

These estrogenic substances can affect the oestrous cycle and cause infertility and weak estrogenic activity that would affect the female reproductive tract of fish, rodents and livestock (Burton & Wells, 2002).

The presence of E2 with concentrations even lower than 1 ng/L in the aquatic environment can cause infertility, reduce estrogenic activity of fish females (Burton & Wells, 2002), induce hermaphroditism of aquatic organisms (Mills & Chichester, 2005) and change detectable habits which may change reproductivity of fish (Denslow & Sepúlveda, 2007).

Estrogens, including E2, are transferred to the aquatic life via direct discharge of wastewater into aquatic systems. The application "Agro-industrial effluent for microalgae cultivation used for feeding fish" (Cheunbarn & Cheunbarn, 2015), is also another method to transfer these pollutants to fish. When E2 reaches the marine environment, it may be bio-transformed, bio-concentrated (Lai et al., 2002a) and/or accumulate in marine animals

(Lai et al., 2002b; Gomes et al., 2004) through the food chain. In fact, feeding fish on algae cultured on wastewater can even increase the content of E2 level in fish by 3% (Meng-Umphan, 2009). Eventually, the environmental safety, the health of both marine living organisms and humans might be threatened. The danger of estradiol encouraged EPA to frequently monitor these hormones to the most accurate levels (EPA, 2012).

The estrogenic effects of E2 appear to be the most compelling mechanism of endocrine disruption and affect fish in both fresh water and marine environments (Matthiessen et al., 2002). It leads to improper expression of egg proteins; vitellogenin, and zonaradiata proteins (Knoebl et al., 2004). Furthermore, semantic modulation of secondary sexual structures may be noticed (Kirby et al., 2003). Several effects were found on aquatic creatures, including reproductive effects on Roach (*Rutilus rutilus*) (Jobling et al., 2002), (*Pomatoschistus minus*) (Kirby et al., 2003; Robinson et al., 2004), and sand goby (Kirby et al., 2003). Moreover, disruption of growth hormone and prolactin mRNA expression in the rainbow trout (Elango et al., 2006) and changing vitellogenin level in adult male zebra fish (Andersen et al., 2006) were also reported.

In fact, vitellogenin and cyp19b gene expression in zebra fish, were found to be affected by E2 during early embryogenesis and organogenesis (Wang et al., 2011). Mussels were also tested *in vivo* by being injected E2 in the presence of Chlorpyrifos (a common pesticide). They were found to be subjective by an internal interaction between both of them in the digestive gland (Canesi et al., 2011).

A laboratory study and field surveys were carried out in UK by Allen et al. (1999) on euryhaline flounder (*Platichthys flesus*) to evaluate the presence of biological responses in fish to the amount of estrogens and their alternatives in water, and to discover whether the effects are likely to be harmful to populations. They used yolk protein vitellogenin as an indicator of exposure to estrogens in. male fish. They revealed that synthetic estrogens had a decreased effect on the vitellogenin response in *Platichthys flesus* than the freshwater species rainbow trout.

In fact, E2 as well as other estrogens are also able to change immune parameters of fish with functional consequences on their ability to cope with pathogens as reported by (Wenger et al., 2011) on his study on rainbow trout (*Oncorhynchus mykiss*).

Actually, (Denslow & Sepúlveda, 2007) reported that exposure of fish to endocrine disrupting chemicals can affect fish in different ways. Sexually differentiated exposure of fish to endocrine disrupting chemicals causes a clear subsidence in the bioavailability of sex hormones and gonadotropins; this consequently leads to impaired gonadal development, altered reproductive behaviors, and decreased fecundity and fertility.

Also in females, reduction of production of E2 leads to modifications in vitellogenesis. This ends up by detrimental effects on oogenesis and egg quality, ultimately leading to developmental abnormalities, increased embryo and sac fry mortality, and even spawning inhibition. These reproductive changes are usually reversible (activational) with animals. So animals are capable of returning back to normal after the exposure finishes. But exposure during the period of sex differentiation can result in irreversible structural (organizational) changes leading to modified reproductive produce and permanent (irreversible) masculinization or feminization of fish. This effect is expected to be magnified with recycling contaminated water in fish farms. The recycling appears to be appealing owing to its economic efficiency, but unfortunately, (Elnwishy, 2008). It may also lead to higher concentrations of E2 into second rounds of water recycling in farming.

In 1999, a study on Mersey estuary fish revealed that approximately 20% of male fish contained oocytes in their testes, but it was not seen elsewhere in UK (Allen et al., 1999). Actually, in Denmark, investigations on intersex showed that male fish in some areas were affected by contaminated wastewater discharges had intersex percentages between 5 and 11% (Jobling et al., 2002; Bjerregaard et al., 2006).

Therefore, this intersex feature and Vitellogenin changes in fish have been used as biomarkers for estrogenic exposure in the environment (Bjerregaard, 2012).

Also, (Brown et al., 2004) conducted a study, in which they found that E2 disrupts growth hormone and plasma insulin-like growth factor (IGF-I) concentrations in salmon. These effects are ecologically significant for survival of wild salmon even if unveiled to a little concentration. However, regardless of how small the amount of E2 is, aquatic organism would be able to survive though their internal function is not fully normal (Elnwishy et al., 2007). Thus, it was also suggested that it reduces hepatic sensitivity to growth hormone, peripheral production of insulin-like growth factor (Norbeck & Sheridan, 2011) in rainbow fish and vitellogenin induction, and a physiological response consistent with exposure to estrogenic compounds (Barber et al., 2011).

On the other hand, very few reports indicate that treatment of fish with E2 did not cause any change in growth or survival of sword tail fish *Xiphophorus hellerii* (Saeed et al., 2012a) or Green Tiger Barb Fish *Puntius tetrazona* (Saeed et al., 2012b).

Even algae were found to be a host environment for these compounds in higher trophic levels (Mason et al., 1996; Pflugmacher et al., 1999) via bio-concentration and bio-magnification, due to their substantial biomass, extensive range of habitat.

5. Impact of E2 and Estrogens on Humans

Estradiol can act as natural estrogens in human bodies by binding to estrogen receptors in the endocrine system (Ra et al., 2011). Some of these chemicals are suspected to cause human infertility or influence the development of children, or harm the reproductive processes (Guillette & Gunderson, 2001). In addition, some of these compounds can accumulate in sediment from where they can exert negative impacts on aquatic food webs (MPCA, 2008).

It is believed that the accumulation of estrogens in a human body has a role in breast carcinogenesis (Yaghjyan & Colditz, 2011) and prostate cancer (Giese, 2003; Eliassen & Hankinson, 2008). Estradiol and other estrogens were also reported to increase proliferation of breast epithelium and stroma. Consequently, the chances of mutation increase in rapidly proliferating epithelium (Russo et al., 2002; Russo et al., 2006; Pattarozzi et al., 2008).

High amount of E2 is produced in female bodies via secretion result in production of enormous follicles in women and sudden decline in estrogen level because E2 is secreted from cumulus oophorus, resulting in poor-quality oocytes and embryos, lower fertilization, and higher miscarriage rates (Colakoglu et al., 2011).

It seems that the levels of E2, progesterone, cortisol and DHEA were found to be significantly high in patient of burning mouth syndrome (Kim et al., 2012).

6. Degradation and Removal of E2 and Estrogens

The natural estrogens including 17β-estradiol are not a persistent compound and could be degraded by sewage bacteria in aerobic and anaerobic conditions (Lee & Liu, 2002). Therefore, attention was directed towards developing possible methodologies for purification of water via phytoremediation (Imai et al., 2007) and biodegradation (Joss et al., 2004; Takeshi et al., 2004; Scherr et al., 2009) to eliminate these compounds from the water bodies.

Bacteria were the most recommended bioremediation method. In 2002, Fujii et al. (2002) isolated a new *Novosphingobium* species as a 17β-estradiol degrading bacterium from activated sludge in a sewage treatment plant in Japan, and proved that no toxic products were reproduced and accumulated in the medium.

Aerobic and anaerobic experiments made by (Lee & Liu, 2002) on the persistency of 17β-estradiol and its metabolites were intriguing. They used bacteria from activated sludge from a local sewage treatment plant in Canada. Their study depicted that E2 and the metabolites were not enduring and could be rapidly degraded by the used sewage bacteria. No other stable degradation products were noticed as shown in Figure 2.

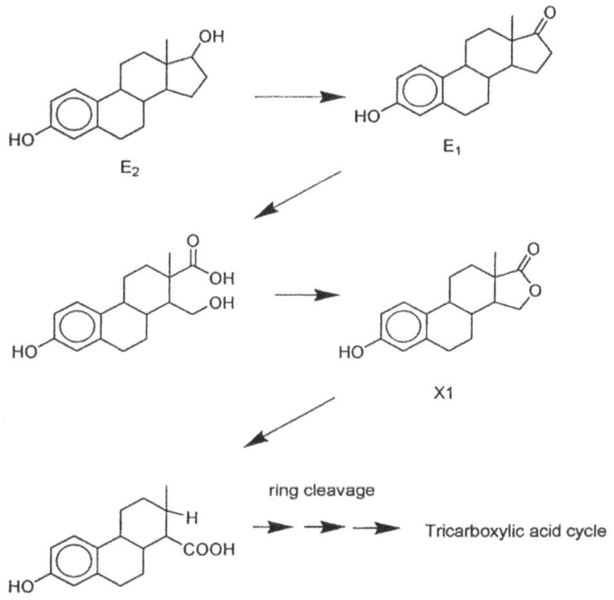

Figure 2. Proposed degradation pathway of estrone (Lee & Liu, 2002)

Different bacteria were discovered to be able to degrade 17β-estradiol successfully. For instance, *Enterobacter* sp. and *Bacillus* sp. were found successful estradiol degrading bacteria, by 57% and 37%, respectively, within 12 hours (Elnwishy et al., 2012). *Rhodococcus zopfii*, could completely degrade 100 mg/L of E2 plus estrone, estriol, and ethinyl estradiol (Yoshimoto et al., 2004). Actually, *Rhodococcus equi*, and *Rhodococcus zopfii* were both suggested as E2 degrading capable strains which degrade estradiol within 24 hours (Yoshimoto et al., 2004). Another bacterium (not named) was reported by (Lee & Liu, 2002) to be able to completely degrade 20 mg of E2 within 18 hours. Many other species were also found efficient such as *Aminobacter*, *Brevundimonas*, *Escherichia*, *Flavobacterium*, *Microbacterium*, *Nocardioides*, *Rhodococcus, and Sphingomonas* (Yu et al., 2007), *Novosphingobium* species (Fujii et al., 2002), ARI-1 and KC8 strains, as indicated by (Roh & Chu, 2010) *Sphingomonas* sp. and *Rhodococcus* (Kurisu et al., 2010). In fact, some bacteria are capable of degrading E2 as well as estrone, estriol, 16α-hydroxyestrone, 2-methoxy-estrone, and 2-methoxyestradiol (Lee & Liu, 2002). *Bacillus* spp., isolated from activated sludge, was also found effective in degrading E2 (Jiang et al., 2010).

In 2004, it was possible to remove estrogens, estrone, 17β-estradiol, and estriol, from the water by nitrifying activated sludge and ammonia-oxidizing bacterium *Nitrosomonaseuropaea*. In fact, there was a suggestion referring that although E2 can be degraded to estrone in the presence of nitrifying activated sludge and ammonia-oxidizing bacterium, but this might have been because of other heterotrophic bacteria, not by ammonia-oxidizing bacteria. This suggestion is due to absence of estrone during the degradation intervals (Shi et al., 2004). Later, the same compounds were successfully removed from water when they were treated with continuous flow algae and duckweed ponds (Shi et al., 2010).

Generally, it is not yet clear how E2 is degraded. Yet, oxidation of E2 to estrone is accepted to be the first step of the degradation pathway (Christoph & Juliane, 2009). This oxidation was suggested to be initiated in the biodegradation process at the hydroxyl group at C-17 (ring D) of E2, leading to the formation of the major metabolite (Lee & Liu, 2002).

It is repeatedly mentioned in studies that oxidation of E2 to estrone occurs both in complex culture systems, such as activated sludge (Ternes et al., 1999) and also can occur in purified bacterial cultures (Chang-Ping et al., 2007). Nevertheless, the pathways of bacteria-mediated degradation are not yet fully well understood.

It was reported earlier that during the very early stage (1-5 hr.) of E2 degradation by the culture, unknown metabolite was observed. The frequency of its occurrence was not really as detectable as that of estrone (Lee & Liu, 2002). Later, (Kurisu et al., 2010) conducted a study to identify E2 metabolites; they incubated resting ED8 cells with E2 and then put the meta-cleavage inhibitor "3-chlorocatechol" to inhibit benzene ring meta-cleavage, resulting in the accumulation of intermediate degradation products. In 3 hours, they detected the trimethylsilyl derivatives of E2 and five other metabolites.

In the natural environment, different factors may be involved in the degradation process by bacteria, such as: degradation in the anaerobic river sediments which is more rapid than in the anaerobic marine sediments (Lopez et al., 2002; Tyler et al., 2005; Czajka & Londry, 2006; Christoph & Juliane, 2009), salinity factor that makes E2 more resistant to degradation, or failing of marine microorganism to involve in biodegradation of E2 are all possible factors.

Anaerobic biodegradation is less energy efficient than aerobic biodegradation, so the aerobic degradation of E2 is much faster (Lee & Liu, 2002). After a 22 hours contact of E2 with a culture containing sewage treatment plants, two-thirds of the spiked E2 (200 μg L^{-1}) was oxidized to estrone. The spiked 17β-estradiol was completely removed within 18 hours.

Activated sludge from wastewater treatment plants were used to isolate two strains of *Rhodococcus* as estrogens degraders (Yoshimoto et al., 2004). Strain *Rhodococcus zopfii* completely and rapidly degraded 100 mg of 17β-estradiol, estrone, estriol, and ethinylestradiol/liter, the other was *Rhodococcus equi*, showed degradation activities comparable with *Rhodococcus zopfii. Rhodococcus zopfii* showed the highest activity, because it selectively degraded 100 mg/L of E2, even when glucose was used as a readily utilizable carbon source in the culture medium after 24 h.

Later, genera *Aminobacter*, *Brevundimonas*, *Escherichia*, *Flavobacterium*, *Microbacterium*, *Nocardioides*, *Rhodococcus*, and *Sphingomonas* were isolated from activated sludge of a wastewater treatment plant as well (Yu et al., 2007). These bacteria converted E2 to estrone, but only *Brevundimonas* and *Sphingomonas* showed the ability to degrade estrone.

Actually *Sphingomonas* seems to be an effective E2 degrader, thus it was examined and even tested for its genome sequence (Anyi et al., 2011). This step can lead to identifying the gene responsible for E2 degrading or resistance in the future. Also, *Novosphingobium* sp., (ARI-1) and KC8 were able to remove testosterone, 17β-estradiol and estrone at the same time from wastewater rapidly when it is grown on complex nutrients containing 17β-estradiol (Roh & Chu, 2010).

However, *Novosphingobium* sp. is a common estrogen-degrader which has no ability to degrade estrone after it is grown on a nutrient-estrogen-free medium for 7 days, while strain KC8 would still be able to degrade both 17β-estradiol and estrone after growing on the same medium for 15 days.

They also detected concentrations of strain KC8 2-3 times higher than those of strain *Novosphingobium* sp. in the waste water treatment plants, which led them to suggest that strain KC8 is an ever-present strain in waste water treatment plants and might be very significant in estrogen removal.

7. Conclusion

The emerging water pollution with 17β-estradiol—endocrine disrupting chemical—is present in most of organism in the environment leading to higher concentrations in human bodies. The impact of this pollutant on the environmental ecosystems and human welfare are equally dangerous. Though efforts on removal 17β-estradiol have been carried out, further investigations on environmentally friendly alternative component should be considered.

References

Allen, Y., Scott, A. P., Matthiessen, P., Haworth, S., Thain, J. E., & Feist, S. (1999). Survey of estrogenic activity in United Kingdom estuarine and coastal waters and its effects on gonadal development of the flounder Platichthys flesus. *Environmental Toxicology and Chemistry, 18*(8), 1791-1800. http://dx.doi.org/10.1002/etc.5620180827

Andaluri, G., Suri, R., & Kumar, K. (2011). Occurrence of estrogen hormones in biosolids, animal manure and mushroom compost. *Environmental Monitoring and Assessment*, 1-9.

Andersen, L., Goto-Kazeto, R., Trant, J. M., Nash, J. P., Korsgaard, B., & Bjerregaard, P. (2006). Short-term exposure to low concentrations of the synthetic androgen methyltestosterone affects vitellogenin and steroid levels in adult male zebrafish (*Danio rerio*). *Aquatic Toxicology, 76*(3-4), 343-352. http://dx.doi.org/10.1016/j.aquatox.2005.10.008

Anyi, H., Jibing, H., Kung-Hui, C., & Chang-Ping, Y. (2011). Genome Sequence of the 17β-Estradiol-Utilizing Bacterium Sphingomonas Strain KC8. *Journal of Bacteriology, 193*(16), 4266-4267. http://dx.doi.org/10.1128/JB.05356-11

Barber, L. B., Brown, G. K., Nettesheim, T. G., Murphy, E. W., Bartell, S. E., & Schoenfuss, H. L. (2011). Effects of biologically-active chemical mixtures on fish in a wastewater-impacted urban stream. *Science of The Total Environment, 409*(22), 4720-4728. http://dx.doi.org/10.1016/j.scitotenv.2011.06.039

Barel-Cohen, K., Shore, L. S., Shemesh, M., Wenzel, A., Mueller, J., & Kronfeld-Schor, N. (2006). Monitoring of natural and synthetic hormones in a polluted river. *Journal of Environmental Management, 78*(1), 16-23. http://dx.doi.org/10.1016/j.jenvman.2005.04.006

Belfroid, A. C., Schrap, S. M., & De Voogt, P. (2006). Occurrence of estrogenic hormones, bisphenol-A and phthalates in the aquatic environment of The Netherlands. In A. D. Vethaak, et al. (Ed.), *Estrogens and xenoestrogens in the aquatic environment: An integrated approach for field monitoring and effect assessment* (pp. 53-75).

Biddle, S., Teale, P., Robinson, A., Bowman, J., & Houghton, E. (2007). Gas chromatography-mass spectrometry/mass spectrometry analysis to determine natural and post-administration levels of oestrogens in bovine serum and urine. *Analytica Chimica Acta, 586*(1-2), 115-121. http://dx.doi.org/10.1016/j.aca.2006.10.044

Birkett, J. W., & Lester, J. N. (2003). *Endocrine Disrupters in Wastewater and Sludge Treatment Processes* (1st ed., February 28, 2003). IWA Publishing (Intl. Water Assoc.).

Bjerregaard, L., Korsgaard, B., & Bjerregaard, P. (2006). Intersex in wild roach (*Rutilis rutilis*) from Danish sewage effluent-receiving streams. *Ecotoxicology Environmental Safety, 64*, 321-328. http://dx.doi.org/10.1016/j.ecoenv.2005.05.018

Bjerregaard, P. (2012). Estrogen mimicking effects of xenobiotics in fish. *Acta Veterinaria Scandinavica, 54*(Supl. 1), 1-2. http://dx.doi.org/10.1186/1751-0147-54-S1-S12

Boxall, A. B. A. (2004). The environmental side effects of medication. *EMBO Rep, 5*(12), 1110-1116. http://dx.doi.org/10.1038/sj.embor.7400307

Brix, R., Postigo, C., Gonzà¡Lez, S., Villagrasa, M., Navarro, A., Kuster, M., ... Barcelà, D. (2010). Analysis and occurrence of alkylphenolic compounds and estrogens in a European river basin and an evaluation of their importance as priority pollutants. *Analytical and Bioanalytical Chemistry, 396*(3), 1301-1309. http://dx.doi.org/10.1007/s00216-009-3358-8

Brown, M., Robinson, C., Davies, I. M., Moffat, C. F., Redshaw, J., & Craft, J. A. (2004). Temporal changes in gene expression in the liver of male plaice (*Pleuronectes platessa*) in response to exposure to ethynyl oestradiol analysed by macroarray and Real-Time PCR. *Mutation Research/Fundamental and Molecular Mechanisms of Mutagenesis, 552*(1-2), 35-49. http://dx.doi.org/10.1016/j.mrfmmm.2004.06.002

Burton, J. L., & Wells, M. (2002). The effect of phytoestrogens on the female genital tract. *Journal of Clinical Pathology, 55*(6), 401-407. http://dx.doi.org/10.1136/jcp.55.6.401

Caldwell, D. J., Mastrocco, F., Nowak, E., Johnston, J., Yekel, H., Pfeiffer, D., ... Anderson, P. D. (2009). An Assessment of Potential Exposure and Risk from Estrogens in Drinking Water. *Environmental Health Perspective, 118*(3). http://dx.doi.org/10.1289/ehp.0900654

Canesi, L., Negri, A., Barmo, C., Banni, M., Gallo, G., Viarengo, A., & Dondero, F. (2011). The organophosphate Chlorpyrifos interferes with the responses to 17beta-estradiol in the digestive gland of the marine mussel Mytilus galloprovincialis. *PLoS One, 6*(5), e19803. http://dx.doi.org/10.1371/journal.pone.0019803

CAPMAS. (2016). *Central Agency for Public Mobilization and Statistics: Egypt in Figures, 3.*

Chang-Ping, Y., Hyungkeun, R., & Kung-Hui, C. (2007). 17β-Estradiol-Degrading Bacteria Isolated from Activated Sludge. *Environmental Science & Technology, 41*(2), 486-492. http://dx.doi.org/10.1021/es060923f

Chetrite, G. S., Cortes-Prieto, J., Philippe, J. C., Wright, F., & Pasqualini, J. R. (2000). Comparison of estrogen concentrations, estrone sulfatase and aromatase activities in normal, and in cancerous, human breast tissues. *Journal of Steroid Biochemistry and Molecular Biology, 72*(1-2), 23-27. http://dx.doi.org/10.1016/S0960-0760(00)00040-6

Cheunbarn, T., & Cheunbarn, S. (2015). Cultivation of Algae in Vegetable and Fruit Canning Industrial Wastewater Treatment Effluent for Tilapia (*Oreochromis niloticus*) Feed. *Survival, 1*(F2), 100. http://dx.doi.org/10.17957/ijab/17.3.14.502

Chimento, A., Sirianni, R., Casaburi, I., & Pezzi, V. (2014). Role of Estrogen Receptors and G Protein-Coupled Estrogen Receptor in Regulation of Hypothalamus-Pituitary-Testis Axis and Spermatogenesis. *Front Endocrinol, 5*, 1-5. http://dx.doi.org/10.3389/fendo.2014.00001

Christoph, M., & Juliane, H. (2009). *Microbial Degradation of Steroid Hormones in the Environment and Technical Systems*. Lecture. Retrieved from http://www.ibp.ethz.ch/research/aquaticchemistry/teaching/archive_past_lectures/term_paper_08_09/HS08_CHRISTOPH_MOSCHET_rev_termpaper.pdf

Colakoglu, M., Toy, H., Icen, M. S., Vural, M., Mahmoud, A. S., & Yazici, F. (2011). *The impact of estrogen supplementation on IVF outcome in patients with polycystic ovary syndrome*. Abstracts of the 27th Annual Meeting of the European Society of Human Reproduction and Embryology, Stockholm, Sweden, 3-6 July.

Collier, A., Wang, J., Diamond, D., & Dempsey, E. (2005). Microchip micellar electrokinetic chromatography coupled with electrochemical detection for analysis of synthetic oestrogen mimicking compounds. *Analytica Chimica Acta, 550*(1-2), 107-115. http://dx.doi.org/10.1016/j.aca.2005.06.053

Cook, J. W., Dodds, E. C., Hewett, C. L., & Lawson, W. (1934). The oestrogenic activity of some condensed-ring compounds in relation to their biological activities. *Proc. Roy. Soc. Lond., 114*, 272-286. http://dx.doi.org/10.1098/rspb.1934.0006

Czajka, C. P., & Londry, K. L. (2006). Anaerobic biotransformation of estrogens. *Science of the Total Environment, 367*(2-3), 932-941. http://dx.doi.org/10.1016/j.scitotenv.2006.01.021

Delvoux, B., Husen, B., Aldenhoff, Y., Koole, L., Dunselman, G., Thole, H., & Groothuis, P. (2007). A sensitive HPLC method for the assessment of metabolic conversion of estrogens. *The Journal of Steroid Biochemistry and Molecular Biology, 104*(3-5), 246-251. http://dx.doi.org/10.1016/j.jsbmb.2007.03.006

Denslow, N. D., Chow, M. C., Kroll, K. J., & Green, L. (1999). Vitellogenin as a Biomarker of Exposure for Estrogen or Estrogen Mimics. *Ecotoxicology, 8*(5), 385-398. http://dx.doi.org/10.1023/A:1008986522208

Denslow, N., & Sepúlveda, M. (2007). Ecotoxicological effects of endocrine disrupting compounds on fish reproduction. In P. J. Babin, J. Cerdà & E. Lubzens (Eds.), *The Fish Oocyte* (pp. 255-322). Springer Netherlands. http://dx.doi.org/10.1007/978-1-4020-6235-3_10

Desbrow, C., Routledge, E. J., Brighty, G. C., Sumpter, J. P., & Waldock, M. (1998). Identification of Estrogenic Chemicals in STW Effluent. 1. Chemical Fractionation and in Vitro Biological Screening. *Environmental Science & Technology, 32*(11), 1549-1558. http://dx.doi.org/10.1021/es9707973

Elango, A., Shepherd, B., & Chen, T. T. (2006). Effects of endocrine disrupters on the expression of growth hormone and prolactin mRNA in the rainbow trout pituitary. *General and Comparative Endocrinology, 145*(2), 116-127. http://dx.doi.org/10.1016/j.ygcen.2005.08.003

Eliassen, A., & Hankinson, S. (2008). Endogenous hormone levels and risk of breast, endometrial and ovarian cancers: prospective studies. *Advances in Experimental Medicine and Biology, 630*, 148-165. http://dx.doi.org/10.1007/978-0-387-78818-0_10

Elnwishy, N. H. (2008). Effectiveness of fish culture implementation to improve irrigation water quality. *International Journal of Agriculture and Biology, 10*(5), 591-592.

Elnwishy, N. H., Ahmed, M. T., El-Sherif, M. S., & El-Hameed, M. A. (2007). The effect of Diazinon on glutathine and Acetylcholinesterase in Tilapia (*Oreochromis niloticus*). *Journal of Agriculture and Social Sciences*.

Elnwishy, N., Hanora, A., Afifi, R., Omran, H., & Matiasson, B. (2012). A Potential 17β-Estradiol degrader Bacterium Isolated from Sewage water. *Egyptian Academic Journal of Biological Sciences, 4*(1), 27-34.

EPA. (2009). Drinking Water Contaminant Candidate List 3-Final. *Federal Register, Doc No: E9-24287, 74*(194), 51850-51862.

EPA. (2012). Environmental Protection Agency: Revisions to the Unregulated Contaminant Monitoring Regulation (UCMR 3) for Public Water Systems. *Federal Register, Doc No: 2012-9978, 77*(85).

Erdal, S., & Dumlupinar, R. (2011). Mammalian sex hormones stimulate antioxidant system and enhance growth of chickpea plants. *Acta Physiologiae Plantarum, 33*(3), 1011-1017. http://dx.doi.org/10.1007/s11738-010-0634-3

Fu, S.-B., Riaz, H., Khan, M. K., Zhang, H.-L., Chen, J.-G., & Yang, L.-G. (2014). Influence of Different Doses of Equine Chorionic Gonadotropin on Follicular Population and Plasma Estradiol Concentration in Chinese Holstein Dairy Cows. *International Journal of Agriculture & Biology, 16*(2).

Fujii, K., Kikuchi, S., Satomi, M., Ushio-Sata, N., & Morita, N. (2002). Degradation of 17B-Estradiol by a Gram-Negative Bacterium Isolated from Activated Sludge in a Sewage Treatment Plant in Tokyo, Japan. *Applied and Environmental Microbiology, 68*(4). http://dx.doi.org/10.1128/AEM.68.4.2057-2060.2002

Giese, R. W. (2003). Measurement of endogenous estrogens: Analytical challenges and recent advances. *Journal of Chromatography A, 1000*(1-2), 401-412. http://dx.doi.org/10.1016/S0021-9673(03)00306-6

Gomes, R. L., Deacon, H. E., Lai, K. M., Birkett, J. W., Scrimshaw, M. D., & Lester, J. N. (2004). An assessment of the bioaccumulation of estrone in Daphnia magna. *Environmental Toxicology and Chemistry, 23*(1), 105-108. http://dx.doi.org/10.1897/02-613

Gong, J., Ran, Y., Chen, D.-Y., & Yang, Y. (2011). Occurrence of endocrine-disrupting chemicals in riverine sediments from the Pearl River Delta, China. *Marine Pollution Bulletin, 63*(5-12), 556-563.

Gracia-Lor, E., Sancho, J., Serrano, R., & Hernández, F. (2012). Occurrence and removal of pharmaceuticals in wastewater treatment plants at the Spanish Mediterranean area of Valencia. *Chemosphere, 87*(5), 453-462. http://dx.doi.org/10.1016/j.chemosphere.2011.12.025

Guillette, L. J. Jr., & Gunderson, M. P. (2001). Alterations in development of reproductive and endocrine systems of wildlife populations exposed to endocrine-disrupting contaminants. *Reproduction, 122*(6), 857-864. http://dx.doi.org/10.1530/rep.0.1220857

Hahlbeck, E., Katsiadaki, I., Mayer, I., Adolfsson-Erici, M., James, J., & Bengtsson, B.-E. (2004). The juvenile three-spined stickleback (*Gasterosteus aculeatus* L.) as a model organism for endocrine disruption IIâ€"kidney hypertrophy, vitellogenin and spiggin induction. *Aquatic Toxicology, 70*(4), 311-326. http://dx.doi.org/10.1016/j.aquatox.2004.10.004

Halm, S., Martinez-Rodriguez, G., Rodriguez, L., Prat, F., Mylonas, C., Carrillo, M., & Zanuy, S. (2004). Cloning, characterisation, and expression of three oestrogen receptors (ER alpha, ER beta 1 and ER beta 2) in the European sea bass, Dicentrarchus labrax. *Molecular and Cellular Endocrinology, 223*(1-2), 63-75. http://dx.doi.org/10.1016/j.mce.2004.05.009

Hansen, P. D., Dizer, H., Hock, B., Marx, A., Sherry, J., Mcmaster, M., & Blaise, C. (1998). Vitellogenin—A biomarker for endocrine disruptors. *TrAC Trends in Analytical Chemistry, 17*(7), 448-451. http://dx.doi.org/10.1016/S0165-9936(98)00020-X

Haring, R., Xanthakis, V., Coviello, A., Sullivan, L., Bhasin, S., Wallaschofski, H., ... Vasan, R. S. (2012). Clinical correlates of sex steroids and gonadotropins in men over the late adulthood: the Framingham Heart Study. *International Journal of Andrology, 35*(6), 775-782. http://dx.doi.org/10.1111/j.1365-2605.2012.01285.x

Harries, J., Jobling, S., & Sumpter, J. (1997). Estrogenic activity in five United Kingdom rivers detected by measurement of vitellogenesis in caged male trout. *Environmental Toxicology and Chemistry, 15*, 534-542. http://dx.doi.org/10.1002/etc.5620160320

Harries, J., Sheahan, D., & Matthiessen, P. (1996). A survey of estrogenic activity in United Kingdom inland waters. *Environmental Toxicology and Chemistry, 15*, 1993-2002. http://dx.doi.org/10.1002/etc.5620151118

Havlíková, L., Nováková, L., Matysová, L., Sícha, J., & Solich, P. (2006). Determination of estradiol and its degradation products by liquid chromatography. *Journal of Chromatography A, 1119*(1-2), 216-223. http://dx.doi.org/10.1016/j.chroma.2006.01.085

Hintemann, T., Schneider, C., Schöler, H. F., & Schneider, R. J. (2006). Field study using two immunoassays for the determination of estradiol and ethinylestradiol in the aquatic environment. *Water Research, 40*(12), 2287-2294. http://dx.doi.org/10.1016/j.watres.2006.04.028

Hirsch, R., Ternes, T., Haberer, K., & Kratz, K. L. (1999). Occurrence of antibiotics in the aquatic environment. *Science of the Total Environment, 225*(1-2), 109-118. http://dx.doi.org/10.1016/S0048-9697(98)00337-4

Holtorf, K. (2009). The bioidentical hormone debate: Are bioidentical hormones (estradiol, estriol, and progesterone) safer or more efficacious than commonly used synthetic versions in hormone replacement therapy? *Postgrad Med, 121*(1), 73-85. http://dx.doi.org/10.3810/pgm.2009.01.1949

Imai, S., Shiraishi, A., Gamo, K., Watanabe, I., Okuhata, H., Miyasaka, H., … Hirata, K. (2007). Removal of phenolic endocrine disruptors by Portulaca oleracea. *Journal of Bioscience and Bioengineering, 103*(5), 420-426. http://dx.doi.org/10.1263/jbb.103.420

Isobe, T., Shiraishi, H., Yasuda, M., Shinoda, A., Suzuki, H., & Morita, M. (2003). Determination of estrogens and their conjugates in water using solid-phase extraction followed by liquid chromatography-tandem mass spectrometry. *Journal of Chromatography A, 984*(2), 195-202. http://dx.doi.org/10.1016/S0021-9673(02)01851-4

Jiang, L., Yang, J., & Chen, J. (2010). Isolation and characteristics of 17β-estradiol-degrading *Bacillus* spp. strains from activated sludge. *Biodegradation, 21*(5), 729-736. http://dx.doi.org/10.1007/s10532-010-9338-z

Jiang, T., Zhao, L., Chu, B., Feng, Q., Yan, W., & Lin, J.-M. (2009). Molecularly imprinted solid-phase extraction for the selective determination of 17[beta]-estradiol in fishery samples with high performance liquid chromatography. *Talanta, 78*(2), 442-447. http://dx.doi.org/10.1016/j.talanta.2008.11.047

Jobling, S., Coey, S., Whitmore, J. G., Kime, D. E., Van Look, K. J. W., Mcallister, B. G., … Sumpter, J. P. (2002). Wild Intersex Roach (*Rutilus rutilus*) Have Reduced Fertility. *Biology of Reproduction, 67*, 515-524. http://dx.doi.org/10.1095/biolreprod67.2.515

Joss, A., Andersen, H., Ternes, T., Richle, P. R., & Siegrist, H. (2004). Removal of Estrogens in Municipal Wastewater Treatment under Aerobic and Anaerobic Conditions: Consequences for Plant Optimization. *Environmental Science & Technology, 38*(11), 3047-3055. http://dx.doi.org/10.1021/es0351488

Katie, A. (2008). *Occurrence in waste water treatment plant and waste disposal site water samples* (pp. 28-34). Retrieved from http://nywea.org/clearwaters/08-3-fall/05-EstrogenInWastewater.pdf

Katsiadaki, I., Scott, A. P., Hurst, M. R., Matthiessen, P., & Mayer, I. (2002). Detection of environmental androgens: a novel method based on enzyme-linked immunosorbent assay of spiggin, the stickleback (*Gasterosteus aculeatus*) glue protein. *Environmental Toxicology and Chemistry, 21*, 1946-1954. http://dx.doi.org/10.1002/etc.5620210924

Kim, H. I., Kim, Y. Y., Chang, J. Y., Ko, J. Y., & Kho, H. S. (2012). *Salivary cortisol, 17beta-estradiol, progesterone, dehydroepiandrosterone, and alpha-amylase in patients with burning mouth syndrome.* Oral presentaion.

Kirby, M. F., Bignell, J., Brown, E., Craft, J. A., Davies, I., Dyer, R., … Robinson, C. (2003). The presence of morphologically intermediate papilla syndrome in United Kingdom populations of sand goby (*Pomatoschistus* spp.): Endocrine disruption? *Environmental Toxicology and Chemistry, 22*(2), 239-251. http://dx.doi.org/10.1897/1551-5028(2003)022%3C0239:tpomip%3E2.0.co;2

Klaschka, U. (2008). The infochemical effec—A new chapter in ecotoxicology. *Environmental Science and Pollution Research, 15*(6), 452-462. http://dx.doi.org/10.1007/s11356-008-0019-y

Knoebl, I., Hemmer, N., & Denslow, N. (2004). Induction of zona radiata and vitellogenin genes in estradiol and nonylphenol exposed male sheepshead minnows (*Cyprinodon variegatus*). *Marine Environmental Researh, 58*, 547-551. http://dx.doi.org/10.1016/j.marenvres.2004.03.043

Kolodziej, E. P., Gray, J. L., & Sedlak, D. L. (2003). Quantification of steroid hormones with pheromonal properties in municipal wastewater effluent. *Environmental Toxicology and Chemistry, 22*(11), 2622-2629. http://dx.doi.org/10.1897/03-42

Kolpin, D., Furlong, E., Meyer, M., Thurman, E., Zaugg, S., Barber, L., & Buxton, H. (2002). Pharmaceuticals, hormones, and other organic wastewater contaminants in U.S. streams, 1999-2000: A national reconnaissance. *Environmental Science and Technology, 15*(6), 1202-1211. http://dx.doi.org/10.1021/es011055j

Körner, W., Bolz, U., Triebskorn, R., Schwaiger, J., Negele, R.-D., Marx, A., & Hagnemaier, H. (2001). Steroid analysis and xenosteroid potentials in two small streams in southwest Germany. *Journal of Aquatic Ecosystem Stress and Recovery, 8*(3), 215-229. http://dx.doi.org/10.1023/A:1012976800922

Kosjek, T., Heath, E., & Kompare, B. (2007). Removal of pharmaceutical residues in a pilot wastewater treatment plant. *Analytical and Bioanalytical Chemistry, 387*(4), 1379-1387. http://dx.doi.org/10.1007/s00216-006-0969-1

Kuch, H. M., & Ballschmiter, K. (2000). Determination of endogenous and exogenous estrogens in effluents from sewage treatment plants at the ng/L-level. *Fresenius' Journal of Analitical Chemistry, 366*(4), 392-395. http://dx.doi.org/10.1007/s002160050080

Kurisu, F., Ogura, M., Saitoh, S., Yamazoe, A., & Yagi, O. (2010). Degradation of natural estrogen and identification of the metabolites produced by soil isolates of *Rhodococcus* sp. and *Sphingomonas* sp. *Journal of Bioscience and Bioengineering, 109*(6), 576-582. http://dx.doi.org/10.1016/j.jbiosc.2009.11.006

Lagadic, L., Coutellec, M.-A., & Caquet, T. (2007). Endocrine disruption in aquatic pulmonate molluscs: few evidences, many challenges. *Ecotoxicology, 16*(1), 45-59. http://dx.doi.org/10.1007/s10646-006-0114-0

Lai, K. M., Scrimshaw, M. D., & Lester, J. N. (2002a). Prediction of the bioaccumulation factors and body burden of natural and synthetic estrogens in aquatic organisms in the river systems. *The Science of the Total Environment, 289*(1-3), 159-168. http://dx.doi.org/10.1016/S0048-9697(01)01036-1

Lai, K. M., Scrimshaw, M. D., & Lester, J. N. (2002b). Biotransformation and Bioconcentration of Steroid Estrogens by *Chlorella vulgaris*. *Applied Environmental Microbiology, 68*(2), 859-864. http://dx.doi.org/10.1128/AEM.68.2.859-864.2002

Larsson, D. G. J., De Pedro, C., & Paxeus, N. (2007). Effluent from drug manufactures contains extremely high levels of pharmaceuticals. *Journal of Hazardous Materials, 148*(3), 751-755. http://dx.doi.org/10.1016/j.jhazmat.2007.07.008

Le Noir, M., Lepeuple, A.-S., Guieysse, B., & Mattiasson, B. (2007). Selective removal of 17[beta]-estradiol at trace concentration using a molecularly imprinted polymer. *Water Research, 41*(12), 2825-2831. http://dx.doi.org/10.1016/j.watres.2007.03.023

Lee, H., & Liu, D. (2002). Degradation of 17β-Estradiol and Its Metabolites by Sewage Bacteria. *Water, Air, and Soil Pollution, 134*, 353-368. http://dx.doi.org/10.1023/A:1014117329403

Liu, S., Ying, G.-G., Zhao, J.-L., Chen, F., Yang, B., Zhou, L.-J., & Lai, H.-J. (2011). Trace analysis of 28 steroids in surface water, wastewater and sludge samples by rapid resolution liquid chromatography-electrospray ionization tandem mass spectrometry. *Journal of Chromatography A, 1218*(10), 1367-1378. http://dx.doi.org/10.1016/j.chroma.2011.01.014

Lopez, D. A., Maria, J., Gil, A., Paz, E., & Barcelo, D. (2002). Occurrence and analysis of estrogens and progestogens in river sediments by liquid chromatography-electrospray-mass spectrometry. *Analyst, 127*(10), 1299-1304. http://dx.doi.org/10.1039/B207658F

Lu, G., Yan, Z., Wang, Y., & Chen, W. (2011). *Assessment of estrogenic contamination and biological effects in Lake Taihu.*

Luna, T. O., Plautz, S. C., & Salice, C. J. (2013). Effects of 17alpha-ethynylestradiol, fluoxetine, and the mixture on life history traits and population growth rates in a freshwater gastropod. *Environ Toxicol Chem, 32*(12), 2771-2778. http://dx.doi.org/10.1002/etc.2372

Mason, R. P., Reinfelder, J. R., & Morel, F. O. M. M. (1996). Uptake, Toxicity, and Trophic Transfer of Mercury in a Coastal Diatom. *Environmental Science & Technology, 30*(6), 1835-1845. http://dx.doi.org/10.1021/es950373d

Matsumoto, Y., Kuramitz, H., Itoh, S., & Tanaka S. (2005). Quantitative analysis of 17beta-estradiol in river water by fluorometric enzyme immunoassay using biotinylated estradiol. *Analitical Science, 21*(3), 219-224. http://dx.doi.org/10.2116/analsci.21.219

Matthiessen, P., Allen, Y., Bamber, S., Craft, J., Hurst, M., Hutchinson, T., ... Thomas, K. (2002). The impact of oestrogenic and androgenic contamination on marine organisms in the United Kingdom - Summary of the EDMAR programme. Endocrine Disruption in the Marine Environment. *Marine Environmental Research, 54*(3-5), 645-649. http://dx.doi.org/10.1016/s0141-1136(02)00135-6

Mehinto, A. C., Hill, E. M., & Tyler, C. R. (2010). Uptake and biological effects of environmentally relevant concentrations of the nonsteroidal anti-inflammatory pharmaceutical diclofenac in rainbow trout (*Oncorhynchus mykiss*). *Environmental Science & Technology, 44*(6), 2176-2182. http://dx.doi.org/10.1021/es903702m

Meng-Umphan, K. (2009). Growth performance, sex hormone levels and maturation ability of Pla Pho (*Pangasius bocourti*) fed with Spirulina supplementary pellet and hormone application. *International Journal of Agriculture and Biology, 11*(4), 458-462.

Mills, L. J., & Chichester, C. (2005). Review of evidence: Are endocrine-disrupting chemicals in the aquatic environment impacting fish populations? *Science of the Total Environment, 343*(1), 1-34. http://dx.doi.org/10.1016/j.scitotenv.2004.12.070

MPCA (Minnesota Pollution Control Agency). (2008). Minnesota's nonpoint source management. *Water Quality in Minnesota, 2010.* Retrieved from http://wrc.umn.edu/prod/groups/cfans/@pub/@cfans/@wrc/documents/asset/cfans_asset_213099.pdf

Narender, K. N., & Cindy, L. M. (2009). Water Quality Guidelines for Pharmaceutically-active-Compounds (PhACs): 17α-ethinylestradiol (EE2). *Report by Ministry of Environment province of British Columbia.* Retrieved from http://www.env.gov.bc.ca/wat/wq/BCguidelines/PhACs-EE2/PhACs-EE2-tech.pdf

Navas, J. M., & Segner, H. (2006). Vitellogenin synthesis in primary cultures of fish liver cells as endpoint for in vitro screening of the (anti)estrogenic activity of chemical substances. *Aquatic Toxicology, 80*(1), 1-22. http://dx.doi.org/10.1016/j.aquatox.2006.07.013

Norbeck, L. A., & Sheridan, M. A. (2011). An in vitro model for evaluating peripheral regulation of growth in fish: Effects of 17 β-estradiol and testosterone on the expression of growth hormone receptors, insulin-like growth factors, and insulin-like growth factor type 1 receptors in rainbow trout (*Oncorhynchus mykiss*). *General and Comparative Endocrinology, 173*(2), 270-280. http://dx.doi.org/10.1016/j.ygcen.2011.06.009

Ogunleye, A. A., & Holmes, M. D. (2009). Physical activity and breast cancer survival. *Breast Cancer Research, 11*(5), 106. http://dx.doi.org/10.1186/bcr2351

OSPAR. (2013). Background Document on Clotrimazole (2013 update). *Publication Number: 595/2013.* ISBN 978-1-909159-28-0.

Panter, G. H., Hutchinson, T. H., Hurd, K. S., Bamforth, J., Stanley, R. D., Duffell, S., ... Tyler, C. R. (2006). Development of chronic tests for endocrine active chemicals: Part 1. An extended fish early-life stage test for oestrogenic active chemicals in the fathead minnow (*Pimephales promelas*). *Aquatic Toxicology, 77*(3), 279-290. http://dx.doi.org/10.1016/j.aquatox.2006.01.004

Pattarozzi, A., Gatti, M., Barbieri, F., Wurth, R., Porcile, C., Lunardi, G., ... Florio, T. (2008). 17beta-estradiol promotes breast cancer cell proliferation-inducing stromal cell-derived factor-1-mediated epidermal growth factor receptor transactivation: Reversal by gefitinib pretreatment. *Molecular Pharmacology, 73*(1), 191-202. http://dx.doi.org/10.1124/mol.107.039974

Pedrouzo, M., Borrull, F., Pocurull, E., & Marcé, R. M. (2009). Estrogens and their conjugates: Determination in water samples by solid-phase extraction and liquid chromatography-tandem mass spectrometry. *Talanta, 78*(4-5), 1327-1331. http://dx.doi.org/10.1016/j.talanta.2009.02.005

Petrovic, M., Eljarrat, E., López De Alda, M. J., & Barceló, D. (2001). Analysis and environmental levels of endocrine-disrupting compounds in freshwater sediments. *TrAC Trends in Analytical Chemistry, 20*(11), 637-648. http://dx.doi.org/10.1016/S0165-9936(01)00118-2

Pflugmacher, S., Wiencke, C., & Sandermann, H. (1999). Activity of phase I and phase II detoxication enzymes in Antarctic and Arctic macroalgae. *Marine Environmental Research, 48*(1), 23-36. http://dx.doi.org/10.1016/S0141-1136(99)00030-6

Pojana, G. A., Bonfa, F., Busetti, A., Collarin, A., & Marcomini, A. (2004). Estrogenic potential of the Venice, Italy, lagoon water. *Environmental Toxicology and Chemistry, 23*, 1874-1880. http://dx.doi.org/10.1897/03-222

Pojana, G., Gomiero, A., Jonkers, N., & Marcomini, A. (2007). Natural and synthetic endocrine disrupting compounds (EDCs) in water, sediment and biota of a coastal lagoon. *Environment International, 33*(7), 929-936. http://dx.doi.org/10.1016/j.envint.2007.05.003

Ra, J.-S., Lee, S.-H., Lee, J., Kim, H. Y., Lim, B. J., Kim, S. H., & Kim, S. D. (2011). Occurrence of estrogenic chemicals in South Korean surface waters and municipal wastewaters. *Journal of Environmental Monitoring, 13*(1), 101-109. http://dx.doi.org/10.1039/C0EM00204F

Robinson, C. D., Craft, J. A., Moffat, C. F., Davies, I. M., Brown, E. S., & Megginson, C. (2004). Oestrogenic markers and reduced population fertile egg production in a sand goby partial life-cycle test. *Mar Environ Res, 58*(2-5), 147-150. http://dx.doi.org/10.1016/j.marenvres.2004.03.009

Rodriguez-Mozaz, S., López De Alda, M. J., & Barceló, D. (2004). Monitoring of estrogens, pesticides and bisphenol A in natural waters and drinking water treatment plants by solid-phase extraction-liquid chromatography-mass spectrometry. *Journal of Chromatography A, 1045*(1-2), 85-92. http://dx.doi.org/10.1016/j.chroma.2004.06.040

Roh, H., & Chu, K.-H. (2010). A 17β-Estradiol-utilizing Bacterium, Sphingomonas Strain KC8: Part I - Characterization and Abundance in Wastewater Treatment Plants. *Environmental Science & Technology, 44*(13), 4943-4950. http://dx.doi.org/10.1021/es1001902

Rujiralai, T., Bull, I. D., Llewellyn, N., & Evershed, R. P. (2011). In situ polar organic chemical integrative sampling (POCIS) of steroidal estrogens in sewage treatment works discharge and river water. *Journal of Environmental Monitoring, 13*(5), 1427-1434. http://dx.doi.org/10.1039/c0em00537a

Russo, J., Fernandez, S. V., Russo, P. A., Fernbaugh, R., Sheriff, F. S., Lareef, H. M., ... Russo, I. H. (2006). 17-Beta-estradiol induces transformation and tumorigenesis in human breast epithelial cells. *Journal of the Federation of American Societies for Experimental Biology, 20*(10), 1622-1634. http://dx.doi.org/10.1096/fj.05-5399com

Russo, J., Lareef, M. H., Tahin, Q., Hu, Y. F., Slater, C., Ao, X., & Russo, I. H. (2002). 17Beta-estradiol is carcinogenic in human breast epithelial cells. *Journal of Steroid Biochemistry and Molecular Biology, 80*(2), 149-162. http://dx.doi.org/10.1016/S0960-0760(01)00183-2

Saeed, B., Mohammad, S., Seyed, A., & Hamideh, K. (2012a). Effect of 17beta-Estradiol on Growth and Survival of sword tail (*Xiphophorus hellerii*). *World Journal of Fish and Marine Sciences, 4*(4), 335-339.

Saeed, B., Mohammad, S., Seyed, A., & Hamideh, K. (2012b). Effect of 17 beta Estradiol on Growth and Survival of Green Tiger Barb (*Puntius tetrazona*). *Global Veterinaria, 8*(5), 445-448.

Sandeep, D., Richard, F., Mehul Vora, M., Husam, G. P., Ajay, C., & Paresh, D. (2011). Low Estradiol Concentrations in Men with Subnormal Testosterone Concentrations and Type 2 Diabetes. *Diabetes Care August, 34*(8), 1854-1859. http://dx.doi.org/10.2337/dc11-0208

Sawssan, M., Monia, M., Gauthier. T., Kristell, K., Alain, G., Christophe, M., & Amel, H. (2015). The endocrine-disrupting effect and other physiological responses of municipal effluent on the clam *Ruditapes decussatus*. *Environmental Science and Pollution Research, 22*(24), 19716-19728. http://dx.doi.org/10.1007/s11356-015-5199-7

Scherr, F. F., Sarmah, A. K., Di, H. J., & Cameron, K. C. (2009). Degradation and metabolite formation of 17beta-estradiol-3-sulphate in New Zealand pasture soils. *Environment International, 35*(2), 291-297. http://dx.doi.org/10.1016/j.envint.2008.07.002

Seifert, M., Li, W., Alberti, U., Kausch, M., & Hock, B. (2003). Biomonitoring: Integration of biological endpoints into chemical monitoring. *Pure and Applied Chemistry. IUPAC Workshop 3.3, 75*(Nos. 11-12), 2451-2459. http://dx.doi.org/10.1351/pac200375112451

Shane, S., Cecil, L.-H., Joe, C., Jörg, E., Drewes, A. E., Richard, C. P., & Dan, S. (2011). *Pharmaceuticals in the Water Environment*. Retrieved from http://www.dcwater.com/waterquality/PharmaceuticalsNACWA.pdf

Shi, J., Fujisawa, S., Nakai, S., & Hosomi, M. (2004). Biodegradation of natural and synthetic estrogens by nitrifying activated sludge and ammonia-oxidizing bacterium *Nitrosomonas europaea*. *Water Research, 38*(9), 2323-2330. http://dx.doi.org/10.1016/j.watres.2004.02.022

Shi, W., Wang, L., Rousseau, D., & Lens, P. (2010). Removal of estrone, 17α-ethinylestradiol, and 17ß-estradiol in algae and duckweed-based wastewater treatment systems. *Environmental Science and Pollution Research, 17*(4), 824-833. http://dx.doi.org/10.1007/s11356-010-0301-7

Sultan, M. K., Bakr, I., Ismail, N. A., & Arafa, N. (2010). Prevalence of unmet contraceptive need among Egyptian women: A community-based study. *Journal of Preventive Medicine and Hygiene, 51*(2).

Takeshi, Y., Fumiko, N., Junji, F., Koichi, W., Harumi, M., Takashi, M., ... Hiroshi, O. (2004). Degradation of estrogens by *Rhodococcus zopfii* and *Rhodococcus equi* isolates from activated sludge in wastewater treatment plants. *Applied and Environmental Microbioloy, 70*(9), 5283-5289. http://dx.doi.org/10.1128/AEM.70.9.5283-5289.2004

Tanaka, T., Takeda, H., Ueki, F., Obata, K., Tajima, H., Takeyama, H., ... Matsunaga, T. (2004). Rapid and sensitive detection of 17[beta]-estradiol in environmental water using automated immunoassay system with bacterial magnetic particles. *Journal of Biotechnology, 108*(2), 153-159. http://dx.doi.org/10.1016/j.jbiotec.2003.11.010

Tawfic, M. (2006). Presistent Organic Pollutants in Egyrpt - An Overview. In I. Twardowska, H. E. Allen, M. M. Häggblom & S. Stefaniak (Eds.), *Soil and Water Pollution Monitoring, Protection and Remediation* (pp. 25-38). Springer Netherlands.

Ten Kulve, J. S., De Jong, F. H., & De Ronde, W. (2011). The effect of circulating estradiol concentrations on gonadotropin secretion in young and old castrated male-to-female transsexuals. *The Aging Male, 14*(3), 155-161. http://dx.doi.org/10.3109/13685538.2010.511328

Ternes, T. A., Kreckel, P., & Mueller, J. (1999). Behaviour and occurrence of estrogens in municipal sewage treatment plants-II. Aerobic batch experiments with activated sludge. *The Science of the Total Environment, 225*(1), 91-99. http://dx.doi.org/10.1016/S0048-9697(98)00335-0

Tyler, C. R., Spary, C., Gibson, R., Santos, E. M., Shears, J., & Hill, E. M. (2005). Accounting for differences in estrogenic responses in rainbow trout (*Oncorhynchus mykiss*: Salmonidae) and roach (*Rutilus rutilus*: Cyprinidae) exposed to effluents from wastewater treatment works. *Environmental Science and Technology, 39*(8), 2599-2607. http://dx.doi.org/10.1021/es0488939

Versteeg, D. J., Alder, A. C., Cunningham, V. L., Kolpin, D. W., Murray-Smith, R., & Ternes, T. (2005). Environmental Exposure Modeling and Monitoring of Human Pharmaceutical Concentrations in the. *Human pharmaceuticals: Assessing the impacts on aquatic ecosystems* (Vol. 71).

Vethaak, A. D., Lahr, J., Schrap, S. M., Belfroid, A. L. C., Rijs, G. B. J., Gerritsen, A., ... De Voogt, P. (2005). An integrated assessment of estrogenic contamination and biological effects in the aquatic environment of The Netherlands. *Chemosphere, 59*(4), 511-524. http://dx.doi.org/10.1016/j.chemosphere.2004.12.053

Wagner, M., & Oehlmann, J. (2009). Endocrine disruptors in bottled mineral water: total estrogenic burden and migration from plastic bottles. *Environmental Science and Pollution Research International, 16*(3), 278-286. http://dx.doi.org/10.1007/s11356-009-0107-7

Wang, J., Shi, X., Du, Y., & Zhou, B. (2011). Effects of xenoestrogens on the expression of vitellogenin (vtg) and cytochrome P450 aromatase (cyp19a and b) genes in zebrafish (*Danio rerio*) larvae. *Journal of Environmental Science and Health, Part A. Toxic/Hazardous Substances and Environmental Engineering, 46*(9), 960-967. http://dx.doi.org/10.1080/10934529.2011.586253

Wang, L., Cai, Y.-Q., He, B., Yuan, C.-G., Shen, D.-Z., Shao, J., & Jiang, G.-B. (2006). Determination of estrogens in water by HPLC-UV using cloud point extraction. *Talanta, 70*(1), 47-51. http://dx.doi.org/10.1016/j.talanta.2006.01.013

Wei, H., Yan-Xia, L., Ming, Y., & Wei, L. (2011). Presence and Determination of Manure-borne Estrogens from Dairy and Beef Cattle Feeding Operations in Northeast China. *Bulletin of Environmental Contamination and Toxicology, 86*(5), 465-469. http://dx.doi.org/10.1007/s00128-011-0247-6

Wenger, M., Sattler, U., Goldschmidt-Clermont, E., & Segner, H. (2011). 17Beta-estradiol affects the response of complement components and survival of rainbow trout (*Oncorhynchus mykiss*) challenged by bacterial infection. *Fish Shellfish Immunology, 31*(1), 90-97. http://dx.doi.org/10.1016/j.fsi.2011.04.007

Xu, J. F., Xia, L.-L., Xia, L. F., Song, N., Song, N. F., Chen, S.-D., ... Wang, G. (2016). Testosterone, Estradiol, and Sex Hormone-Binding Globulin in Alzheimer's Disease: A Meta-Analysis. *Curr Alzheimer Res., 13*(3), 215-222. http://dx.doi.org/10.2174/1567205013666151218145752

Yaghjyan, L., & Colditz, G. (2011). Estrogens in the breast tissue: A systematic review. *Cancer Causes and Control, 22*(4), 529-540. http://dx.doi.org/10.1007/s10552-011-9729-4

Yoshimoto, T., Nagai, F., Fujimoto, J., Watanabe, K., Mizukoshi, H., Makino, T., ... Omura, H. (2004). Degradation of Estrogens by *Rhodococcus zopfii* and *Rhodococcus equi* Isolates from Activated Sludge in Wastewater Treatment Plants. *Applied and Environmental Microbiology, 70*(9), 5283-5289. http://dx.doi.org/10.1128/AEM.70.9.5283-5289.2004

Yu, C.-P., Roh, H., & Chu, K.-H. (2007). 17β-Estradiol-Degrading Bacteria Isolated from Activated Sludge. *Environmental Science & Technology, 41*(2), 486-492. http://dx.doi.org/10.1021/es060923f

Zhang, S., Zhang, Q., Darisaw, S., Ehie, O., & Wang, G. (2007). Simultaneous quantification of polycyclic aromatic hydrocarbons (PAHs), polychlorinated biphenyls (PCBs), and pharmaceuticals and personal care products (PPCPs) in Mississippi river water, in New Orleans, Louisiana, USA. *Chemosphere, 66*(6), 1057-1069. http://dx.doi.org/10.1016/j.chemosphere.2006.06.067

Zhou, Y., Zha, J., Xu, Y., Lei, B., & Wang, Z. (2011). Occurrences of six steroid estrogens from different effluents in Beijing, China. *Environmental Monitoring and Assessment, 184*(3), 1719-1729. http://dx.doi.org/10.1007/s10661-011-2073-z

Zou, L., Lin, H., & Jiang, J. (2007). Determination of [beta]-Estradiol Residues in Fish/Shellfish Muscle by Gas Chromatography-Mass Spectrometry. *Chinese Journal of Analytical Chemistry, 35*(7), 983-987. http://dx.doi.org/10.1016/S1872-2040(07)60063-2

Yield and Quality Traits of Some Flax Cultivars as Influenced by Different Irrigation Intervals

Emad Rashwan[1], Ahmed Mousa[2], Ayman EL-Sabagh[3] & Celaleddin Barutçular[4]

[1] Department of Agronomy, Faculty of Agriculture, Tanta University, Egypt

[2] Fiber Crops Research Section, Field Crops Research Institute, Egypt

[3] Department of Agronomy, Faculty of Agriculture, Kafrelsheikh University, Egypt

[4] Department of Field Crops, Faculty of Agriculture, Cukurova University, Turkey

Correspondence: Ayman EL-Sabagh, Department of Agronomy, Faculty of Agriculture, Kafrelsheikh University, Egypt. E-mail: ayman.elsabagh@agr.kfs.edu.eg

Abstract

Flax is a potential winter crop for Egypt that can be grown for both seed and fiber. The study was conducted during two successive winter seasons of 2013/14 and 2014/15 in the experimental farm of El-Gemmeiza Agricultural Research Station, Agriculture Research Centre, Egypt. The objective of this work was to evaluate the effect of irrigation intervals (25, 35 and 45) on the straw, seed, oil, fiber yields and quality of flax cultivars (Sakha1, Giza9 and Giza10). Irrigation intervals significantly influenced all studied traits except oil percentage. Irrigated flax plants every 35 days gave the maximum values for all traits, while irrigation every 45 days gave the minimum values. In respect to cultivars, significant differences were found in most yield and quality characters. Furthermore, the performance of Sakha 1 cultivar was superior in main stem diameter, biological, straw yields per faddan, seed index, seed, oil yields per faddan and oil percentage. Meanwhile, Giza 10 cultivar highly significantly out yielded Giza9 and Sakha1 in plant height, fiber fineness, fiber length, total fiber percentage and fiber yield per faddan. The interactions between irrigation intervals and flax cultivars were highly significant for all traits. Based on the results, Sakha1 cultivar recorded the maximum values for main stem diameter, biological, straw yields per faddan, seed, oil yields per faddan and oil percentage and Giza 10 recorded the maximum values for plant height, fiber fineness, fiber length, total fiber percentage and fiber yield per faddan under irrigation of plants every 35 days.

Keywords: flax, fiber, irrigation intervals, oil, straw, seed yield

1. Introduction

Flax (*Linum usitatissimum* L.), is one of the most versatile and useful crops and has been grown for thousands of years (Genser & Morris, 2003). Its oil types known as linseed considered one of the most important oil crops for the extraction of oil. About 80% of the linseed oil goes for industrial purpose and the remaining 20% is used for edible purposes (Asgharipour & Rafiei, 2010). It is a multi-purpose crop with benefits extending to both human and animal nutrition (Szilgyi, 2003). This reflects its very high content of essential fatty acids (EFAs), a high percentage of dietary fiber, and the highest level of “lignans” from any plant or seed products used for human food (Pandey & Agarwal, 1998). Flax is considered one of the most important dual purpose crops for oil and fiber production in Egypt and also in the world, rich in oil (41%), protein (20%), and dietary fiber (28%), (Bakry et al., 2012). Flax is an important economic crop which plays a role in our policy through its local fabrication as well as exportation. Although, the cultivated area in Egypt is relatively small and decreased dramatically in last decade, great reduction had happened in flax cultivated area (Aermae, 2007).

Many researchers found significant differences among the fiber, dual purpose and oil types of flax such as: El-Refaey et al. (2015), where they found that, Giza 10 cultivar (fiber type) gave the highest values for plant height, technical stem length, fiber fineness, fiber length, total fiber percentage and fiber yield per faddan compared with other dual purpose and oil types cultivars. El-Seidy et al. (2015) also inferred that, Line 22 (oil type) gave the highest values for number of fruiting branches per plant, number of capsules per plant, number of seeds per capsule, fertility percentage, seed yield per faddan, oil percentage and oil yield per faddan compared with other dual purpose and fiber types cultivars. In addition, Bauer et al. (2015) mentioned that, the higher fiber contents along with the higher straw yields resulted in the fiber-type cultivars yielding 60-70% more fiber than the seed-type cultivar. However, the seed-type cultivar had higher seed weight and seed yield than the fiber-type cultivar. Flax cultivars significantly differed in yield and its attributes (El-Kady et al., 1995; Abo-Zaid, 1997). Many investigators reported significant differences among flax cultivars concerning straw, seed, oil and fiber yields and their components (El-Hariri et al., 2004; El-Hariri et al., 2012; Barky et al., 2014).

Concerning to irrigation intervals, Chorumale et al. (2001) and Yenpreddiwar et al. (2007) mentioned that, two irrigations applied at flowering and capsule filling stages significantly increased the yield attributes, yield, oil content and oil yield of flax compared with no irrigation and irrigation at flowering stage only. In addition, Sharma et al. (2012) indicated that, irrigation at both 30 and 60 days after sowing (DAS) produced the highest values of growth characters compared with irrigation at 30 DAS only. Lisson and Mendham (2000) found that, irrigation increased flax straw and seed yield when precipitation was low and with poor distribution. Bauer and Frederick (1997) conducted a two-year study on flax in adjacent irrigated and rainfed areas and found the irrigated flax had approximately 1000 kg ha^{-1} higher straw yield.

It is necessary to increase flax productivity per unit area which could be achieved by using high yielding cultivars and improving the agricultural treatments (El-Sahrawi et al., 2008; Barky et al., 2013; El-Hariri et al., 2012). Therefore, in the recent years many efforts were devoted to increase the productivity of the flax through improving genetic traits and use of improved cultivars which have high yields and high water use efficiency. The cultivars and irrigation intervals were considered two of the main factors that affecting directly the growth and productivity of flax plants. With keeping the above points in view, the objective of this research were to evaluate the performance and response of some flax cultivars under different irrigation intervals.

2. Material and Methods

2.1 Plant Material and Procedures

The present investigation was carried out at El-Gemmeiza Agriculture Research Station, Gharbiua Governorate, Egypt during the two successive winter seasons of 2013/14 and 2014/15 to study the performance and response of some flax cultivars to different irrigation intervals and their influences on straw, seed, oil and fiber yields and its components and quality attributes for these cultivars. The preceding crop was maize (*Zea mays*) in both seasons. The soil texture was clay loam, pH 7.7 and EC 0.40 in both seasons. Precipitation in the experiments area-generally didn't exceed potential evapotranspiration for most crops. As a result, all crops in this area get its needs of water by irrigation, especially surface irrigation. Worthy to mention that, average of precipitation for this experimental location was about 42 mm/m^2 in the winter seasons.

Field experiment was carried out each season using a split-plot design with three replications, and the irrigation intervals were plotted in the main plots. Three irrigation intervals were applied as follows: (25, 35 and 45 days) started after the first irrigation, which means (5, 4 and 3 irrigation times) taking into consideration the sowing and first irrigations. These irrigation intervals are frequently used by farmers under surface irrigation system in the fields of the surrounding areas. However, flax cultivars (Sakha 1, Giza 9 and Giza 10) were arranged in the sub-plots, where each sub-plot size was 6 m^2 (2 m × 3 m) or (1/700 fad.). The two experiments were sown on 27 and 29 October in the first and second season, respectively. Calcium super phosphate was added at the rate of 100 kg/faddan (15.5% P_2O_5) before sowing and potassium sulphate (48% K_2O) was applied in one dose at the rate of 50 kg/faddan before sowing in both seasons. Nitrogen was added to the sub-plots at the rate of 60 kg N/faddan in the form of urea (46% N) in two equal doses, the first half was added before the first irrigation and the second one was added before the second irrigation. Other recommended cultural practices for growing flax were done as usual in the area.

2.2 Sampling and Measurement

Ten individual plants were chosen randomly at full maturity for each sub-plot to record plant height (cm), main stem diameter (mm) and seed index (gm), while other traits were taken from whole plants of the plot.

2.2.1 Straw Yield and Its Related Characters

Plant height (cm): the distance from the cotyledonary node to the top of plant, main stem diameter (mm): at the middle region to the nearest 0.1 mm by using biocles, biological yield per faddan (ton) (faddan = 4200 m^2): estimated all plants from the sub-plots and converted to record biological yield per faddan before removing the capsules, and straw yield per faddan (ton): estimated from the sub-plots and converted to record straw yield per faddan after removing the capsules.

2.2.2 Seed Yield and Its Related Characters

Seed index (gm): It was determined as the average weight of 1000 seeds obtained from each sub-plot, seed yield per faddan (kg): seed yield of an area of each plot (6 m^2) was estimated and transformed to kg per faddan, oil percentage (%): was determined as described by the (AOAC, 1990) methods, using petroleum ether (40-60 °C) in Soxhlet apparatus and oil yield per faddan: was calculated from the following formula: Oil yield per faddan = oil percentage × seed yield per faddan

2.2.3 Fiber Yield and Its Technological Characters

Fiber Fineness (N.m) determined using Radwan and Momtaz (1966) method according to the following equation:

$$N.m = (N \times L)/W \tag{1}$$

Where, N.m = Metrical number, N = Number of fibers (20 fibers and the length for each one = 10 cm), L = Length of fibers in mm (10 cm), and W = Weight of fibers in mg.

Fiber yield per faddan (Kg): calculated from plot fiber yield, Fiber length (cm): Ten fiber ribbons from each treatment were spreaded out and each ribbon was measured then the average fiber length was recorded, Total fiber percentage: It was calculated from the following formula:

$$\text{Total fiber percentage} = \frac{\text{Fiber yield}}{\text{Straw yield after retting}} \times 100 \tag{2}$$

2.3 Statistical Analysis

Using Michigan State University Computer Statistical Package (MSTATC), the analysis of variance was used for the two experiments according to Snedecor and Cochran (1982). The data was statistically analyzed for each season and the homogeneity of experimental error in both seasons was tested, then the combined analysis of data was performed for the characters over two seasons (Le Clerg et al., 1962) to present the main factors and its first and second order interactions. The least significant difference (LSD) test at 0.05 and 0.01 levels of significance was used to indicate mean comparison.

3. Results and Discussions

3.1 Straw Yield and Its Related Characters

The analysis of variance for the combined data with regard to plant height, main stem diameter, biological and straw yields per faddan showed highly significant differences among the three tested irrigation intervals as has presented in Table 1. It was observed that, irrigation every 35 days (4 irrigation times during the growth season) increased plant height, main stem diameter and straw yield per faddan and recorded the highest values for these traits followed by the irrigation every 25 days (5 irrigation times) which ranked the second. While the lowest values for all above mentioned traits were recorded by irrigation every 45 days (3 irrigation times). The only noticeable change in the data direction was irrigation every 25 days (5 irrigation times) recorded the highest values of biological yield per faddan with no significant difference with irrigation every 35 days (4 irrigation times). These results may be due to the reduction of plant growth which occurs as results for some responses to water deficit (irrigation every 45 days) from one side, and excess moisture (irrigation every 25 days) from the other side.

Table 1. Means of straw yield and yield traits as affected by irrigation intervals (combined data)

Irrigation intervals	Plant height (cm)	Main stem diameter (mm)	Biological yield (ton)/faddan	Straw yield (ton)/faddan
I1 every 25 days	96.74	0.9150	4.710	3.848
I2 every 35 days	101.4	0.9672	4.700	4.244
I3 every 45 days	85.11	0.7744	3.821	3.342
$LSD_{0.01}$	0.3203	0.001118	0.07909	0.06126

It is known that, there is a decrease in the photochemical activity of photosynthesis, rubisco enzyme activity and the accumulation of secondary metabolites when plants are under water deficit stress as well as abscisic acid and solutes increase due to water reduction. These factors reduce stomatal conductance and consequently photosynthetic activity which ultimately resulted in a reduction in the synthesis of proteins and cell walls, as well as a decrease in the rate of cell expansion (Chavarria & dos Santos, 2012). On the other hand, flax plant has a shallow rooting system and needs adequate water in the 0-10 cm soil layer (Wood, 1997), and under excess moisture conditions, oxygen was often in short supply, normal exchange of gases from roots to soil was frequently disturbed and altered plant metabolism. Moreover, lodging was more pronounced under excess water conditions especially with high sowing rates. The sum of these responses contributes to explain the reduction of plant growth under excess moisture conditions. These results are in harmony with those obtained by Gabiana (2005) when he reported that, straw and total dry matter production was increased by irrigation compared with rainfed plants. Islam et al. (2011), Ahmadizadeh (2013), El sabagh et al. (2015a, 2015b), Abd El-Wahed et al. (2015a), and Barutçular et al. (2016) revealed that, drought stress causes a reduction in plant growth. De Carvalho et al. (2005) observed an increase in the plants height with increasing the water availability. Taiz and Zeiger (2006) stated that, lesser water availability tends to present lesser plant height, because the water restriction can affect the metabolic processes of growth. Flax is sensitive to water shortage and plant height reduced due to lack of water (Ahlawat & Gangaiah, 2009), Mirshekari et al. (2012) mentioned that, the highest plant height, biological yield and harvest index of flax were obtained from control irrigation treatment, while limited irrigation stress resulted in the lowest these traits.

In this study, the analysis of variance for the combined data showed highly significant differences among the three tested flax cultivars Sakha 1, Giza 9 and Giza 10 (Table 2). It was observed that, Giza 10 cultivar gave the tallest plants and recorded the highest values of plant height. However, Sakha 1 cultivar ranked the first and recorded the highest values of main stem diameter, biological and straw yields per faddan, followed by Giza 10 only in main stem diameter trait, while Giza 9 ranked the second in biological and straw yields per faddan. The present results are mainly due to the genetic differences and potentiality between the fiber and dual-purpose cultivars of flax. Sakha 1 (dual-purpose type) recorded the thicker stem diameter and surpassed other cultivars in seed yield and consequently surpassed it in biological and straw yields. However, Giza 10 (fiber type) gave the tallest plants.

Table 2. Means of straw yield and yield traits as affected by flax cultivars (combined data)

Cultivars	Plant height (cm)	Main stem diameter (mm)	Biological yield (ton)/faddan	Straw yield (ton)/faddan
Sakha 1	90.68	1.047	4.960	4.294
Giza 9	91.44	0.8017	4.289	3.827
Giza 10	101.1	0.8078	3.981	3.312
$LSD_{0.01}$	0.2064	0.0009323	0.05896	0.05106

These results are in good agreement with those obtained by Assar (2008), El-Refaey et al. (2010) and El-Refaey et al. (2015), where they indicated that, fiber types were superior in total height compared with dual-purpose and oil types. However, dual and oil types surpassed fiber types in main stem diameter, biological and straw yields per faddan. El-Refaey et al. (2009) also reported that, ideal fiber types are characterized by the thin stem diameter.

The effect of the first order interactions between (seasons × irrigation intervals), (seasons × flax cultivars) and (irrigation intervals × flax cultivars) on straw yield and its traits are presented in Figure 1. The interaction between seasons and irrigation intervals had only significant effect on main stem diameter and straw yield per faddan. Moreover, the interaction between seasons and flax cultivars had highly significant effect on main stem diameter. While, the effect of the interaction between irrigation intervals and flax cultivars on plant height, main stem diameter biological and straw yield per faddan was highly significant and Giza 10 cultivar recorded the highest plant height under irrigation every 35 days. Whereas, Sakha 1 cultivar recorded the highest main stem diameter, biological and straw yields per faddan under irrigation every 35 days.

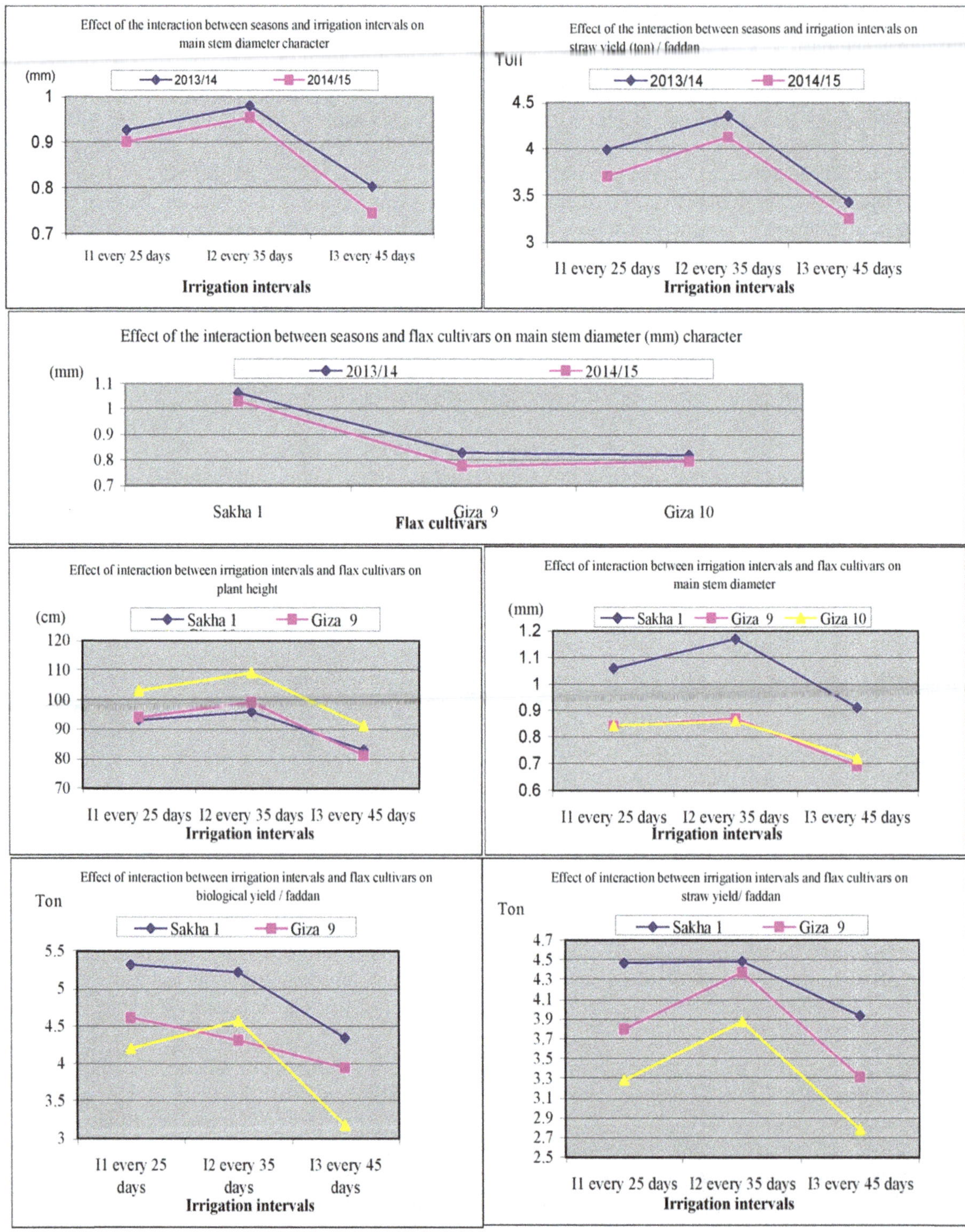

Figure 1. Effects of the first order interactions (seasons × irrigation intervals), (seasons × flax cultivars) and (irrigation intervals × flax cultivars) on straw yield and its traits

The effect of the second order interaction (seasons × irrigation intervals × flax cultivars) on plant height, main stem diameter and straw yield per faddan was highly significant (Table 3), and Giza 10 cultivar recorded the highest plant height under irrigation every 35 days in the first season. Sakha 1 cultivar recorded the highest main stem diameter and straw yield per faddan under irrigation every 35 days in the first season.

These results clearly indicated that, the cultivars significantly differed in their responses to the irrigation intervals. Results revealed that, the variability among tested flax cultivars which may be expected due to the differences of these cultivars in origin, growth habit and genetic constituent and the environmental conditions.

Table 3. Interaction effects among seasons, irrigation intervals and flax cultivars on straw yield and is traits

Seasons	Irrigation intervals	Flax cultivars		
		Sakha 1	Giza 9	Giza 10
Plant height (cm)				
2013/14	I1 every 25 days	93.60	94.70	103.6
	I2 every 35 days	96.50	99.22	109.7
	I3 every 45 days	83.35	81.60	91.60
2014/15	I1 every 25 days	92.60	93.36	102.6
	I2 every 35 days	95.41	99.02	108.7
	I3 every 45 days	82.62	80.77	90.70
$LSD_{0.01}$		0.5055		
Main stem diameter (mm)				
2013/14	I1 every 25 days	1.070	0.8633	0.8500
	I2 every 35 days	1.190	1.190	0.8600
	I3 every 45 days	0.9300	0.7300	0.7500
2014/15	I1 every 25 days	1.050	0.8200	0.8367
	I2 every 35 days	1.150	0.8500	0.8633
	I3 every 45 days	0.8933	0.6567	0.6867
$LSD_{0.01}$		0.02284		
Straw yield per faddan (ton)				
2013/14	I1 every 25 days	4.588	3.931	3.451
	I2 every 35 days	4.693	4.465	3.923
	I3 every 45 days	3.945	3.421	2.923
2014/15	I1 every 25 days	4.344	3.664	3.112
	I2 every 35 days	4.274	4.280	3.828
	I3 every 45 days	3.922	3.203	2.636
$LSD_{0.01}$		0.1251		

3.2 Seed Yield and Its Related Characters

The analysis of variance for the combined data with regard to seed index (gm), seed yield per faddan, oil percentage and oil yield per faddan showed highly significant differences among the three irrigation intervals (Table 4). Irrigation every 35 days (4 irrigation times during the growth season) was achieved the highest values for all previous mentioned traits without significant differences with both irrigation every 25 days (5 irrigation times) for seed index trait and irrigation every 45 days (3 irrigation times) for oil percentage. In addition, irrigation every 25 days (5 irrigation times) ranked the second, while irrigation every 45 days (3 irrigation times) recorded the lowest values and ranked the last for seed yield and oil yield per faddan. The best result for all these traits were obtained from irrigation every 35 days (4 irrigation times). It could be ascribed to favorable moisture conditions resulting from irrigations at critical physiological stages of initiation of flowering and seed filling. Adequate moisture may also increase photosynthesis which is responsible for carbohydrate formation, seed filling and final seed yield. On the other hand, it appears that water stress hampered flowering and reduced the probability of developing flower to capsule and its occurrence during flowering and capsule formation resulted in capsule abortion.

Table 4. Means of seed yield and yield traits as affected by irrigation intervals (combined data)

Irrigation intervals	Seed index (gm)	Seed yield per faddan (Kg)	Oil percentage (%)	Oil yield per faddan (Kg)
I1 every 25 days	6.744	364.4	32.55	119.3
I2 every 35 days	6.746	406.1	33.04	134.4
I3 every 45 days	5.611	327.1	32.90	107.7
$LSD_{0.01}$	0.0353	3.916	0.2694	1.207

These results are in agreement with those obtained by Gabiana (2005), when he found a significant positive effect on seed yield attributed to higher capsule numbers and more seeds per capsule under adequate water conditions. Meanwhile, flax plants subjected to water stress performed poorly in biomass production and yield. Mirshekari et al. (2012) mentioned that, the highest number of primary branches per plant, capsules number per plant, seed numbers per capsule, seed and oil yields per faddan and oil percentage of flax seed were obtained from control irrigation treatment, while limited water stress during flowering and seed filling stages resulted in lowest these traits. Bauer et al. (2015) inferred that, irrigation significantly increased seed weight, although it did not significantly impact on flax seed yield and the average of seed yield under irrigation was higher than rainfed condition.

The analysis of variance for the combined data with regard to seed index (gm), seed yield per faddan, oil percentage and oil yield per faddan as affected by the three tested flax cultivars Sakha 1, Giza 9 and Giza 10 are presented in Table 5. A highly significant variation was found among cultivars, and Sakha 1 cultivar recorded the highest values for all these traits, followed by Giza 9 cultivar with no significant differences between them only for oil percentage. However, the lowest values were recorded by Giza 10. These differences among the tested cultivars could be mainly attributed to the differences in their genetical construction and their response to the environmental conditions. The marked differences in seed index are mainly due to the differences in genetical make up of the tested cultivars. In addition, the variations among cultivars in seed yield per faddan and oil percentage are mainly due to the genetical variation possessed by the tested cultivars, and the increment in seed yield per faddan for Sakha 1 is attributed to the increase of number of capsules per plant, number of seeds per capsule (unpublished data) and seed index. Consequently, the increment in oil yield per faddan for Sakha 1 is attributed to the increment in seed yield per faddan and oil percentage.

Table 5. Means of seed yield and yield traits as affected by flax cultivars (combined data)

Cultivars	Seed index (gm)	Seed yield per faddan (Kg)	Oil percentage (%)	Oil yield per faddan (Kg)
Sakha 1	7.741	545.7	33.19	181.1
Giza 9	5.824	284.7	33.15	94.34
Giza 10	5.536	267.1	32.15	86.00
$LSD_{0.01}$	0.02948	5.719	0.07800	2.003

Such results are in harmony with those reported by Zahana et al. (2003), El-Azzouni et al. (2003), Kineber (2004) and Abou-Zaied and Mousa (2007), where they found that, Sakha 1 flax cultivar significantly produced the highest values of capsules number per plant, seed index, seed yield per faddan, oil percentage and oil yield per faddan compared with other tested dual-purpose and fiber cultivars. El-Seidy et al. (2010) and El-Seidy et al. (2015) reported that, flax oil types realized the highest seed yield per faddan, oil percentage and oil yield per faddan followed by dual purpose types, while the lowest values were recorded by fiber types. Bauer et al. (2015) concluded that, the seed-type cultivar had higher seed weight and seed yield than the fiber-type cultivar. The differences between the tested cultivars could mainly be attributed to the differences in their genetical constitution and their response to the environmental conditions. In this connections, many investigators obtained higher levels of varietal differences in yield and its components in many regions (Kineber et al., 2006; El-Kady Eman & Abd El-Fatah, 2009; Abd El-Wahed et al., 2015b; EL Sabagh et al., 2016a, 2016b).

The effect of the first order interactions (seasons × irrigation intervals) and (irrigation intervals × flax cultivars) on seed yield and its traits are presented in Figure 2. The interaction between seasons and irrigation intervals had only significant effect on seed yield per faddan and oil yield per faddan. While, the effect of the interaction between irrigation intervals and flax cultivars had highly significant effect on all seed traits, and Sakha 1 cultivar recorded the highest value for seed index under irrigation every 25 days. Moreover, the same cultivar recorded the highest values for oil percentage, seed yield per faddan and oil yield per faddan under irrigation every 35 days, with no significant differences with irrigation every 25 days for the last two mentioned traits.

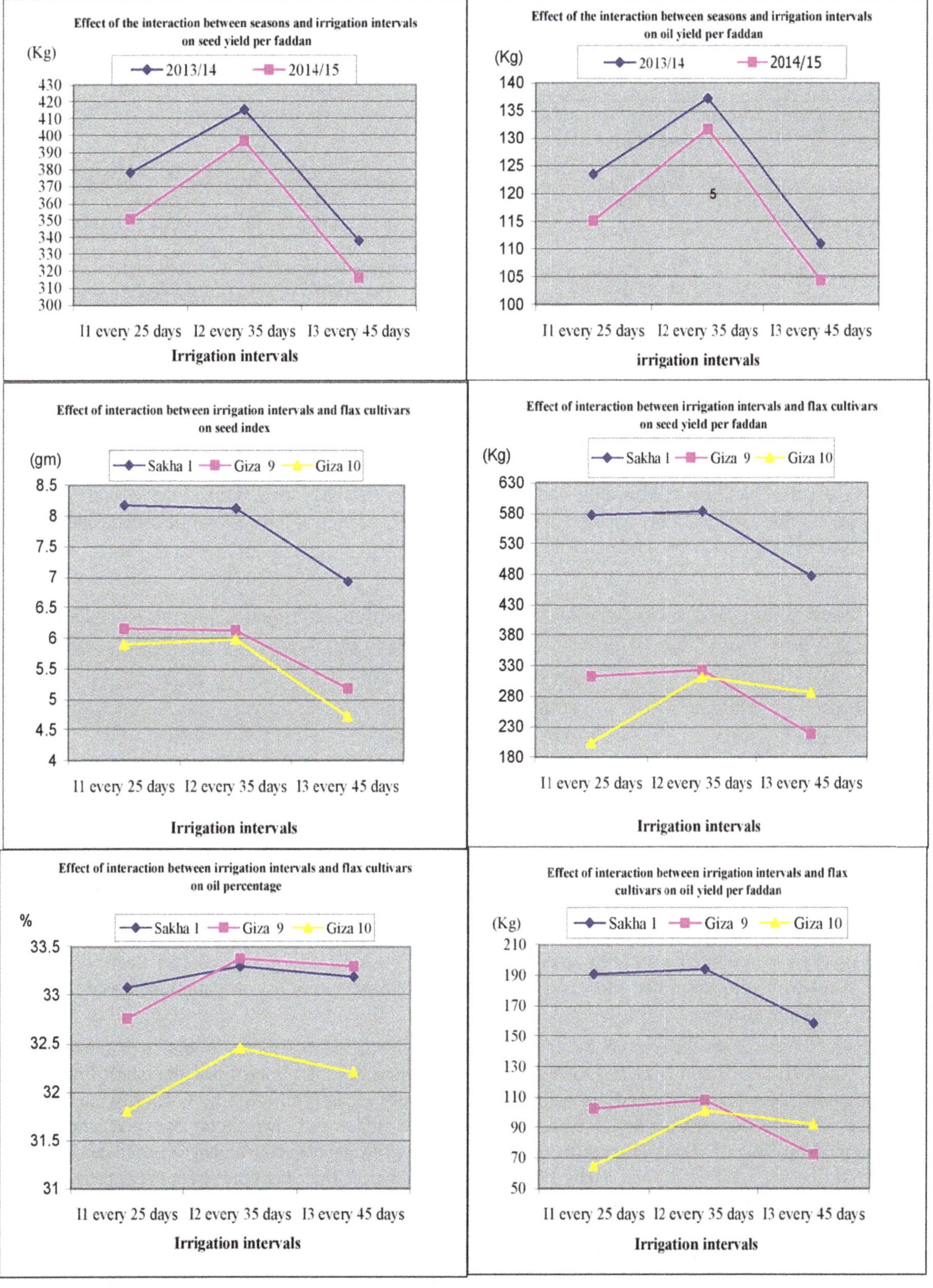

Figure 2. Effects of the first order interactions (seasons × irrigation intervals) and (irrigation intervals X flax cultivars) on seed yield and its traits

The effect of the second order interaction (seasons × irrigation intervals × flax cultivars) on seed yield per faddan was only significant (Table 6), while Sakha 1 cultivar recorded the highest value under both irrigation every 35 days and every 25 days in the first season.

Table 6. Interaction effects among seasons, irrigation intervals and flax cultivars on seed yield (per faddan trait)

Seasons	Irrigation intervals	Flax cultivars		
		Sakha 1	Giza 9	Giza 10
Seed yield per faddan (Kg)				
2013/14	I1 every 25 days	588.0	333.0	213.0
	I2 every 35 days	594.3	329.7	322.3
	I3 every 45 days	489.3	228.3	296.7
2014/15	I1 every 25 days	566.0	293.0	193.3
	I2 every 35 days	571.3	317.0	301.7
	I3 every 45 days	465.3	207.3	275.7
$LSD_{0.05}$		10.34		

3.3 Fiber Yield and Its Technological Characters

The analysis of variance for the combined data with regard to fiber fineness, fiber length, total fiber percentage and fiber yield per faddan showed highly significant differences among the three tested irrigation intervals (Table 7). Irrigation every 35 days gave the finest fiber, the tallest fiber length, the highest total fiber percentage and the highest fiber yield per faddan, followed by irrigation every 25 days which ranked the second. While the most coarseness fiber, shortest fiber length, lowest fiber yield per faddan and lowest total fiber percentage values were obtained by irrigation every 45 days (3 irrigation times).

Table 7. Means of fiber yield and its technological characters as affected by irrigation intervals (combined data)

Irrigation intervals	Fiber fineness (N.m)	Fiber length (cm)	Total fiber percentage (%)	Fiber yield per faddan (Kg)
I1 every 25 days	319.6	84.05	16.18	452.4
I2 every 35 days	329.2	87.36	17.04	515.6
I3 every 45 days	302.2	73.57	14.66	332.6
$LSD_{0.01}$	2.285	1.532	0.08664	9.345

The trend of these data was similar to the trend of straw yield and its related characters, which emphasize that, straw yield and plant height especially technical stem length (a long unbranched stem which contains the most high quality fiber) define the fiber biomass (Sankari, 2000). Similar trends were also stated by Elhaak et al. (1999) as a high percentage of valuable long fibers were a function of high soil moisture and fertility especially of N, P, K and Mg. Gabiana (2005) also mentioned that, the fiber yield per plant was three times more in irrigated plots than in rainfed plants. In other words, water stressed plants produced less fiber compared to irrigated plants with adequate water. Moreover, Bauer et al. (2015) reported that, irrigation increased fiber yield compared to rainfed plants.

The analysis of variance for the combined data showed highly significant differences among the three tested flax cultivars: Sakha 1, Giza 9 and Giza 10 (Table 8). Giza 10 cultivar gave the finest fiber, the tallest fiber length, the highest total fiber percentage and the highest fiber yield per faddan, followed by Giza 9, while the most coarseness fiber, the shortest fiber length, lowest fiber yield per faddan and lowest total fiber percentage values were obtained by Sakha 1. Such differences could be attributed to genetic constituents of cultivars, whereas, Giza 10 is a fiber type and Giza 9 and Sakha 1 are dual-purpose types. The fiber types gave the highest values of fiber percentage which were proportionally with the loss in straw yield per faddan after retting. The loss in straw yield per faddan after retting for dual-purpose (Giza 9 and Sakha 1) was higher than the other fiber type. These results are in agreement with these obtained by Abd El-Fatah (2007) and El-Refaey et al. (2010) and they reported that, fiber types exceeded dual and oil types in fiber fineness and it had superiority for fiber length character and recorded the highest fiber percentage and fiber yield per faddan. El-Azzouni and Zedan (2009) opined that, Giza 10 surpassed all other tested cultivars in fiber fineness. These results are in the same line with these obtained by Bauer et al. (2015) as they mentioned that, the higher fiber contents along with the higher straw yields resulted in the fiber-type cultivars yielding 60-70% more fiber than the seed-type cultivar.

Table 8. Means of fiber yield and its technological characters as affected by flax cultivars (combined data)

Cultivars	Fiber fineness (N.m)	Fiber length (Cm)	Total fiber percentage (%)	Fiber yield per faddan (Kg)
Sakha 1	280.4	78.97	10.95	359.2
Giza 9	322.9	81.93	17.99	444.2
Giza 10	347.6	84.08	18.94	497.3
L.S.D $_{0.01}$	2.132	1.208	0.08339	6.611

The effect of the first order interactions (seasons × irrigation intervals), (seasons × flax cultivars) and (irrigation intervals × flax cultivars) on fiber yield and its technological characters are presented in Figure 3. The interaction between seasons and irrigation intervals had highly significant effect on total fiber percentage, while it had only significant effect on fiber yield per faddan. Moreover, the interaction between seasons and flax cultivars had highly significant effect on total fiber percentage and fiber yield per faddan. In addition, the effect of the interaction between irrigation intervals and flax cultivars was highly significant on all fiber traits and Giza 10 cultivar recorded the highest values for all these traits under irrigation every 35 days.

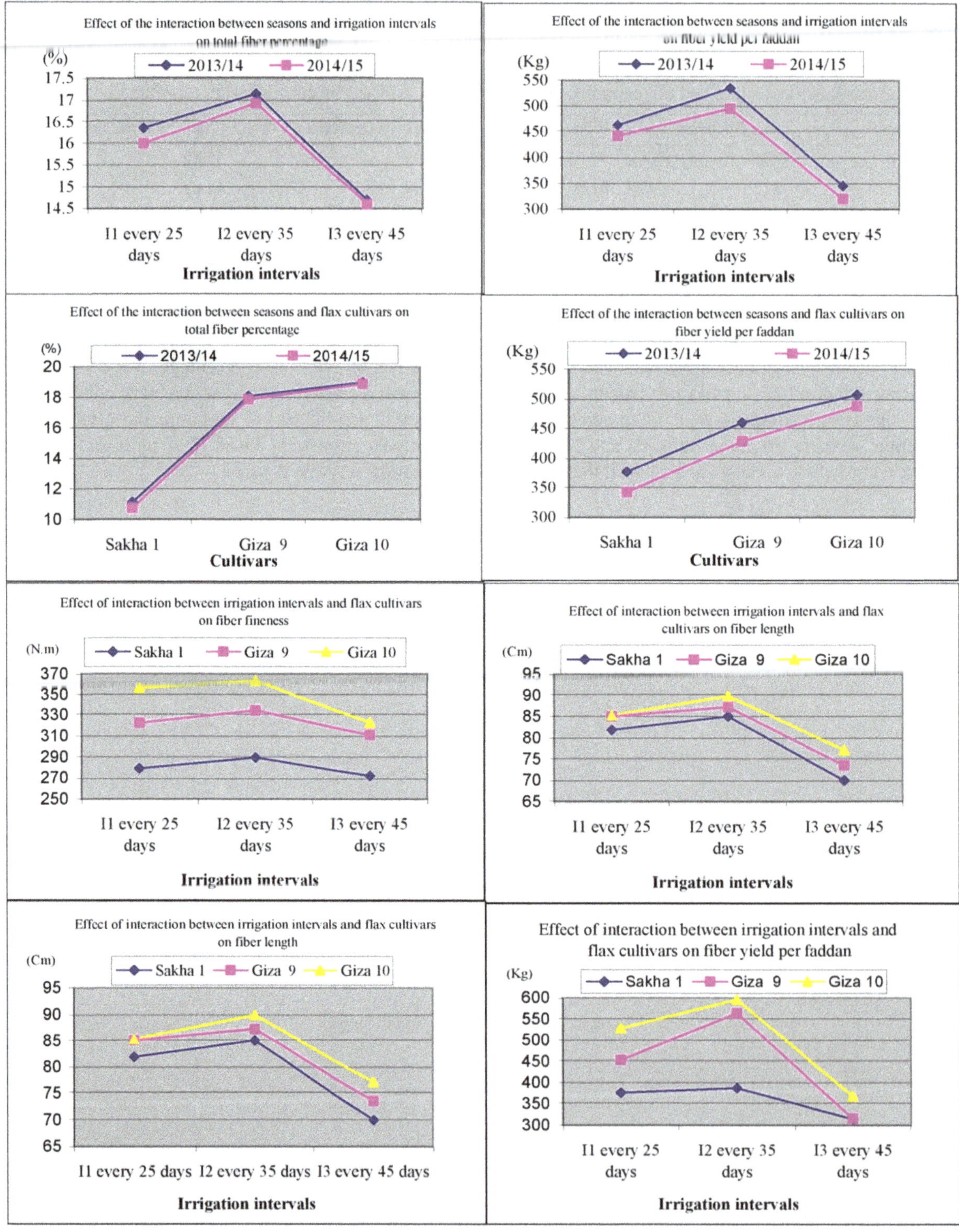

Figure 3. Effects of the first order interactions (seasons × irrigation intervals), (seasons × flax cultivars) and (irrigation intervals × flax cultivars) on fiber yield and its technological characters

The effect of the second order interaction (seasons × irrigation intervals × flax cultivars) on fiber fineness, total fiber percentage and fiber yield per faddan was highly significant and Giza 10 cultivar recorded the highest values for all these traits under irrigation every 35 days in the first season (Table 9).

These results clearly indicated that, the cultivars significantly differed in their responses to the irrigation intervals and the variability among tested flax cultivars which may be expected due to the differences of these cultivars in genetic constituent and the environmental conditions (Khalifa et al., 2011; El-Hariri et al., 2012).

Table 9. Interaction effects among seasons, irrigation intervals and flax cultivars on fiber yield and its technological characters

Seasons	Irrigation intervals	Flax cultivars		
		Sakha 1	Giza 9	Giza 10
Fiber fineness (N.m)				
2013/14	I1 every 25 days	280.5	324.9	357.9
	I2 every 35 days	292.5	332.2	363.9
	I3 every 45 days	273.5	312.5	320.6
2014/15	I1 every 25 days	278.1	320.9	355.1
	I2 every 35 days	286.8	337.4	362.6
	I3 every 45 days	271.1	309.8	325.4
$LSD_{0.01}$		5.222		
Total fiber percentage (%)				
2013/14	I1 every 25 days	11.39	18.21	19.50
	I2 every 35 days	11.64	19.69	20.10
	I3 every 45 days	10.36	16.38	17.36
2014/15	I1 every 25 days	10.84	17.73	19.44
	I2 every 35 days	11.23	19.52	20.04
	I3 every 45 days	10.22	16.39	17.23
$LSD_{0.01}$		0.2043		
Fiber yield per faddan (Kg)				
2013/14	I1 every 25 days	391.1	459.4	538.5
	I2 every 35 days	411.1	593.9	600.8
	I3 every 45 days	327.2	327.2	381.4
2014/15	I1 every 25 days	361.1	446.5	517.9
	I2 every 35 days	363.6	533.6	590.4
	I3 every 45 days	300.9	304.3	354.6
$LSD_{0.01}$		16.19		

4. Conclusion

From this study, it can be concluded that, irrigation intervals have a significant role on flax production and irrigation every 35 days (4 irrigation times during the growing season) performed the best traits. Among the cultivars, Giza 10 cultivar was superior in plant height, fiber fineness, fiber length, total fiber percentage and fiber yield per faddan traits. Meanwhile, Sakha 1 cultivar highly significantly surpassed other cultivars in main stem diameter, biological, straw yields per faddan, seed index, seed, oil yields per faddan and oil percentage.

References

Abd El-Fatah, A. A. E. (2007). Comparative study on some flax cultivars. *J. Agric. Sci. Mansoura Univ., 71*(9), 11-19.

Abd El-Fatah, A. A. E. (2007). Comparative study on some flax cultivars. *J. Agric. Sci. Mansoura Univ., 32*(9), 7111-7119.

Abd El-Wahed, M. H., El-Sabagh, A., Zayed, A., Sanussi, A., Saneoka, A., & Barutçular, C. (2015a). Improving yield and water productivity of maize grown under deficit-irrigated in dry area conditions. *Azarian Journal of Agriculture, 2*(5), 123-132.

Abd El-Wahed, M., El-Sabagh A., Saneoka, H., Abdelkhalek, A., & Barutçular, C. (2015b). Sprinkler irrigation uniformity and crop water productivity of barley in arid region. *Emirates J. of Food and Agriculture, 27*(10), 770-775.

Abo Zaied, T. A. (1997). *Comparative study of yield technological characters of some flax varieties* (PhD Thesis). Fac. of Agric. Mansoura Univ., Egypt.

Abou-Zaied, T. A., & Mousa, A. M. (2007). Effect of different NPK treatments on yield and yield components of two flax varieties. *J. Agric. Sci., Mansoura Univ., 32*(10), 8057-8064.

Aermae. (2007). *Agriculture Economic Report*. Ministry of Agriculture, Egypt.

Ahlawat, I. P. S., & Gangaiah, B. (2009). Effect of configuration and irrigation on sole and linseed (*Linum usitatissimum*) intercropped chickpea (*Cicre artietinum*). *Indian J. Agril. Sci., 80*(3), 250-253.

Ahmadizadeh, M. (2013). Physiological and Agro-Morphological Response to Drought Stress. *Middle-East J. Scientific Res., 13*(8), 998-1009.

Alessi, J., & Power, J. F. (1970). Influence of row spacing, irrigation, and weeds on dryland flax yield, quality, and water use. *Agron. J., 62*, 635-637. http://dx.doi.org/10.2134/agronj1970.00021962006200050026x

AOAC. (1990). *Official Methods of Analysis of the Association of Official Analytical Chemists* (15th ed.). Published by Association of Official Analytical Chemists, Arlington, Virginia, U.S.A.

Asgharipour, M., & Rafiei, M. (2010). Intercropping of Isabgol (*Plantago ovata* L.) and Lentil as Influenced by Drought Stress. *American-Eurasian Journal of Sustainable Agriculture, 4*(3), 341-348.

Assar, M. B. (2008). *Evaluation of some flax genotypes under different conditions* (M.Sc. thesis, p. 86). Fac. Agric., Tanta Univ., Egypt.

Bakry, A. B., Elewa, T. A., & Ali, O. A. M. (2012). Effect of Fe foliar application on yield and quality traits of some flax varieties grown under newly reclaimed sandy soil. *Australian J. Basic and Applied Sci., 6*(7), 532-536.

Barky, A. B., El-Hariri, D. M., Mervat Sh, S., & El-Bassiouny, H. M. S. (2012). Drought stress mitigation by foliar application of salicylic acid in two linseed varieties grown under newly reclaimed sandy soils. *Journal of Applied Sciences Research, 8*(7), 3503-3514.

Barky, A. B., Ibrahim, O. M., Elewa, T. A., & El-Karamany, M. F. (2014). Performance Assessment of Some Flax (*Linum usitatissimum* L.) Varieties Using Cluster Analysis under Sandy Soil Conditions. *Agricultural Sciences, 5*, 677-686. http://dx.doi.org/10.4236/as.2014.58071

Barutçular, C., Yıldırım, M., Koç M., Akıncı, C., Toptaş, I., Albayrak, O., ... El-Sabagh, A. (2016). Evaluation of SPAD chlorophyll in spring wheat genotypes under different environments. *Fresenius Environmental Bulletin, 25*(4), 1258-1266.

Bauer, P. J., & Frederick, J. R. (1997). Winter crop effect on double-cropped cotton grown with and without irrigation. In R. N. Gallaher & R. McSorley (Eds.), *Proc. of the 20th Ann. Southern Conservation Tillage Conf. for Sustainable Agric., June 24-26, 1997, University of Florida Special Series SS-AGR-60* (pp. 220-222).

Bauer, Ph. J., Stone, K. C., Foulk, J. A., & Dodd, R. B. (2015). Irrigation and cultivar effect on flax fiber and seed yield in the Southeast USA. *Industrial Crops and Products, 67*, 7-10. http://dx.doi.org/10.1016/j.indcrop.2014.12.053

Chavarria, G., & Dos Santos, H. P. (2012). Plant Water Relations: Absorption, Transport and Control Mechanisms, Advances in Selected Plant Physiology Aspects. In G. Montanaro (Ed.), *InTech.*

Chorumale, P. B., Dahatonde, B. N., & Vyas, J. S. (2001). Response of linseed to nitrogen under varied moisture regimes. *Ann. Plant Physiol., 13*(2), 192-194.

De Carvalho, G., De Sousa, B., Dos Santos, U. M., Fernandes, A. V., Barbosa, S., & Buckeridge, M. S. (2005). Growth, photosynthesis and stress indicators in young rosewood plants (*Aniba rosaeodora* Ducke) under different light intensities. *Braz. J. Plant Physiol., 17*, 325-334.

El-Azzouni, A. M. A., & Zedan, S. Z. (2009). Effect of sowing dates, foliar x as balanced compound fertilizer on yield and its components some flax varieties in sandy soil. *J. Agric. Sci., Mansora Univ., 20*(3), 2071-2084.

El-Azzouni, A. M. A., Moawed, E. A., & Salama, S. M. (2003). Effect of seeding rate, potassin fertilizer on some genotypes of flax (*Limun usitatissimum* L.). *J. Agric. Sci., Mansoura Univ., 28*(8), 5887-5902.

Elhaak, M. A., El-Shourbagy, M. N., & El-Nagar, R. (1999). Effect of edaphic factors on technical properties of flax fibre. *J. Agron. Crop Sci., 182*, 113-120. http://dx.doi.org/10.1046/j.1439-037x.1999.00273.x

El-Hariri, D. M., Barky, A. B., Elewa, T. A., & Ibrahim, O. M. (2012). Evaluation of some flax (*Linum usitatissimum* L.) varieties under newly reclaimed sandy soils conditions. *International Journal of Academic Research, 4*(1). 98-102.

El-Hariri, D. M., Hassanein, M. S., & El-Sweify, A. H. (2004). Evaluation of some flax genotypes, straw yield, yield components and technological characters. *J. of Natural Fibers, 1*(2), 697-715. http://dx.doi.org/10.1300/J395v01n02_01

El-Kady Eman, A. E., & Abd El-Fatah, A. A. E. (2009). Comparison of yield, its components physical properties and chemical composition of twelve flax genotypes. *J. Agric. Res. Kafr El-Sheikh Univ., 35*(1), 69-85.

EL-Kady, E. F. A., Shafshak, S. E., Gab-Allah, F. L., & Kineber, M. E. A. (1995). Effect of seeding rates on yield and its components for six promising flax genotypes under saline conditions. *J. Agric. Sci. Mansoura Univ., 20*(2), 593-602.

El-Refaey, R. A., El-Seidy, E. H., Abou-Zaied, T. A., & Ewes, A. A. (2009). *Breeding for the ideal plant type of fiber in flax through line × tester analysis* (pp. 69-84). The Fifth Inter. Conf. Sustain. Agric. Develop., Fac. Agric., Fayoum Univ.

El-Refaey, R. A., El-Seidy, E. H., Abou-Zaied, T. A., & Rashwan, E. A. (2010). *Evaluation of some genotypes of flax (Linum usitatissimum L.) for fiber and its related characters under different plant densities and retting methods* (pp. 165-187). The 12th Conference of Agronomy, Suez Canal Univ., Fac., Environ., Agric. Sci., September 20-22, 2010, El-Arish, Egypt.

El-Refaey, R. A., El-Seidy, E. H., Abou-Zaied, T. A., Abd El-Razek, U. A., & Rashwan, E. A. (2015). Effect of Different Mineral and Biological Nitrogenous Fertilizers Combinations on Straw Yield and Fiber Quality of Some Flax '*Linum usitatissimum* L.' Genotypes Glob. *J. Agric. Food Safety Sci., 2*(3), 346-364.

El-Sabagh, A., Barutçular, C., & Saneoka, H. (2015a). Assessment of drought tolerance maize hybrids at grain growth stage in mediterranean area. *International Journal of Biological, Biomolecular, Agricultural, Food and Biotechnological Engineering, 9*(9), 962-965.

El-Sabagh, A., Omar, A., Barutçular, C., & Saneoka, H. (2016b). Role of integrated use of nitrogen fertilizer sources in improving seed quality of canola (*Brassica napus* L.). *Turkish Journal of Agriculture - Food Science and Technology, 4*(2), 73-78.

El-Sabagh, A., Sorour, S., Morsi, A., Islam, M. S., Ueda, A., Barutcular, C., ... Saneoka, H. (2016a). Role of Osmoprotectants and compost application in improving water stress tolerance in soybean (*Glycine max* L.). *International Journal of Current Research, 8*(2), 25949-25954.

El-Sabagh, A., Sorour, S., Omar, A., Islam, M. S., Ueda, A., Saneoka, H., & Barutçular, C. (2015b). *Soybean (Glycine max L.) Growth Enhancement under Water Stress Conditions* (pp. 148-152). International Conference on Chemical, Agricultural and Biological Sciences, (CABS-2015), Sept. 4-5, 2015, Istanbul, Turkey.

El-Seidy, E. H., El-Refaey, R. A., Abou-Zaied, T. A., & Rashwan, E. A. (2010). *Evaluation of some flax genotypes for seed yield and its related characters under different plant densities* (pp. 691-703). The 12th Conference of Agronomy, Suez Canal Univ., Fac. Environ. Agric. Sci., September 20-22, 2010, El-Arish, Egypt.

El-Seidy, E. H., El-Refaey, R. A., Abou-Zaied, T. A., Abd El-Razek, U. A., & Rashwan, E. A. (2015). Effect of Different Mineral and Biological Nitrogenous Fertilizers Combinations on Seed Yield and its components of Some Flax '*Linum usitatissimum* L.' Genotypes. *Glob. J. Agric. Food Safety Sci., 2*(3), 365-383.

El-Shahaway, T. A., Rookiek, K. G., Balbaa, L. K., & Abbes, S. M. (2008). Micronutrients, B-vitamins and yeast in relation to increasing flax (*Linum usitatissimun* L.). Growth, yield productivity and controlling associated weeds. *Asian Journal of Agricultural Research, 2*, 1-14. http://dx.doi.org/10.3923/ajar.2008.1.14

El-Sweify Amna, H. H., Abd El-Daim, M. A., & Hussein, M. M. M. (2006). Response of some flax genotypes to pulling date under newly reclaimed sandy soil and sprinkler irrigation conditions. *Egypt. J. Agric. Res., 84*(4), 1103-1115.

Gabiana, C. P. (2005). *Response of linseed (Linum usitatissimum L.) to irrigation, nitrogen and plant population* (M.Sc. thesis, p. 85). Lincoln Univ., New Zealand.

Genser, A. D., & Morris, N. D. (2003). History of cultivation and uses of flaxseed. In A. D. Muir & N. D. Westcott (Eds.), *Flax - The genus Linum*. Taylor and Francis. London.

Islam, M. S., Akhter, M. M., El-Sabagh, A., Liu, L. Y., Nguyen, N. T., Ueda, A., et al. (2011). Comparative studies on growth and physiological responses to saline and alkaline stresses of Foxtail millet (*Setaria italica* L.) and Proso millet (*Panicum miliaceum* L.). *Aust. J. Crop Sci., 5*, 1269-1277.

Khalifa, R. Kh. M., Manal, F. M., Bakry, A. B., & Zeidan, M. S. (2011). Response of some flax varieties to micronutrients foliar application under newly reclaimed sandy soil. *Australian Journal of Basic and Applied Sciences, 5*(8), 1328-1334.

Kinebar, M. E. A. (2004). Effect of environmental conditions on yield and its quality of two flax cultivars. *J. Agric. Sci., Mansoura Univ., 44*(8), 1-12.

Kineber, M. E. A., El-Emary, F. A., & El-Nady, M. F. (2006). Botanical studies on some fibrous flax (*Linum usitatissmum* L.) genotypes grown under Delta conditions. *J. Agric. Sci., Mansoura Univ., 31*(5), 2881-2890.

Le Clerg, E. L., Leonard, W. H., & Clerk, A. G. (1962). *Field plot technique* (p. 300). Burgess Publication Company, Minneapolis, Minnesota.

Lisson, S. N., & Mendham, J. J. (2000). Agronomic studies of flax (*Linum usitatissimum* L.) in south-eastern *Australia. Aust. J. Exp. Agric., 40*, 1101-1112. http://dx.doi.org/10.1071/EA00059

Mirshekari, M., Amiri, R., Nezhad, H., Noori, S. A. S., & Zandvakili, O. R. (2012). Effects of planting date and water deficit on quantitative and qualitative traits of flax seed. *American-Eurasian J. Agric. Environ. Sci., 12*(7), 901-913.

Pandey, R., & Agarwal, R. M. (1998). Water stressinduced changes in praline contents and nitrate reductase activity in rice under light and dark conditions. *Physiol. Mole. Biol. Plants, 4*, 53-57.

Radwan, S. R. H., & Momtaz, A. (1966). The technological properties of flax fibers and the methods of estimating them. *El-Falaha J., 46*(5), 466-476.

Sankari, H. S. (2000). Linseed (*Linum usitatissimum* L.) cultivars and breeding lines as stem biomass producers. *Journal of Agronomy and Crop Science, 184*, 225-231. http://dx.doi.org/10.1046/j.1439-037x.2000.00375.x

Sharma, G., Sutalya, R., Prasad, S., & Sharma, I. (2012). Effect of irrigation and intercropping system on growth, yield and quality of mustered and linseed. *Crop Res. Hisar., 25*(3), 579-581.

Snedecor, G. W., & Cochran, W. G. (1982). *Statistical Methods* (6th ed.). Iowa State Univ. Press, Ames., Iowa, USA.

Szilgyi, L. (2003). Influence of drought on seed yield components in common bean. *Bulg. J. Plant Physiol., Special Issue*, 320-330.

Taiz, L., & Zeiger, E. (2006). *Plant Physiology* (5th ed., p. 20). Sinauer Associates Inc., Publishers Sunderland, Massachusetts, U.S.A.

Wood, I. M. (1997). *Fibre Crops - New opportunities for Australian agriculture* (pp. 18-24). Department of Primary Industries, Brisbane.

Yenpreddiwar, M. D., Nikam, R. R., Thakre, N. G., Harsha, K., & Sharma, S. K. (2007). Effect of irrigation and moisture conservation practices on yield of linseed. *J. Soils Crops, 17*(1), 121-127.

Zahana, A. E. A., Abo-Kaied, H. M. H., & Ashry, N. A. (2003). Effect of different zinc levels and Mycorrhiza fungi and their combinations on flax. *J. Agric. Sci. Mansoura Univ., 28*(1), 67-76.

5

Shade's Benefit: Coffee Production under Shade and Full Sun

Valdir Alves[1,2], Fernando F. Goulart[3], Tamiel Khan B. Jacobson[1,2], Reinaldo J. de Miranda Filho[4] & Clarilton Edzard D. Cardoso Ribas[2]

[1] Programa de Pós Graduação em Meio Ambiente e Desenvolvimento Rural, Faculdade UnB Planaltina, Universidade de Brasília, Planaltina, DF, Brazil

[2] Programa de Pós Graduação em Agroecossistemas, Centro de Ciências Agrárias, Universidade Federal de Santa Catarina, Florianópolis, SC, Brazil

[3] Análise e Modelagem de Sistemas Ambientais/Centro de Sensoriamento Remoto, Dept. Cartografia, Instituto de Geociências, Universidade Federal de Minas Gerais, Belo Horizonte, MG, Brazil

[4] Faculdade UnB Planatina, Universidade de Brasília, RADIS/Fup, Planaltina, DF, Brazil

Correspondence: Fernando Goulart, Análise e Modelagem de Sistemas Ambientais/Centro de Sensoriamento Remoto, Dept. Cartografia, Instituto de Geociências, Universidade Federal de Minas Gerais, Belo Horizonte, MG Cep: 31270-901, Brazil. E-mail: goulart.ff@gmail.com

Abstract

Coffee has major importance in tropical landscapes from agronomic, economic and ecological perspectives. Yet the conversion of shade-coffee into full sun monocultures has deep effect on the potential of those systems to conserve biodiversity and ecosystems services (such as pest control and pollination). Despite of this, effect of shade on production has not been sufficiently addressed, particularly in Brazil, the world major coffee producer. This study compared the performance of shaded coffee and full sun management in terms of productivity and production costs. The survey was conducted in Municipality of Mirante da Serra, in the Brazilian Amazon and eight coffee agroecosystems, four under shade and four under full sun were investigated. The results indicate that shaded systems have lower production costs requiring less working hours than sun plantations. The average production cost of shaded agroecosystems was 49.63%, while in systems under full sun, this value was 82.2%. Shaded and full sun productivity did not differ significantly, with higher variance in the former, showing that shaded systems are more heterogeneous. Shaded coffee agroecosystems presented an economically and environmentally viable alternative. The lower production cost enhances economic viability of these ecosystems in Amazon as well as in the rest of the tropics. Such efficiency may have influenced the persistence of these managements, despite the worldwide agriculture intensification tendency.

Keywords: agroecosystems, agroecology, coffee production cost, full sun coffee, shade-grown coffee

1. Introduction

Coffee plantations covers 10,420,008 thousand hectares (Rudel et al., 2009), playing major role from economic to ecological perspective at global scale. In several coffee producer countries such as Brazil, Colombia, Venezuela, Costa Rica, Panama and Mexico, traditional cultivation of coffee is done under shading trees (Ricci et al., 2006). These shade-coffee plantations are considered biodiversity reservoirs, and yet serving for its productive end (Perfecto & Vandermeer, 2010). Agriculture intensification via tree removal and input use is known to reduce habitat extent and quality for native fauna. Intensification also promotes the decline in the provision of ecosystems services to coffee (such as pest control and pollination) (Goulart et al., 2012; Perfecto et al., 2004). This species loss may, therefore, reduce coffee production at long term (Goulart et al., 2012). The maintenance of those biodiversity rich ecosystems is largely depending upon the economic and productive viability for farmers.

Brazil is the world's major coffee producer, supplying one third of the worlds' coffee gross production (FAOSTAT, 2016). Despite of this, there are many knowledge gaps on many aspects of coffee production in Brazil, particularly regarding the shade influences. In 2014 national coffee production was 2712 million kg of benefited, being 70.5% of the total species *arabica* and 29.5% of the total species *Coffea canephora* (Brazilian Institute of Geography and Statistics [IBGE], 2014). Rondônia occupies the 5th place on the national ranking of coffee producers, being the second largest coffee producer of *Coffea canephora*, which corresponds to approximately 11.27% of the Brazilian production of coffee from group *Coffea canephora* (National Supply Company [CONAD], 2014).

Coffee is the most widespread perennial crop in the state of Rondônia, composing one of the main sources of income for many families in the countryside (Marcolan, 2009). Among agroecological systems, agroforestry stands out as an alternative for reconciling agricultural, social and environmental goals by combining crops with other backbone trees in forest-like agroecosystems (Nair, 1991). The cultivation of shade-grown coffee is an example of such management, which consists in shaded of coffee farms with native and exotics, crops and non-crops trees, having coffee as the main crop (Ferreira, 2005).

The effects of shade on coffee production are controversial and it may exert a positive effect by increasing soil quality, reducing climatic stress, and reduce weed invasion. On the other hand, low light incidence leads to low photosynthetic activity, and consequently less productivity. Therefore, shading should not exceed 30 to 40% of the entire agroecosystems (Mancuso, 2013). Perez et al. (1977) suggested that the removal of trees may increase in up to 30% productivity. According to Baggio et al. (1997), there is no difference between moderate shadow and full sun. Finally, other authors suggest that there is a ramped shaped relationship in which there is an increase of productivity with increase of shading up to a certain level, beyond which, a decline in productivity is observed (Soto-Pinto et al., 2000; Staveret et al., 2001). Therefore, cultivation of coffee may suffer favorable and unfavorable variations depending on shade level, soil and climate characteristics and management.

In Brazil, there is a great demand for assessing shading system in agronomic and economic terms, and there is scant quantitative information on shade effects in coffee systems (Perdoná, 2013). We here compared the productivity and production cost between shade and full sun systems of agroecological coffee *Coffea canephora* in the Brazilian Amazon. Furthermore, we draw the overall implications of our study to sustainability of coffee systems in the tropics.

2. Method

The study site is located in the Amazon Biome, in Padre Ezekiel settlement, in Mirante da Serra, Rondônia state, Brazil. The settlement is placed in the central region of the State of Rondônia, as shown in (Alves, 2010). The assessments were conducted in four properties with shaded coffee agroecosystem and four under full sun agroecosystems (Table 1). The data collection was carried out between June 2013 and June 2014 through visits to the crops accompanied by the farmers. The information were given by the farmers that agreed in participate in the survey and all activities occurred in this period were recorded, even if the researcher was not present. The shaded agroecosystems are formed of native vegetation left when the crops were implemented; therefore, the trees have the same age of the coffee crops. Backbone trees were composed of *Inga edulis*, *Tabebuia* ssp, *Bertholletia excels*, *Orbignya phalerata*, *Hymenaea courbaril*, *Musa paradisiaca* (banana trees), *Annona muricata*, *Caesalpinia equinata* and *Tabebuia cassiniodes*. Many areas were previously composed of pasture, which have been abandoned for deployment of coffee.

Crop management and the dates of the attainment were recorded (such as pruning, manual/mechanical thinning, mowing, harvest, drying and sale) in terms of working hours. The working costs analyzed here, only takes into account the operations in coffee plants without considering the depreciation of machinery, taxes and cost of land. We opt to assess only the costs related to the workforce due to the fact that it represents the major cost in coffee production. Cost of production is here defined as the sum of the values from all resources (inputs) and operations (services) used in a given productive process. For economic analysis purposes, cost of production is the compensation that production factors (land, labor and capital), use to produce determined goods. Production costs have been used to verify if resources employed in a production process are compensating, and enabling to check activity profitability (Viana & Silveira, 2009).

Table 1. Latitude, longitude and altitude of agroecosystems, total area, coffee plants interspacing and number of coffee plants

Agroecosystem	Latitude	Longitude	Altitude (m)	Total Area (ha)	Spacing between plants (m)	Number of plants
Shaded						
I	11°00'57.24"	62°37' 04"	252	1	3.5 × 3	952
II	11°01'44.34"	62°37'37.90"	228	1.5	3 × 3	1666
III	11°01'36.2"	62°37'02.1"	200	1.6	2.5 × 2, 2 × 5	2200
IV	11°01'08.5"	62°36'39.8"	236	0.27	3 × 3	300
Full Sun						
V	11°01'14.5"	62°36'43"	237	2.42	3 × 3	2688
VI	11°01'06.6"	62°36'27.1"	230	1	3 × 3	1111
VII	11°01'01.84"	62°36'41.22"	251	1	3 × 3	1111
VIII	11°01'07.21"	62°36'26.32"	228	1.5	3 × 3	1111

The values paid by labor to carrying out harvesting were calculated using the economic value of kg of processed coffee. There was no variation in labor costs between shaded and under full sun systems, because the values paid to harvest were equal per kg in all agroecosystems. The variations in production costs in this activity occurred only in accordance with the quantity (in kg of coffee per hectare) produced by agroecosystems.

Expenditure necessary for the production and cultivation related to the workforce detached for each activity were evaluated. *F test* (variance comparison) and t test ($\alpha = 0.05$) were performed to compare the average productivity per hectare between shaded and under full sun agroecosystems Kolmogorov-Smirnov tests were carried out to check for data normality distribution ($\alpha = 0.05$). The tests were performed using the Paleontological Statistics Software Package for Education and Date Analysis (PAST 2.04 for Windows).

3. Results

Full sun systems present higher production costs, accounting for more than twice the average of hours per hectare spent on shade systems in pruning and thinning activities (Table 2), while higher maximum average mowing hours were also found in full sun systems (Table 3).

Table 2. Frequency of activity, period and quantity of labor used in pruning and thinning

Agri.	Frequency	Period	Hours/per activity	Average hours/ha	Average hours/ha shaded and under full sun
Shaded					
I	1	August 2013	48	48	
II	2	July 2013	88	58.6	55.2
III	0	00	00	00	
IV	1	January 2014	16	59.2	
Full Sun					
V	2	Jan and FEB. 2014	256	105.7	
VI	2	Aug/Sep 13 Jan/Feb. 14	128	128	
VII	2	Jul/13 Feb 14	112	112	113,07
VIII	2	Jul/13 Jan 14	160	106.6	

Table 3. Frequency of mechanical/manual mowing and amount of labor hours required for the activity

Agroecosystem.	Area (ha)	Frequency	Date	Hours	Average h/ha
Shaded					
I	1	3.0	Nov 2013/Feb and May 2014	72.0	72
II	1.5	1.0	April 2014	12.0	8.0
III	1.6	2.0 Manual	June 2013	96.0	49.2
		1.0 Mechanical	May 2014	24.0	
IV	0.27	1.0	March 2014	8.0	29.6
Full Sun					
V	2.42	2.0	Oct 2013/Mar 2014	128,0	52.8
VI	1	3.0	Jul 2013/nov2013/Feb/2014	96.0	96.0
VII	1	2.0	Oct/13 and March/14	72.0	72.0
VII	1.5	2.0	Sep./13, Feb. and Mar/14	88.0	58.6

The activity of drying were carried in farm and using a drier. Coffee beans were dried in the property, performed outdoors, bare soil or using direct fire dryers placed in the municipality of Mirante da Serra. Coffee grains produced by agroecosystem II the coffee were marketed mature, the sale was conducted without drying. Drying in the dryer was only performed for coffee produced in agroecosystem VI and represented 6.5% of the value obtained with the marketing of the kg of coffee.

The table below (Table 4) shows that the average productivity per hectare in shaded agroecosystems suffered the greatest variation when compared to productivity of under full sun agroecosystems.

Table 4. Harvest period, number of kg per hectare per agroecosystems, value paid for harvesting and total cost.

Agroecos.	Harvest Period	Kg per hectare	Kg per agrieco.	Value of the kg/US$ for harvesting	Cost of harvesting US$
Shaded					
I	May and June 2014	1260,00	1260,00	0,28	355,69
II	April and May 2014	499,80	750.00	0,28	210.00
III	April and May 2014	750.00	1200,00	0,28	336,00
IV	May 2014	2221,80	600.00	0,28	168.00
Full Sun					
V	April and May 2014	618,00	1500.00	0,28	420.00
VI	April and May 2014	870,00	870,00	0,28	242,60
VII	April and May	960,00	960,00	0,28	268,80
VIII	April and May 2014	920,00	1380,00	0,28	386,40

Table 5 provides a synthesis of agroecosystems production, total production by agroecosystems and average per hectare.

Table 5. Total production, production and average production of shaded and full sun agroecosystems

Agri.	Total production (kg)	Production (kg)	Production (kg) average per hectare of shaded and unshaded agroecosystems
Shaded Coffee			
I	1260,00	1260,00	
II	750,00	499,80	1.182,45
III	1200,00	750,00	
IV	600,00	2220,00	
Full Sunlight			
V	1500,00	619,00	
VI	870,00	870,00	842,25
VII	960,00	960,00	
VIII	1380,00	920,00	

The variation of sale values (US$) per kg of coffee occurred in accordance with the period in which this had been marketed. For this reason, the prices of kg of coffee between April and July ranged from US$ 0.882 to US$ 0.952, and between October and December, the price ranged from US$ 0.98 to US$ 1. 064.

Shaded agroecosystems (I, II, III and IV) presented a lower frequency in crop handling reflected in the decrease in required labor quantity (h) for the activities performed in the agroecosystems under full sun. This occurred mainly due to tree species in the agroecosystems, which increased the shade to coffee plants and altered the spontaneous weed populations, reducing competition.

The agroecosystems under full sunlight showed low variation in productivity (standard deviation SD = 153.31 and variation coefficient VC = 18.20 %), while variation was significantly higher in shaded agroecosystems, (standard deviation SD = 760.60 and variation coefficient VC = 64.32%). In agroecosystems under full sun, cultivation also came with small differentiation in total cost of production (88.30% and 78.45% 94.60% and 67.45%). The possible causes of similarity in production may be due to crops being conducted with the same crop handling. The systems under full sun are more homogeneous, presenting low differentiation amongst the agroecosystems themselves. The costs with the workforce in under full sun systems were higher compared to shaded crops (Tables 6 and 7) due to the absence of the woodland component (tree layer) , which inhibits the incidence of direct light in cultivation and consequently reduces the need for time in labor to suppress weed.

Table 6. Production cost per activity (pruning and thinning, mowing and weed suppression, harvest, drying), total cost and net margin in percentage of the cost required per activity

Agrecossytem	Pruning and thinning	Mowing	Harvesting	Drying	Total cost of production	Net Margin
Shaded						
I	6.20%	16.30%	28.50%	4,13%	55.13%	44.87%
II	20.00%	4.8%	30%	00	54.80%	45.20%
III	00	17.11%	26%	2.97%	46.08%	53.86%
IV	4.56%	4.0%	30%	6.84%	45.4%	54.6%
Full Sun						
V	29.10%	25.60%	30%	3.60%	88.30%	11.70%
VI	25%	33.10%	30%	6.5%	94.60%	5.40%
VII	29.41%	12.65%	29.41%	6.98%	78.45%	21.55%
VIII	26.31$	9.56%	26.31%	5.27%	67.45%	32.55%

Table 7. Total Cost, gross income, percentage and net margin per agroecosystem

Agri.	Workforce total cost US$	Total gross income in US$	Net Percentage	Net margin in R$
Shaded				
Agri. I	686.53	1245.30	44.87%	558.76
Agri. II	212.72	470.63	45.20%	257.90
Agri. III	374.58	811.84	53,86%	437.26
Agri. IV	260.55	564.76	54.6%	304.20
Full Sun				
Agri. V	515.17	583.43	11.70%	68.26
Agri. VI	774.67	818.90	5.40%	44.22
Agri. VII	723.06	921.70	21.55%	198.62
Agri. VIII	665.88	987.20	32.55%	321.33

The data on productivity presented a normal distribution ($p = 0.53$). The differences between the variances was significant ($F = 24.6$, $p = 0.02$), and considering unequal variances, we compared the averages using the t test for unequal variances (Welch test), which showed no significant differences between the productivity in the different agroecosystems ($p = 0.44$). Thus, the raw values are higher in shaded agroecosystems, however, differences were not statistically significant, due to the large amplitude in average productivity among shaded agroecosystems (499.8-2222.22 kg/ha).

4. Discussion

The results of this research indicate that the shaded agroecological systems presented higher economical results than cultivation under full sun, because it requires less working hours. Systems under full sun have smaller variation in productivity than shaded intercropping; possibly due to the fact that full sun agroecosystems are more homogeneous among themselves, while shade agroecosystems are much more heterogeneous. Another possible cause of high variation in shaded agroecosystems is area variation in the shade systems. For instance, agroecosystem IV presented a small area (0.27 ha) and presenting the highest productivity (2222.22 kg/ha). The conversion of the value of total production (0.27 ha) in production per hectare may have overestimated the value of average systems productivity. The negative relationship between farm size and productivity is long known in the agricultural sciences, and smaller farms are more easily and effectively management and thus produce more than larger counterparts (Rosset, 1999).

Some studies show an increase of 10% to 30% in coffee productivity with the removal of trees (Perez, 1977). Baggio et al. (1997) show that there is no difference between moderate shade and full sun. Finally, other authors suggest that there is productivity increase with shading up to a certain level, beyond which, a decline in productivity takes place (Soto-Pinto et al., 2000; Staveret et al., 2001). In a revision paper, Damatta (2004) concludes that the productivity in shaded systems is superior in comparison to crops under full sun in situations in which the edaphic conditions are sub-optimal. Shading agroforestry is also known for reducing the impact of frosts (Baggio et al., 1997) and water stress on the events of drought (Damatta, 2004). Moreover, in cultivations under full sun that are more than one or two decades old, the quality of the soil is committed, thus there is lower productivity compared to shaded systems (Damatta, 2004).

We did not find significant statistical differences in the productivity of shaded versus full sun systems, although raw values suggested higher performance of shaded systems. Lower variation in productivity was found in full sun systems, possibility due to complexity and heterogeneity of shaded systems in comparison with the former. The combination of different shade degrees, agriculture design, backbone tree composition, farm size and other management variations may greatly increase shade system variability.

Full sun systems are more labor intensive than shade counterparts, due to the greater frequency and time spend per activity of pruning, thinning and mowing for weed suppression. Due to less light incidence in shaded agroecossytems, weed invasion is reduced and coffee plants show lower sprouting, easing the systems management. For Mancuso (2013), shading can modify the composition of weed species, reducing the number of most competitive plants. Weed invasion drops to very low levels in systems with more than 40% of shade (Mancuso, 2013). Other positive effects of shading are the increase of organic matter, edaphic fauna enrichment, and increase in nutrient cycling, reduction of soil erosion, biodiversity conservation, and attenuation of extreme temperature and wind incidence (Mancuso, 2013). Furthermore, productive lifespan of shaded coffee is longer than plants under full sunlight (Damatta, 2004).

Back-bone trees promote changes in energy distribution, in thermal air conditions, soil and in the coffee plants which presents differentiated growth. Plant´s stress reduction by the improvement of microclimate and soil quality (Lemos, 2007). Additionally, shading with appropriate tree species produces larger fruits, with softer and sweetened tissues, improves vegetative aspect of coffee, increase in the number of primary and secondary branches, increasing reproductive capacity of coffee trees (Lemos, 2007).

Intermediate levels of shade are known to alleviate extreme climatic conditions in *Coffea canephora* providing greater sustainability to the systems (Damatta, 2004). Back bone trees also adds to the productive potential for the shaded systems, providing extra income for growers due to production of wood, fruits, medicinal plants, herbs and essential oils, fuel and fiber (Pezzopane et al., 2007; Coelho, 2010). These additional crops can be harvested during the periods between coffee harvests, securing a more constant income source for famers. Furthermore, shade systems may also contribute for better working conditions, protecting farmers from direct solar radiation, promoting health and wellbeing (Mangabeira, 2009). Also, shade coffee certification can increase coffee price in markets, due to its ecological and social benefits. Fair trade, shade coffee, organic is example of agriculture certification which is growing around the world, increasing the economic return of the agroforestry products (Perfecto et al., 2005).

By and large, the cultivation of shaded coffee *Coffea canephora* presented higher performance than full sun, due to less workforce requirements, enabling extra-income from associated crops, contributing to farms wellbeing and biodiversity conservation, although shade systems present very high variation in productivity in the Amazon. Many other tropical landscapes may share similarities with this environment and thus may present similar trends in terms of shade effect on coffee production. Much of the tropical farmers are small holders, owing properties not larger than two hectares (Lowder et al., 2014). These farmers are mainly economically poor population particularly susceptible to price elasticity of agriculture inputs. Our results corroborates with the high productive efficiency of shaded coffee, being a viable alternative and environmental friendly practice in the tropical region.

Acknowledgements

FFG received a post-doc fellowship from Análise e Modelagem de Sistemas Ambientais-UFMG/CAPES-PNPD. Project "Regularização Ambiental e Diagnóstico dos Sistemas Agrários dos Assentamentos da Região Norte do Estado do Mato Grosso", INCRA-FINATEC n° 5788/2015 funded the publication fees. The Programa de Pós-Graduação em Agroecossistemas-UFSC provided a master scholarship for VA. All the farmers involved in the research for all the support.

References

Alves, V. (2010). *Avanços e limites no processo de cooperação de famílias do assentamento Palmares, Nova União-RO*. 68 f. Monografia (graduação em agronomia) departamento de agronomia, Universidade do Estado de Mato Grosso-UNEMAT. Cáceres, MT.

Baggio, A. J., Caramori, P. H., Androcioli Filho, A., & Montoya, L. (1997). Productivity of southern Brazilian coffee plantations shaded by different stockings of grevillea robusta. *Agroforestry Systems, 37*, 111-120. http://dx.doi.org/10.1023/A:1005814907546

Bolero, J. C., Martinez, H. E. P., & Santos, R. H. S. (2006). *Características do café (Coffea arabica L.) sombreado no norte da América Latina e no Brasil: Análise comparativa* (Vol. 1, No. 2, pp. 94-102). Coffea Science, Lavras.

Coelho, A. R., et al. (2010). Nível de sombreamento, umidade do solo e morfologia do cafeeiro em sistemas agroflorestais. *Revista Ceres, 57*(1), 95-102. http://dx.doi.org/10.1590/S0034-737X2010000100016

CONAB (Companhia Nacional de Abastecimento). (2014). *Acompanhamento da safra brasileira de café, safra 2014* (pp. 1-59). Terceiro Levantamento, Brasília.

Damatta, F. M., & Ramalho, J. D. C. (2006). Impacts of drought and temperature stress on coffee physiology and production: A review. *Braz. J. Plant Physiol., 18*(1), 55-81. http://dx.doi.org/10.1590/S1677-0420200600 0100006

FAOSTAT. (2016). *Database collections*. Food and Agriculture Organization of the United Nations. Rome. Retrieved August, 2016, from http://faostat.fao.org

Ferrão, R. G., et al. (2004). *Biometria aplicada ao melhoramento genético do café Conilon*. 256 f Tese (Doctor Scientiae), Programa de Pós Graduação em Genética e Melhoramento, Universidade Federal de Viçosa.

Ferrão, R. G., et al. (2007). *Café conilon*. Vitória, ES: Incaper.

Ferreira, J. M. L. (2005). *Indicadores de qualidade do solo e de sustentabilidade em cafeeiros arborizados* (p. 90). Florianópolis, SC. Dissertação (Mestrado em Agroecossistemas), Centro de Ciências Agrárias, Universidade Federal de Santa Catarina (USFC).

Gomes, J. C. C., & Assis, W. S. de. (Eds.). (2013). Agroecologia: Princípios e reflexões conceituais. *Coleção Transição Agroecológica* (p. 245). Brasília, DF. Embrapa.

Goulart, F. F., Jacobson, T. K. B., Zimbres, B. Q. C., Machado, R. B., Aguiar, L. M. S., & Fernandes, G. W. (2012). *Agricultural Systems and the Conservation of Biodiversity and Ecosystems in the Tropics*.

IBGE. (n.d.). *CENSO 2006/2013/2014*. Instituto de geografia e estatística.

Lemos, C. L., et al. (2007). *Avaliação do desenvolvimento vegetativo em cafeeiros sombreado e a pleno sol. Revista Brasileira de Agroecologia, 2*(2), 4.

Lowder, S. K., Skoet, J., & Singh, S. (2014). What do we really know about the number and distribution of farms and family farms worldwide? Background paper for The State of Food and Agriculture 2014. *ESA Working Paper No. 14-02*. Rome, FAO.

Mancuso, M. A. C., Soratto, R. P., & Perdoná, M. J. (2013). *Produção de café sombreado* (Vol. 9, No.1, pp. 31-44). Colloquium Agrariae.

Mangabeira, J. A. de C., et al. (2009). *Análise comparativa entre café produzido a pleno Sol e no sistema agroflorestal em Machadinho D'Oeste-RO*. Congresso Brasileiro de Sistemas Agroflorestais 7. Anais. Brasília, DF: SBSAF: Embrapa.

Marcolan, A. L., et al. (2009). *Cultivo dos cafeeiros conilon e robusta para Rondônia* (3rd ed. rev. atual). Porto Velho, Embrapa Rondônia: EMATER-RO.

Nair, P. K. R. (1991). State-of-the-art of agroforestry systems. *Forest Ecology and Management, 45*(1), 5-29. http://dx.doi.org/10.1016/0378-1127(91)90203-8

Perdoná, M. J. (2013). *Cultivo consorciado do cafeeiro (Coffea arabica L.) e cultivares da nogueira-macadâmia (Macadamia integrifolia maiden e betche) sob os regimes sequeiro e irrigado*. Botucatu, 130 f. Tese (Doutor em Agricultura), ciências agronômicas, Universidade Estadual Paulista, Júlio de Mesquita Filho.

Perfecto, I., & Vandermeer, J. (2010). The agroecological matrix as alternative to the land-sparing/agriculture intensification model. *Proceedings of the National Academy of Sciences, 107*(13), 5786-5791. http://dx.doi.org/10.1073/pnas.0905455107

Perfecto, I., Vandermeer, J. H., Bautista, G. L., Nunez, G. I., Greenberg, R., Bichier, P., & Langridge, S. (2004). Greater predation in shaded coffee farms: The role of resident neotropical birds. *Ecology, 85*(10), 2677-2681. http://dx.doi.org/10.1890/03-3145

Pezzopane, J. R. M., Pedro Jr, M. J., & Gallo, P. B. (2007). Caracterização microclimática em cultivo consorciado café/banana. *Revista Brasileira de Engenharia Agrícola e Ambiental, 11*(3), 256-264. http://dx.doi.org/10.1590/S1415-43662007000300003

Ricci, M. dos S. F., Costa, J. R., Pinto, A. N., & Santos, V. L. da S. (2006). Cultivo orgânico de cultivares de café a pleno sol e sombreado. *Pesquisa Agropecuária Brasileira, 41*(4). http://dx.doi.org/10.1590/S0100-204X2006000400004

Rosset, P. (1999). On the benefits of small farms. *Backgrounder, 6*(4).

Rudel, T. K., Schneider, L., Uriarte, M., Turner, B. L., DeFries, R., Lawrence, D., et al. (2009). Agricultural intensification and changes in cultivated areas, 1970-2005. *Proceedings of the National Academy of Sciences, 106*(49), 20675-20680. http://dx.doi.org/10.1073/pnas.0812540106

Soto-Pinto, L., et al. (2000). Shade effect on coffee production at the northern Tzeltal zone of the state of Chiapas, Mexico. *Agriculture, Ecosystems and Environment, 80*, 61-69. http://dx.doi.org/10.1016/S0167-8809(00)00134-1

Staver, C., et al. (2001). Designing pest-suppressive multistrata perennial crop systems: Shade-grown coffee in Central America. *Agroforestry Systems, 53*(2), 151-170. http://dx.doi.org/10.1023/A:1013372403359

Viana, J. G. A., & Silveira, V. C. P. (2009). Análise econômica da ovinocultura: Estudo de caso na Metade Sul do Rio Grande do Sul, Brasil. *Cienc. Rural, 39*(4). http://dx.doi.org/10.1590/s0103-84782009005000030

6

Assessment of Fisherfolk Information Seeking Behaviour with Mobile Phone for Improve Extension and Advisory Services

P. I. Ifejika[1,2]

[1] National Institute for Freshwater Fisheries Research, New Bussa, Niger State, Nigeria

Correspondence: P. I. Ifejika, National Institute for Freshwater Fisheries Research, P.M.B. 6006, New Bussa-912105, Niger State, Nigeria. E-mail: ifejikaphilip@gmail.com; Ifejikaphilip@yahoo.com

Abstract

The study assessed information seeking behaviour of fisherfolk with mobile phone in fishing communities around Kainji Lake basin, Nigeria. Primary data was generated through interview schedule from 165 respondents and analysed with descriptive and factor analysis. Result revealed that mobile phone improved information seeking behaviour of fisherfolk with associates in the fishing communities than with outsiders in government establishment. Kinds of information sought with mobile phone cut across economic, social and health issues in fish market, social gathering, fish catch/gear, health, weather and security related matters. Pattern of information seeking revealed regular use of close associates than extension workers. Close associates mostly contacted with mobile phone for information were fellow fisherfolk (64.2%), family members (64.8%) and community members (55.8%) but weak with fisheries institute (4.8%) and extension agents (0.6%). Fisherfolk rated voice call as the most effective medium in information seeking over flashing, SMS, voice message, video and pictures. Result of factor analysis categorised the challenges into poor financial status, inadequate knowledge and skill, poor network services and lack of energy to recharge mobile phone batteries. It is recommended that extension providers should use the established effective medium and pattern of information behaviour to package and disseminate messages to meet needs of fisherfolk in the lake basin. Interim measure is to provide tool free mobile lines to improve contact and trust to strengthen rapport. Also, mobile network operators should reduce tariff, improve quality of services as well as incorporate training in their social responsibility and promotion strategies for fishing communities to benefit.

Keywords: mobile phone, fisherfolk, fish, information seeking, Africa

1. Introduction

Information has emerged as one of the topmost resources needed for successful economic activities in combination with labour, capital, knowledge, and infrastructures. Fisherfolk that derive their livelihood in artisanal fishery is among those in need of high quality information to take informed decision to work smarter and intelligently. Roopchand (2013) in CRFM Technical & Advisory Document Series define fisherfolk as people who perform different types of work and have different roles in the fishing industry. Hoffmann et al. (2009) view information as processed data that reduce uncertainty at the user level. While Gachie et al. (2006) clarified that "operational information" is information that is practical, concrete and able to help solve existing problems. As such, fisherfolk should seek for operational information from reliable sources to update knowledge and facilitate decision making in the adoption process. Quality information should be credible, relevant, accurate and timely to add value to knowledge in decision process. In this regard, Solano et al. (2003) made known that throughout the phases of the decision-making process, farmers prefer different information sources for problem detection, seeking for problem solutions, seeking for new practices and seeking for opinion.

Wilson (2000) defines information seeking behaviour as the purposive seeking for information as a consequence of a need to satisfy some goal. Vergot et al. (2005) clarify that the source of information was an individual or institution that originated a message. Above clarification shows that information seeking is a personal effort undertaken to find solution to identified needs through a source. Information seeking is a premeditated attempt by the seeker to get operational information that is relevant and useful to take informed decision to resolve complications. To seek is to ask, look, demand and search for information to be better informed on specific issue.

Consequently, information seeking behaviour helps to understand subjects' pattern of information behaviour, facilitate the design of information dissemination within the established sources, reduce the chance of decision made on incomplete or erroneous information (Hill, 2009). Also, it enables subjects obtain reliable information more quickly and easily within short time (Solano et al., 2003). Therefore, extension service providers need to understand the information seeking behaviour of fisherfolk to improve information packaging and delivery.

Information seeking and utilisation through mobile phone technology which exist in infrastructure, services and applications provide opportunities for fisherfolk in fishing communities to unlock their economic vulnerability in fisheries livelihood. According to OECD (2001) globally definition of a fishing community is substantially dependent on, or substantially engaged in, the harvest or processing of fishery resources to meet social and economic needs; and includes fishing vessel owners, operators, crew and fish processors that are based in such community. Clay and Olson (2008) sum fishing communities' vulnerability thus "as populations of many fish species worldwide have declined, the price of input has increased, and coastal development has mushroomed, fishing communities have suffered economic and social vulnerability". Vulnerability of fishing communities is complicated by fisherfolks' information poverty linked to their neglect, lack of access, availability and affordability to services from the public and private extension agents.

Recent GSMA (2015a) report revealed that as at 2015 in Sub-Saharan, there exist 386 million mobile subscribers (41% penetration), 23% mobile internet penetration, 160 million adopted smartphones. Torero (2013) showed that as at 2009 in Nigeria, 88.3% of urban households and 60.3% of rural households had acquired mobile phone, but GSMA (2015b) report showed that current mobile penetration rate stood at 31% in 2015. Interestingly, empowerment of remote riverine fishing communities with mobile networks supports their mobile phone acquisition and usage to enjoy mobile services and applications in health, finance, disaster, agric-news. In view of this trend, Verma et al. (2012) asserted that information and communication technology (ICT) tools especially mobile telephony is the best methods for providing information on agricultural activities. Also, Campaigne et al. (2006) said that mobile phone offers a more reliable and cost effective tool for serving farmers' needs on information compared to internet. As a result, different mobile phone tools has been deployed in agricultural extension mobile services to deliver messages to agricultural communities like SMS, voice call, video and voice recorder (Saravanan, 2010; Fafchamps & Minten, 2012; Ganesan et al., 2015). Labonne and Chase (2009) reported that a World Bank study conducted in the Philippines found strong evidence that purchasing a mobile phone is associated with higher growth rates of incomes in the range of 11%-17% measured through consumption behaviour. Abila et al. (2011) reported that in Lake Victoria, Kenya, mobile phone was used to enhance fish market information service between 2009 and 2010 to stakeholders in fishery value chain through short message sending (SMS) supplied by 165 fish markets and landing sites which was relayed from database 24 hours and 7 days a week. Abila and group added that the visible effect of the multi institutional mobile phone project were increased fish species price by 25%, 91% and 137%; increased income by 30%; reduced post-harvest loss from 5% to 4.5%; generated revenue of $2,550.00 (200,000) Kenya shilling (Ksh) from 20,000 SMS and made profit of 39,700 Ksh. Also, Muthiah et al. (2015) provided another example of mobile agricultural information dissemination to five delta districts in Tamil Nadu State, India, through recorded voice messages delivered to crop farmers' mobiles at zero costs including feedback voice calls. Details of the mobile voice message project relayed between August, 2012 to July, 2013 revealed that a total of 3,833,650 recorded voice messages were disseminated in government scheme (29.53%), best practices (27.12%), fertilizer (23.36%), pesticide application (10.88%), crop insurance (5.80%) and seed varieties (3.31%) which was adopted by 54.0%, yet to adopt (4.00%) and no adoption (42.0%).

In the context of study, a fishing community is a traditional small-scale and family-based fishing village with fishers, processors and marketers in riverine. Previous studies in the area revealed the following characteristics found in the community; low education, early marriage, dominant Muslims, low income, middle age and speak Hausa language (Ifejika, 2012). Fisherfolk in the lake basin comprises of indigenous and migrants from northern and southern states in the country. Also, fisherfolk remain one of the marginalised rural dwellers in agricultural communities due to negligence by the three tiers of government in infrastructure provision like schools, health, road and extension delivery services by change agents. For instance, the Nigeria Federal Ministry of Agriculture and Rural Development (FMARD) in 2012 used mobile phone platform called "E-wallet System or Paper Vouchers" to distribute seedlings and fertilizers to 1.2 million crop farmers within 120 days through SMS alert out of 4.5million national farmers' in the database (Ifejika, 2015). As such, fisherfolk suffer information poverty, denial of productive assets and wallop in abject poverty. Inspite of the odds, these fisherfoks produce over 65% of domestic fish production from inland water bodies consumed in the country. Above scenario prompted the study to verify how mobile phone is aiding fisherfolk in information seeking in riverine communities around the lake basin. The specific objectives were to;

- Determine the kind of information they sought.
- Ascertain the people they contact for information.
- Verify the effectiveness of mobile phone tool in information seeking.
- Ascertain challenges in the use of mobile phone.

2. Study Area

Kainji lake is one of the popular inland freshwater sites for capture fishing in the past forty-five years in northern Nigeria. The lake covers an area of 1250 km^2 and is located at longitude 9°50′-10°55′N and latitude 4°23′-45′E in Niger and Kebbi States, northern Nigeria (Okoye, 1992). In 1996, Kainji lake fish production was 38,246 tonnes which accounted for 12.36% of capture fishery production in the country. Kainji lake fish is dominated by clupeids, tilapinnes and citharinus (du Feu & Abayomi, 1996). Tonnes of these fish caught are processed by smoking and transported weekly from New Bussa (Niger) and Yauri (Kebbi) respectively to some cities in the country like Onitsha and Enugu in south east, Lagos in south west, Ilorin and Kaduna in the north central, Nigeria. The lake has twenty fish taxon for fishing which provides livelihood to 286 fishing villages, 5,499 fishing entrepreneurs and 12,449 fishing assistants (Abayomi & du Feu, 1996). Also, 115 of the fishing villages and 1,975 fishing entrepreneurs were found in the western side of the lake. On gear, they reported that the lake had 9,278 fishing canoes, 18,655 gill nets, 1,560 drift nets, 753 beach seine, 5,548 cast nets, 7,400 longlines, and 36,979 fishing traps. It shows the volume of fishing activities fishers, fish processors, marketers, and other auxiliaries in the value chain engage and value of fish catch in lake basin put at ₦1.147 billion Nigeria Naira. According to Ifejika (2012) mobile network providers found in the area are MTN, Glo, Airtel and recently is Etisalat making it four.

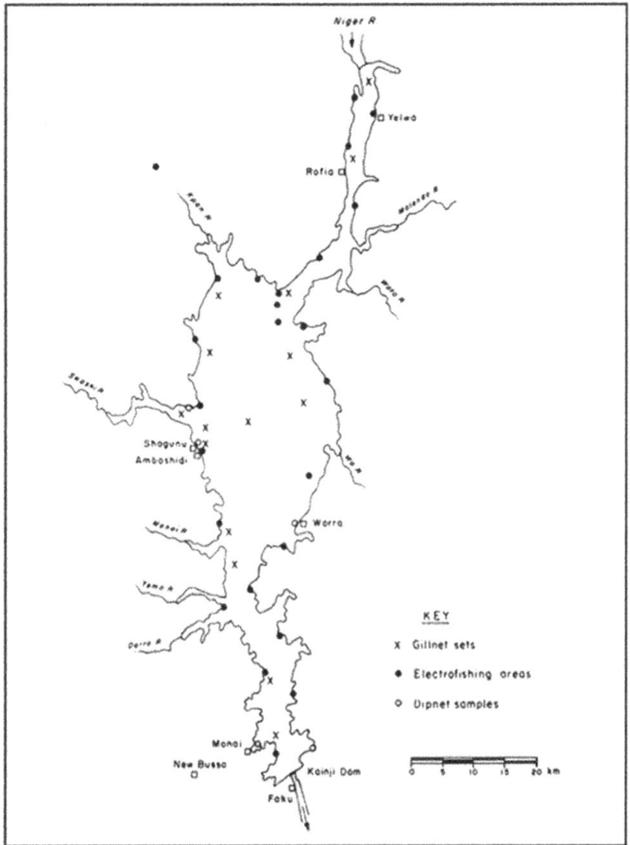

Figure 1. Study area

3. Methodology

Multistage sampling technique was adopted in the study which was carried out around Kainji lake basin. First was stratification of the lake into east and west stratum and western stratum was purposively chosen due to presence of telecommunication service providers' network in some communities. Next was random selection of eleven (11) communities out of 40 identified with telecommunication networks for the study from 115 fishing villages in the western stratum of the lake. The population of the study comprised of all fisherfolk in the selected eleven fishing communities whereas the sample size were mobile phone owners in the communities estimated at 280. From the sample size of 280, respondents were determined using the formulae; n = N/1+N (e) by Israel (1992). Based on the formulae, the respondents for the study were 165 fisherfolk which was randomly selected from the villages thus; Mallale (18), Bussawa (16), Musawa (14), Monai (17), Gwatanwara (9), Kaya (27), Sakajinka (12), Yunawa (5), Tunga Angulu (9), and Tunga Alhaji Ibrahim (18). Primary data was elicited with interview schedule through interview by enumerators' fluent in 2010. Collected data was analysed with descriptive and inferential tools of frequency, percentage and factor analysis as presented in tables.

4. Results and Discussions

Table 1 displays kind of information sought with mobile phone. As shown, respondents sought multi-dimensional information which revolves around fisheries livelihood and non-fishery issues. Top request was fish market information (70.3%) followed by social gathering information (51.5%), household items information (43.0%), fish catch/gear and health information's (24.2%) respectively whereas the least sought information was on weather (23.7%) and security (18.8%). Market information sought was on fish demand, price, supply and new markets whereas financial information sought was on loan, bank alerts and debtors for payment. Fishing information sought revolve around rich fishing ground, gear price, usage of ban gears, while social information was on meeting and weddings. Corroborating the finding on kinds of information sought with mobile phone were studies by Abila et al. (2011), Ifejika et al. (2009), and Jesen (2007). Labonne and Chase (2009) assert that farmers equipped with information have a stronger bargaining position within existing trade relationships in addition to being able to seek out other markets. The finding exposes the diversity of information needs as well as the pattern of information packages for fisherfolk in the riverine community to improve their economic, health and social activities in the area. Moreover, provision of market information will facilitate easier access to market prices, possibly network with fellow fisherfolks and negotiate better prices.

Table 1. Kinds of information sought with mobile phone

	Yes (%)	No (%)
Fish market information	116(70.3)	49(29.7)
Financial information	49(23.9)	116(70.3)
Social gathering information	85(51.5)	80(48.5)
Fish catch and gear information	40(24.2)	125(73.6)
Health information	40(24.2)	125(75.8)
Household items information	71(43.0)	94(57.0)
Weather information	49(23.7)	116(70.3)
Security information	31(18.8)	134(81.2)

Source: Responses from field survey (2010).

Entries in Table 2 are on the people contacted for information through the mobile phone. As shown, mobile phone enabled respondents' to reach out to diversified people within and outside the community to seek for information. It confirmed Hill (2009) finding that farmers sought information from three to nine sources to understand the process for adoption before taking decision. Close associates in fisheries were popular people contacted for information dominated by fellow fisherfolk (64.2%), family members (64.8%) and people in the fishing communities (55.8%). Confirming the result on the use of close associates in work place and community to seek for information were Leckie et al. (1996); Solano et al. (2003); Verma et al. (2012). In the words of Leckie et al. (1996), professionals, such as engineers, nurses, physicians and dentists rely on co-workers and knowledgeable colleagues in their search for work-related information. Probably, they trust and rely on information sought from known and experienced sources than unfamiliar source among associates. However, external people in government with technical information had poor contacted with fisherfolk through mobile phone for information sharing such as extension agents (0.6%) and Fisheries Research staff (4.8%). Observed weak information seeking from government agencies portrays poor extension contact with fishing communities prevalent in public extension services in the country. Patel et al. (2010) found that small farmers in rural India preferred to obtain information from known and trusted experts rather than from other farmers in a field study of Interactive Voice Forum. This is buttressed by Ganesan et al. (2015) finding that 92.50 per cent of the farmers felt the information received could be trusted compared to contrary view by 7.50 per cent on recorded mobile voice messages sent by government extension agency in India. Public and private extension service providers should emulate exemplary mobile phone packages seen in Lake Victoria, Kenya and delta districts of Tamil Nadu, India to design appropriate advisory services to empowered fisherfolk with technical and social information needs in the lake basin.

Table 2. People Seek information with mobile phone

	Yes (%)	No (%)
Fellow fisherfolk (fishers, friends, marketers, processors)	106(64.2)	59(35.8)
Family members (spouse, children, relatives)	107(64.8)	59(35.2)
People in fishing communities	92(55.8)	73(44.2)
Health workers	40(24.2)	125(75.8)
Money lenders	51(30.9)	114(69.1)
Fisheries research staff	8 (4.8)	153(95.2)
Extension agents	1 (0.6)	164(99.4)

Source: Responses from field survey (2010).

Table 3 shows response on mobile phone tools effectiveness in information seeking among respondents. Pooled score discloses that the most effective mobile phone tool for seeking information was voice call (44.26%), followed by flashing (32.26%) and SMS (22.58%) whereas the least was multimedia (0.89%). High usage of voice call tool to seek information attests to respondents' competency to use verbal communication tools and underutilise non-verbal and picture communication mediums such as SMS, video and voice messages. High usage of flashing (call me back) suggests their inability to buy credit or top up due to lack of money which is a sign of poverty. Poor usage of media tools like video, camera and voice message were indication of lack of skill and ignorance. For instance, Donner (2007), confirmed the practice of giving deliberates 'missed calls' or 'flashing' or 'paging' to others has long been a cost reducing measure whereas Ifejika (2012) established 98.8% use of voice call by fisherfolk. Abila et al. (2011) confirmed effective use of SMS by fish workers to send messages in fisheries innovative platform. Also, effectiveness of mobile phone voice call and SMS was confirmed among fisherfolk by Ifejika and Oladosu (2011). Muthiah et al. (2015) confirmed that crop farmers found recorded mobile phone voice messages very satisfactory (99%), better (62.50%) and usefulness (52%). Above evidences attest to using the right mobile phone medium to share the right message to the right audience to make positive impact.

Table 3. Response on effectiveness of mobile phone tools in seeking information

Mobile phone tools	Good (%)	Better (%)	Best (%)	Pooled Score (%)
Adequate content				
SMS	42(25.5)	25(15.4)	17(10.3)	84(22.52)
Voice call	3(1.8)	13(7.9)	149(90.3)	165(44.24)
Flashing	80(48.5)	20(12.0)	20(12.0)	120(32.17)
Video/picture/voice message	4(2.4)	0(0.0)	0(0.0)	4(1.07)
Timely contact				
SMS	41(24.8)	27(16.4)	14(8.5)	82(22.10)
Voice call	2(1.2)	15(9.1)	148(89.7)	165(44.47)
Flashing	85(51.5)	15(9.1)	21(12.7)	121(32.62)
Multimedia tools	3(1.8)	0(0.0)	0(0.0)	3(0.81)
Clear message				
SMS	42(25.5)	27(16.4)	17(10.5)	86(23.12)
Voice call	5(3.0)	16(9.7)	143(86.7)	164(44.08)
Flashing	82(49.7)	16(9.7)	21(12.7)	119(31.99)
Multimedia tools	3(1.8)	0(0.0)	0(0.0)	3(0.81)

Source: Responses from field survey (2010).

Table 4 shows factor analysis of challenges experienced in the use of mobile phone in fishing communities. As revealed, the factors were categorized into four components namely; financial challenges, energy challenges, human capability challenges and quality of services challenges. Factors under financial challenges have high loading on consume money (.827) and high tariff (.619) with severe consequences on mobile phone users in the fishing communities. Second category of factors is on lack of energy supply with high loading on lack of power (.613) and difficulty to recharge battery (.612). Absence of energy in the fishing communities add to financial burden of mobile phone owners and users that pay (₦30.00 to ₦50.00) to recharge battery as well as waste man-hour, time and energy to trek distances to recharge phone battery. Observed lack of energy infrastructure is critical for recharging of phone battery to enable fisherfolk communicate. The third category of factors is human incapability with four challenges; lack of knowledge to operate some phone functions with loading of (-.372) is the most sever followed by phone pilfering (-.038), waste of time (.756) and lack of technician (.323). Low knowledge on written communication is responsible for low use of text message and picture to communicate by respondents. The fourth categories of challenge dwell on quality of services provided by mobile network operator. Prominent and most severe among them is poor network services (-.479), trailed by low quality handset (.587) and fake recharge card (.521). To overcome the identified challenges, there is need for collaboration among key actors; beneficiaries, service providers, mobile phone operators, mobile phone software developers and government. Fisherfolks need to acquire cost saving mobile phones and training to improve capability and skill on the use of multimedia tools and applications. Mobile phone operators should improve quality of service to reduce financial wastage suffered by subscribers, checkmate the incidence of fake recharge cards and reduce the tariff. Present democratic government should tackle the problem of power and road network in fishing communities around the lake basin.

Table 4. Result of factor analysis on challenges encountered in the use of mobile phone

Variables	Poverty	Lack of Energy to recharge battery	Lack of skill & knowledge	Poor network service
High Tariff	.619			
Increased expenditure	.827			
Lack of power		.613		
Difficulty to recharge phone battery		.612		
Lack of technicians			.323	
Waste of time			.756	
Lack of knowledge			-.372	
Phone pilfering			-.038	
Low quality handset				.587
Fake recharge card				.521
Bad network				-.479

Source: Field survey (2010).

5. Conclusion and Recommendation

The study provided empirical evidence that fisherfolk are in desperate need of varieties of information to support fishery livelihood activities in fishing communities. Mobile phone access in the riverine communities has improved information seeking behaviour of fisherfolk among peers but found to be low with extension and fisheries experts outside their domain. Therefore, opportunities provided by mobile phone mediums are enormous to design effective fisheries information dissemination based on established pattern of seeking information. Both fisherfolk and mobile network providers have responsibility to improve the identified challenges in capacity, infrastructure and quality of service. Mobile phone operators should restructure their promotion and corporate responsibility strategy to benefit fisherfolk through training and low tariff and improve service. Concerned change agent in fisheries extension in the lake basin should step-up extension contact with fishing communities as well as establish tool-free mobile phone units for fisheries information sharing with the fisherfolk and others.

References

Abayomi, S. O., & du Feu, F. A. (1996). Frame Survey of Kainji Lake. *Nigeria-German (GTZ) Kainji Lake Fisheries Promotion Project* (pp. 217-219). Annual Report of National Institute for Freshwater Fisheries Research, New Bussa, Niger State.

Abila, R., Ojwang, W., Othina, A., Lwenya, C., Oketch, R., & Okeyo, R. (2011). Enhancing Fish Marketing Through ICT: Experiences of EFMIS project in lake Victoria. *Proceedings of Third Workshop on Fish Technology, Utilisation and Quality Assurance in Africa. Victoria, Mahe, Seychelles, November 22-25, 2011. FAO Fisheries and Aquaculture Report No. 990* (pp. 243-252).

Campaigne, J., Maruti, J., Odhiambo, W., Ashrf, N., Karlan, D., & Gine, X. (2006). *Drumnet: Final Consolidated Technical Report.* Nairobi, Kenya: Pride Africa.

Clay, P. M., & Olson, J. (2008). Defining "Fishing Communities": Vulnerability and the Magnuson- Stevens Fishery Conservation and Management Act1. *Human Ecology Review, 15*(2), 143-160.

Donner, J. (2007). The Rules of Beeping: Exchanging Messages via Intentional Missed Calls on Mobile Phones. *Journal of Computer-Mediated Communication, 13*(1). http://dx.doi.org/10.1111/j.1083-6101.2007.00383.x

GSMA. (2015a). The Mobile Economy. *Sub-Saharan Africa 2015.* Retrieved from file:///C:/Users/USER/Documents/2015-10-08-721eb3d4b80a36451202d0473b3c4a63.pdf

GSMA. (2015b). *Global Mobile Economy Report.* Retrieved from file:///C:/Users/USER/Documents/GSMA_Global_Mobile_Economy_Report_2015.pdf

Hill, M. (2009). Using farmer's information seeking behaviour to inform the design of extension. *Extension Farming Systems Journal, 5*(2), 121-126.

Hoffmann, V., Gerster-Bentaya, M., Christinck, A., & Lemma, M. (2009). *Rural extension volume 1: Basic issues and concepts* (pp. 115-116). Weikersheim, Margraf Publishers.

Ifejika, P. I. (2012). Personal Profile of Mobile Phone Owners among Fisher Folk in Fishing Communities of Kainji Lake Basin, Nigeria. *International Journal of Rural Studies, 19*(2). Retrieved from http://www.vri-online.org.uk/ijrs

Ifejika, P. I., & Oladosu, I. O. (2011). Capability of Fisherfolk to Use Mobile Phone Facilities for Effective Extension Advisory Services around Kainji Lake Basin, Nigeria. *Journal of Rural Research and Information, 6*(2).

Ifejika, P. I., Nwabeze, G. O., Ayanda, J. O., & Asadu, A. N. (2009). Utilization of Mobile Phones as a Communication Channel in Fish Marketing Enterprise among Fishmongers in Western Kainji Lake Basin, Nigeria. *Journal of Information Technology Impact, 9*(2), 107-114.

Israel, G. D. (1992). *Determining Sample Size.* Fact Sheet PEOD-6. Florida Cooperative Extension Service. Institute of Food and Agricultural Sciences, University of Florida.

Jensen, R. T. (2007). The Digital Provide: Information (Technology), Market Performance and Welfare in the South Indian Fisheries Sector. *Quarterly Journal of Economics, 122*(3), 879-924. http://dx.doi.org/10.1162/qjec.122.3.879

Labonne, J., & Chase, R. S. (2009). The Power of Information: The Impact of Mobile Phones on Farmers' Welfare in the Philippines. *Policy Research Working Paper No. 4996*. Washington, DC: World Bank.

Leckie, G. J., Pettigrew, K. E., & Sylvain, C. (1996). Modelling the information seeking of professionals: A general model derived from research on engineers, health care professionals, and lawyers. *Library Quarterly, 66*(2), 161-193. http://dx.doi.org/10.1086/602864

Muthiah, G. (2015). Assessment of Mobile Voice Agricultural Messages Given to Farmers of Cauvery Delta Zone of Tamil Nadu, India. *The Journal of Community Informatics, North America, 11*. Retrieved August 31, 2016, from http://ci-journal.net/index.php/ciej/article/view/1067/1133

OECD. (2001). *Glossary of Statistical Terms: Fishing Community*. Retrieved from https://stats.oecd.org/glossary/detail.asp?ID=993

Okoye, F. C. (1992). Problems and prospects of fish pond culture around the kainji lake basin. *Proceedings of national conference on two decades of research on Lake Kainji held from 29th Nov. to 1st Dec. 1989* (pp. 212-214). NIFFR Publication.

Patel, N., Chittamuru, D., Jain, A., Dave, P., & Parikh, T. P. (2010). Avaaj Otalo - A Field Study of an In-teractive Voice Forum for Small Farmers in Rural India. *In Proc. CHI 2010*, 733-742.

Roopchand, A. Z. (2013). Caribbean Network of Fisherfolk Organizations (CNFO): Advocacy Strategy and Plan for Fisherfolk's Positions on Critical Issues Concerning the Implementation of Regional Fisheries Policies in the Caribbean. *CRFM Technical & Advisory Document* (p. 71, Number 2013/06).

Sane, I., & Traore, M. B. (2002). Mobile phones in time of modernity: The quest for increased self-sufficiency among women fishmongers and fish processors in Dakar. *International Development Research Centre, Document 14. 22.*

Solano, C., Leon, H., Perez, E., & Herrero, M. (2003). The role of personal information sources on the decision making process of Costa Rican dairy farmers. *Agricultural Systems, 76*, 3-18. http://dx.doi.org/10.1016/S0308-521X(02)00074-4

Torero, M. (2013). *Information and communication technology*. Power point presentation at Kigali, Rwanda, November 6, 2013.

Vergot, III P., Israel, G., & Mayo, D. E. (2005). Sources and channels of information used by beef cattle producers in 12 counties of the Northwest Florida extension district. *Journal of Extension, 43*(2).

Verma, A. K., Meena, H. R., Singh, Y. P., Chander, M., & Narayan, R. (2012). Information seeking and sharing behaviour of the farmers-A case study of Uttar Pradesh State, India. *Journal of Recent Advances in Agriculture, 1*(2), 50-55.

Wilson, T. D. (2000). Human Information Behaviour. *Information Science Research, 3*(2), 1-7.

7

The Tilapia Agrifood-Chain from a Sociopoietic Territorial Approach

Verónica Lango-Reynoso[1], Juan L. Reta-Mendiola[1], Felipe Gallardo López[1], Fabiola Lango-Reynoso[3], Katia A. Figueroa-Rodríguez[2] & Alberto Asiain-Hoyos[1]

[1] Colegio de Postgraduados, Campus Veracruz, Manlio Fabio Altamirano, Veracruz, México

[2] Colegio de Postgraduados, Campus Córdoba, Amatlán de los Reyes, Veracruz, México

[3] Instituto Tecnológico de Boca del Río, Boca del Río, Veracruz, México

Correspondence: Juan L. Reta-Mendiola, Colegio de Postgraduados, Campus Veracruz, Manlio Fabio Altamirano, Apartado Postal 421, Veracruz, México. E-mail: jretam@colpos.mx

The research is financed by Consejo Nacional de Ciencia y Tecnología, Colegio de Postgraduados, Campus Veracruz, Instituto de Desarrollo Social & Instituto Nacional de la Pesca.

Abstract

In the state of Veracruz, Mexico, the performance of the Tilapia (*Oreochromis* spp.) production system in the domestic market has been declining. Recent production results are lower than those presented in 1999, revealing that the production model adopted and used since 2001 is ineffective as a development strategy. The reason for the failure is that the model considers the technological production process as the central element of aquacultural competitiveness, without considering that production practices, marketing and consumption of goods are performed by individuals who decide and control their actions and are motivated by the values shared with their social group. This interpretation reveals the need for a new complementary conceptual framework, considering the system of production and consumption as a social self-referencing system. Thus, in this article, a model of an agricultural food-chain with a sociopoietic territorial focus on the development of the aquaculture subsector is outlined. The model is based on constructs having the following theoretical dichotomies: territorial agrifood, neo-institutional business, sociopoiesis and individual motivation.

Keywords: aquaculture, production chain, innovation, motivational values, product system

1. Introduction

The performance of the Tilapia (*Oreochromis* spp.) production system in the state of Veracruz, Mexico, has been declining in the domestic market, despite the investment of 3877.02 million MXN during 2014 and 2015 (Note 1) whose aim was to strengthen the physical, human and technological capital that would support this particular economic system (SHCP, 2016).

The production data show a declining annual growth rate of -0.5% in 2014, a loss of 3.98% during the period 2011 to 2013, and a market deficit of 37% in 2013 (Note 2) (CONAPESCA, 2011, 2012, 2013, 2014; Rangel-López et al., 2014). This information reveals that the contemporary aquacultural model did not develop competitive tilapia farmers capable of meeting the commercial opportunities in Mexico made since the late twentieth century (Note 3) in terms of imports of frozen product (Ceja, 2009).

The Supply Chain model promoted by the Mexican Government considers this primary subsector as an exporter of agro-industrial goods, and which coordinates production and marketing through a structural and decision-making group called Tilapia Product System Committee (Note 4) (Diario Oficial de la Federación, 2001). Although this organization is presented at two levels in Mexico, national and state, the state of Veracruz, similar to other tilapia-producing states, cannot be inserted homogeneously into international agrifood economic dynamics (Gutman & Gorenstein, 2003). Consequently, tilapia producers choose to act as individual agents in their quest for profit maximization (Kreps, 1990) coordinated through free market fluctuation (Gandlgruber, 2004).

The regulatory requirement for tilapia farmers, is established by a State policy (Morales, 2000) which adopted an exogenous economic logic having particular values (Schwartz, 1994) that prevented them from leveraging their social structure (Penrose, 2003) and acting as a cooperative organization capable of taking collective action (Ostrom, Ahn, & Olivares, 2003).

From the various models of agricultural production chains (Davis & Goldberg, 1957), recent research has focused on different technological aspects of their performance (Leonardo, Bijman, & Slingerland, 2015), unfortunately forgetting that production practices, marketing and consumption of agricultural products are made by people who control their actions based on their desires and preferences from memory (Note 5) of the society in which they lived (Bustillo et al., 2009; Casanova-Pérez et al., 2015a; Martínez-Dávila & Bustillo-García, 2010).

To understand these phenomena, a complementary framework founded on the methodological basis of food-chain models and their theoretical frameworks is necessary (Cuevas, 2010; José Muchnik, Requier-Desjardins, Sautier, & Touzard, 2007), as well as the use of third generation systems theories (Martínez-Dávila & Bustillo-García, 2010). This approach is consistent with the historical complexity of agriculture and the prevalence of social reproduction in the origin of agricultural production organization (Casanova-Pérez et al., 2015a).

In this article, we present a representative conceptual model of the territorial production-consumption system of Tilapia, based on the review, analysis and synthesis of bibliographic information and empirical evidence. For its construction, the theoretical framework of the theories and concepts of food-chain models (Cuevas, 2010), agrifood systems (José Muchnik et al., 2007) the theory of neo-institutional business (Penrose, 2003), the sociopoietic systems theory (Arnold-Cathalifaud, 2008), and the universal system of motivational values are included (Schwartz, 2006).

This viewpoint identifies the origin of the dysfunctions in Tilapia aquaculture production system in a diferent dimension compared to the traditional use (Note 6) (Vivanco, Martínez, & Taddei, 2010), visualizing the chain as a complex social phenomenon self-referenced by cultural aspects (Luhmann, 1998), such as motivational values.

1.1 Agricultural Business Context

Logical positivism has influenced economic thinkers of the twentieth century (Landreth, Colander, & Rabasco, 2006), establishing neo-classical capitalism using empirical-deductive postulates (Note 7). Based on econometrics and general laws, we can explain agricultural economic characteristics (Ballestero, 2003) from the end of the twentieth century until today (Note 8) (Bergalli, 2005).

Under this economic rhetoric, industrialized countries have established neo-liberal economic policies (Friedman, 2006), mimicked with few innovative components from other countries (Fajnzylber, 1995) where structural adjustments were made to ensure modernization (Liverman & Vilas, 2006). This structural transformation increased agricultural risk (Note 9) to global society (Costas, 2013). The neo-capitalist critics showed that the neo-liberal assumptions did not reflect agricultural reality (Lawrence, 2009) due to the complexity of the interrelations (Luhmann, 1998) (Table 1). Thus, analytical models and interpretations were included from institutions to counter the risk created by opening economies to a global environment (Luhmann, 1998).

Table 1. Assumptions of neo-liberal and institutionalist economics

Neo-liberal	Neo-institutionalist
Absence of transaction costs in perfect markets.	Institutions (cooperative, firms or organizations) were created to reduce transaction costs from market failures.
To achieve maximum benefits, neo-classical firms behave as black box systems where inputs are converted to products for market.	Institutions explain how they are organized to transform inputs into finished products for maximum benefits.
Explains how pricing is conducted.	Explains the basis of individual purchase decisions, and the infrastructure that keeps the market and companies generating economic activity.

Sources: Veblen (1945), Demsetz (1983), Coase (1996), Sykuta and Chaddad (1999).

Advanced industrial economies of the European Union's economic policy redefined economic policy and produced a model of agricultural development with fewer institutions (North, 1994). Neo-institutionalism proposed a regulatory and organizational framework for a better distribution of the benefits and risks of market integration through rational social management and reduction of organizational burden of productive activity of the State (DiMaggio & Powell, 1991).

On this basis the desectorialización of agriculture was redefined, providing greater involvement of multiple actors of agricultural development (Luhmann, 1998) under strategic, commercial and information guidelines, as well as technological innovation and resource management (Costas, 2013). In sum, agribusiness in the twenty-first century is a complex global phenomenon, oriented towards obtaining economic benefits, which responds to the evolution of the worldview of the dominant society that developed it. Therefore, it is necessary to expand agriculture´s vision itself by involving non-traditional aspects, but inherent to their nature, and to build its own representative and influential institutions in the sectorial politics of economic development and wealth generation for people engaged in this activity using their own resources.

1.2 Evolution of the Agrifood-Chain Approach

1.2.1 Origin

Production chains, as an abstraction of commercial agriculture, began developing more complex models. Historically, the focus on chains as applied to agriculture was promoted from the post-war period to ensure the supplies of raw materials and inputs from agribusiness in developed countries (Cook & Chaddad, 2000).

Initially, the organization of agricultural production was studied under North American agricultural capitalist perspectives. Davis and Goldberg (1957) broadened the vision on agriculture with the "Agribusiness Commodity System" model, referred to as an aggregate of subsystems involving all production-commercialization operations for an agricultural product industry, and the "Subsector Approach", used to analytically describe the productive subsystems (Cook & Chaddad, 2000).

Given the limitations of "Agribusiness" at explaining the forces influencing agricultural structure and dynamics, and their static and deductive approaches (Da Silva, 1994), models were generated from other economic schools to better explain this phenomenon (Caldentey, 2003). The fundamental French agrarian perspective is based the chain concept of Louis Malassis (1968) ("Systèmes Agro-alimentaires and Fillière Alimentaire") as a tool for the analysis of physical flows and interrelationships in agrifood products throughout all their production activities. This instrument evolved into the "Fillière de Production" when it included market satisfaction and its complex relationships (Lesage, 1984).

Early chain models were built on different visions of agricultural business from the neo-liberal, neo-classical, and institutional schools. All were destined to visualize the implicit interdisciplinary complexity in agroindustrial systems established to meet the rapid increase in demand for food in the dominant countries during the post-war period. Subsequent models evolved according to change in economic theory and these industrial production models were diffused to peripheral countries.

1.2.2 Evolution

In the 1980s, the US industrial sector coined the term "Supply Chain" to describe the process providing supplies focused on maximizing agroindustrial assembly (Shukla et al., 2011). Porter (1985), a business/economics researcher, introduced the model "Value Chain" to describe the process of generating product value through the relationships among industrial operational activities. When this model was applied to agricultural production, it referred to activities aligned through strategic alliances among companies, a necessity for products to flow from production to distribution (Peña et al., 2008).

During the 1990s in France, the Agrifood System (AS) was reinterpreted as an institution. In this system, the chains interacted as a logistics distribution unit covering all phases using networks of companies with technical and organizational efficiency (Morales, 2000); agency (Note 10) and convention (Note 11) theories explained their business operation and hierarchical coordination (Caldentey, 2003). The AS was distinguished as a productive process with high technological development that begins with raw material production and ends with the final consumer, thus requiring greater specialization and complex inter-organization (Morales, 2000).

In 1996, under the crisis in rural societies, the fundamentals of Local Food Systems (LFS) and "Milieu" (Local Innovation Systems) from the French school were proposed (Muchnik et al., 2007). This multidisciplinary system was integrated with production organizations and associated territorial services aspiring to adaptively compete through innovation and socioeconomic communication focused on local organizations having quality and social organization (Muchnik & Sautier, 1998).

The agrifood-chain model was rescued to distinguish agricultural food production within agribusiness (Cuevas, 2011), and incorporated the Value Chain to identify competitive advantages among the links. When necessary to compete in the market, value was created through joint operations, transformations, agents and markets sustained by physical and economic flows (Tallec & Bockel, 2005).

The diversity of conceptual models that exist on the production and consumption of agricultural goods is preserved as an axis in open systemic models attempting to describe the phenomenon from the interests of the dominant economic schools. This diversity allows for choices that better represent the characteristics and purposes of capital accumulation, and growth of business organizations or territorial development, even including some models having features that can enrich other compatible models.

Neo-institutional economics focuses on the organization of food and agribusiness subsystems that counteract market forces which restrict trade, including agricultural actors with limited resources (Flores, 2013). The agrifood-chain model represents the production structure and localized food systems through competitiveness of the chain based on the resources within a defined territory. A selected model must respond to historical societal developments as they spread.

1.3 Agrifood-Chain Trends Following Business/Economic Theory

A food-chain is primarily considered as a hierarchical institution (Williamson, 2009) organized as a larger system articulated by subsystems with their own elements. The chain is composed of productive links that bring together companies specializing in carrying out each stage of production of a good (Note 12).

If we consider that a group of companies form the chain, then one company is a fraction of the chain containing a link (Mandelbrot, 1997). Therefore, the behavior of a company is reflected in the behavior of the chain, yet chain performance, understood as a system, is not measured by the performance of any one company. The performance is a retroactive emergent quality, a product of the synergies arising from chain organization acting on businesses (Morin, 1995). Given the chain is a unit of observation and analysis of the production and consumption of an agricultural good, then the theory of institutional business/economics can be employed to understand how the management of company resources (Table 2) influences the competitiveness of the chain.

Table 2. Agrifood-chain resources

Origin	Types	Category	Resources
Internal	Tangible	Products	Human resource managers
			Administrative resources
			Economic resources
			Physical resources
			Productive human resources
			Technological resources
			Market resources
	Intangible	Relational	Organizational coordination
			Production coordination
			Market coordination
		Capacities	Cognitive capacity
			Learning capacity
External	Tangible	Production factors	Economic
			Human resources
			Technology
	Intangible	Outreach	Professionalizing
			Virtual support network
			Unions

Sources: modified from Penrose (1995) and (Nooteboom, 2009).

The company integrates production agents under one contract (Powell, 2003) as an option for coordination in relation to the market and its pricing mechanism (Coase, 1994). Integration depends on the relationship between the costs of buying or producing a good to reduce transaction costs and increase profits. An integrated enterprise is transformed into a system of relations that require a management body (Coase, 1996) (Note 13) responsible for the production and development of adaptive capacities and efficiency in transaction costs (Gandlgruber, 2010).

The company is engaged in the market through the coordination of prices or production (Coase, 1996). The company is competitive and unique *via* the productive coordination that exists when management uses and combines its resources (Penrose, 2003) to synchronize environmental needs with responsiveness (García & Taboada, 2012). Relational resources or inter-organizational collaborative links are part of the intangible internal resources of the company that can focus as a strategy for growth or maintenance (Williamson, 2009).

A command structure or organizational framework aimed at reducing transaction costs (Williamson, 1994) promotes the integration of work teams formed by agents or companies willing to sign an agreement to collaborate (contracts, licenses, franchises, mergers or strategic alliances) in the production by the company (Demsetz, 1997).

The company as a command structure facilitates coordination, monitoring, and dispute resolution to assist in decision-making and stability of contractual relations for production (Williamson, 2009). Its capacity for adjustment depends on the alignment of the structure with transactions (Note 14) that, together with their other accumulated capacities, distinguish it. Such capacities are related to skills, knowledge, strategies, human resources and materials, cooperative behavior and learning (Teece & Pisano, 1994).

The organization of the company (Note 15) depends on its command structures and the behaviors of its individuals defined by the specificity of their actions, the frequency of transactions, and uncertainty (Williamson, 2009). Companies differ in their direction and ability of the employer to use its idiosyncratic knowledge in solving organizational problems. Its limit on growth depends on its ability to develop collaborative partnerships within and outside the organization (García & Taboada, 2012). To increase income by satisfying the need to achieve, gather productive resources, and address inadequate capacities, the company directives use inter-business agreements as a hybrid form of management (Williamson, 2009) defined in terms of the capacity, resources, strategy, and vision of the cooperating companies (Penrose, 2003).

In short, from the perspective presented above, a chain is a productive socio-economic institution linking a group of companies. This view allows for systematization of the chain as a large company that brings together various organizations specialized in one phase of the production process of a commodity, and this is accomplished through cooperative inter-company agreements focused on the coordination of resources to reduce economic transaction costs.

As with any large company, the chain requires a governing body to be aware of what happens in agricultural markets, and to have an internal structure that promotes the alignment of production activities with the requirements of the same. In this perspective, to reduce market uncertainty and promote actions aimed at achieving competitiveness, it is indispensable to guide individual and group behavior of the chain to form an environment for strategic planning. To achieve this corporate vision, chains require an internal environment that fosters team-work, which is achieved when people who run businesses share a culture of cooperative work (Alchian & Demsetz, 2009).

1.4 Motivational Trends in an Agrifood-Chain

The main argument about the importance of social representation for organizations (Note 16) (Arnold-Cathalifaud, 2008) lies in determining their attitude towards the adoption or rejection of organizational strategies and related development with its growth or decline (Parales-Quenza & Vizcaíno-Gutiérrez, 2007). The way they see the world and act in it is stablish by the share beliefs of the society in which the organization evolves (Abric, 1984) is reflected in values, expressed as preferences and expectations (Parales-Quenza & Vizcaíno-Gutiérrez, 2007) that guide their habits, rules, patterns of behavior and learning, individual behavior, idiosyncratic knowledge, and business insight (García & Taboada, 2012).

The placement of values into the lives of individuals or groups is explained from its system requirements (Parsons, 2013). At the base, the physiological needs are followed by social needs and ultimately personal fulfillment (Maslow, 1970). Considering that values are the framework of individual behavior leading towards the satisfaction of needs, and that it is possible to differentiate individuals or social groups (Castro & Nader, 2006), then values serve as predictive frameworks for variations in behavior, and individual interests and attitudes (Luhmann, 2006).

In the model presented by Schwartz, (2006), "Portrait Values Questionnaire", the values are trans-situational goals, variables with degrees of importance, guiding the life and behavior of an individual or a social group. Values exist in this system underlying all cultures, grouped in contrasting dimensions oriented toward promotion of personal interests in collective function (Self-transcendence), prioritizing self-interests (Self-enhancement), safety and order (Conservation), and independence of actions and thoughts (Openness to change). Together, these groups of values define collectivist (Note 17), individualistic (Note 18), or mixed associations of social and cultural aspects (Figure 1).

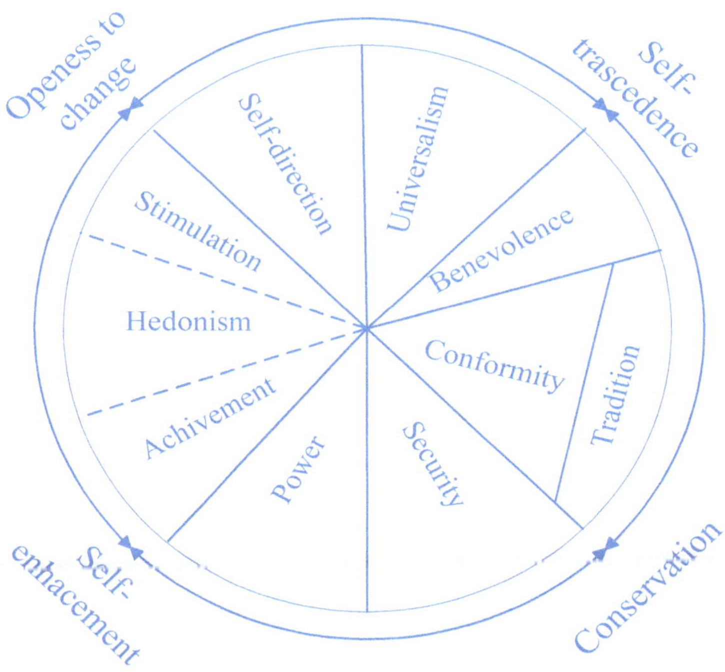

Figure 1. Theoretical relational model of motivational values

Source: Schwartz (2003).

For Schwartz, (2006), the values are personal and group responses to biological needs, coordination of social actions, and group function and survival. Thus, in an agrifood-chain motivational values can be identified that guide actors toward establishing chain organization with the companies as components (Alexander & Colomy, 1990). Based on the identified values, the social group to which the actors belong designs motivating incentives to change individual attitudes and behaviors towards a cooperative organization, such as strategic alliances.

In a culture that values individuality, coinciding with the values of the predominant economic system, power, achievement and hedonism also are valued, making it difficult for people to accept a business growth strategy based on joint work (Schwartz, 2006). Therefore, to encourage cooperation, actors must offer members additional rewards to benefit the chain, and these achievements will allow them to acquire what they value as individuals (Olson, 1965).

Motivational values are a guide for actions occurring within the chain that reflect individual influence on system behavior. To change actions of the actors towards business growth, we must recognize the motives that drive changes in attitude and behavior towards an intervention that capitalizes on its social capital. If the chain does not value team-work, then the introduction of new values will modify knowledge on partnerships from within their cultural environment.

1.5 Social Behavior in an Agrifood-Chain

The research of social and cultural systems developed by Luhmann (1998) focuses on systems that involve human interactions organized by sense (Arnold, 1989). Here, evolution occurs through historical reflexive processes, including beliefs and values, nonexistent in natural systems (Hadis et al., 1976). Among these systems are the societal, organizational, interactional and group (Luhmann, 1998).

This theoretical development is a comparative approach away from the traditional systemic notions applied to the study of society and culture (Arnold-Cathalifaud, 2008), which visualize the problems of contemporary society with the concepts of functional equivalents (Merton, 1968) and contingent functional social structures (Arnold, 1989).

Under this view, society is historically located in a territory defined by its linguistic and cultural system, and is interpreted as a complex self-referencing system that is naturally closed (Martínez-Dávila & Bustillo-García, 2010). Such conditions reduce the complexity created by the range of their possible responses to environmental pressures, and meet their basic needs of self-reproduction with uninterrupted historical differentiation of autonomous hierarchical functional subsystems (Figure 2), but interdependent on the scale of global society (Luhmann, 1998).

Among the functional systems, the economic system is regulated by money exchange for goods obtained from specialized production systems regulated by capital. This system exists in geographical areas oriented toward agrifood systems (Luhmann, 2013), where they are responsible for the generation of food from production to consumption (Thorpe & Bennett, 2004). The process is distributed among several specific subsystems identified by the agricultural production system and interpreted as an agroecosystem (AES) to the consumer system (Figure 2) (Martínez-Dávila & Bustillo-García, 2010).

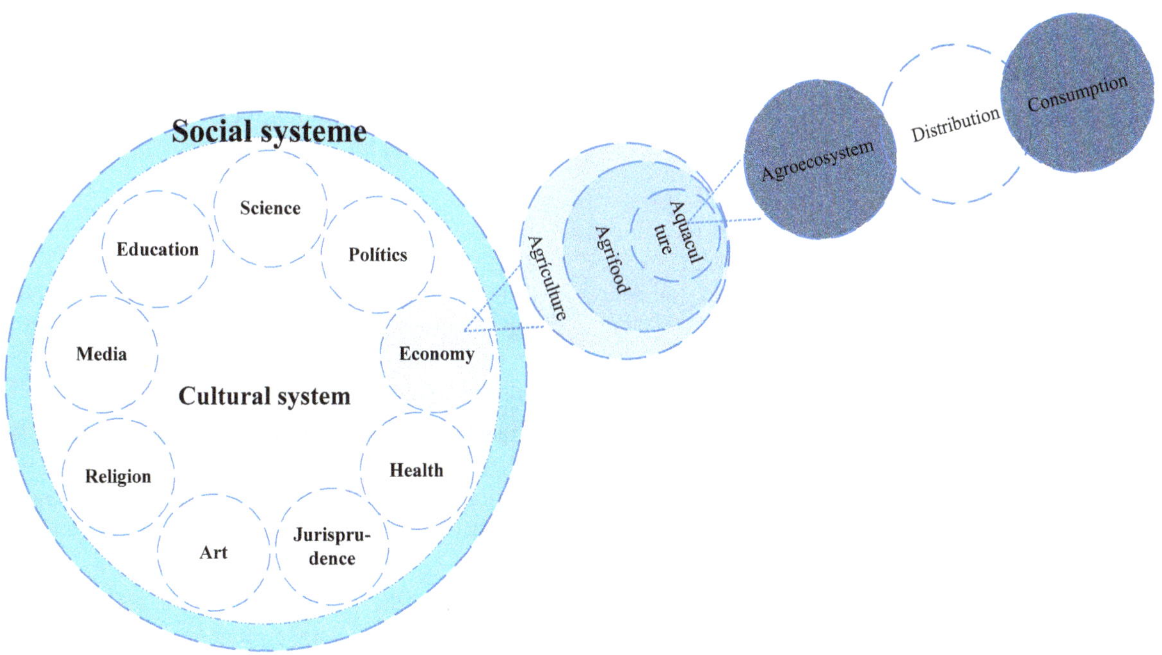

Figure 2. Functional societal systems

Note. Modified from Martínez-Dávila and Bustillo-García (2010).

The autopoiesis applicable to social organizations is known as sociopoiesis. Out of the functional subsystems arose organizational systems that performed production differentiation through the mobilization, orientation and integration of activities necessary to achieve the goal of the functional subsystem (Arnold-Cathalifaud, 2008). Society organized the agrifood system and its subsystems to maintain and reproduce itself by meeting the basic human need for food through its availability and accessibility which is managed by organizations related to the production and consumption of agricultural products (Casanova-Pérez et al., 2015a).

Sociopoiesis permits an understanding of the social reproduction that forms functional systems that meet the basic needs of society. As societal complexity increases, these systems reproduce in an effort to reduce the complexity (Luhmann, 2013). The economic system satisfies the need for food by a food product from a production process regulated by capital. From the agricultural ecosystem, through other systems for transport, marketing and consumption of an aquaculture product, access and availability can be acquired which are regulated by societal demand. Thus, an aquaculture food-chain can be analyzed as a sociopoietic system.

1.6 The Chain as a Model for Agricultural Development in México

In Mexico, the concept of a Production Chain arose in the mid-1970s in work performed for the analysis of agribusiness systems (Vigorito, 1977). Under the influence of the United Nations Economic Commission for Latin America and the Caribbean (ECLAC), the Agrifood System was described as the unit of analysis for economic reproduction and was characterized in order to process agroindustrial transformation and property structures (Morales, 2000).

During the 1980s, researchers disseminated and adapted the model according to their political-economic trend and the type of end-product from the chain. Agroindustrial chains, supply chains, agroindustrial systems, production chains, or production structures were designed (Casar & Ros, 1983; Loría, 2009). Their links were identified with the production, processing, distribution and consumption of an agricultural good, including credit support services, marketing, and production technology (Arroyo, 1986).

The globalizing economy attracted the interest of social scientists to analyze the agroindustrial structure in Mexico. This work was performed to determine the status of important agricultural sectors (Maclas, 1999), where the structure of the agroindustrial chain was modified and its operation focused on quality control of the product for export without considering distribution and consumption (Note 19). In this approach, they only included the links of production and processing, although they acknowledged relations with input from producers and assemblers (Rodríguez, 1999). After the year 2000, the Production Chain model was used regularly to understand the various industries involved in the development of an agricultural product (González, 2002).

The Production Chain model spread within the agricultural sector as a tool of observation and analysis of production-consumption, and became widespread as a tool of production organization for the globalist economic development model implemented in Mexico under the influence of its main trading partners.

1.6.1 Regulatory Base

Contemporary national economic policy is recognized in the production chain as an element for analysis and management of agricultural production structure, and is aimed at increasing productivity, profitability and competitiveness. The Sustainable Rural Development Law (2001) refers to this as the production-consumption chain including the production process of an agricultural or agroindustrial good in rural areas (Diario Oficial de la Federación, 2001). This model is intended to create the conditions for introducing to the world market competitive agricultural products by integrating agents and additional value to agroindustry.

The neo-classical capitalist model of Agribusiness is perceived as the predecessor to the Production Systems (PS) model, referred by Goldberg (1979) as Product Systems, which aims to meet the nutritional needs of consumers based on their lifestyle. In PS, participants in production, marketing, farming supply and storage, and entities and organizations that affect and coordinate goods production and distribution are integrated (Diario Oficial de la Federación, 2001).

The PS is a legal entity created to legally recognize the existence of the production chain and is a form of organization of agricultural production. It is conceived as a set of elements and concurrent agents of production processes for agricultural products, which maintain the overall structure of the supply chain model (Diario Oficial de la Federación, 2001). For operational purposes, the Secretary of Agriculture, Livestock, Rural Development, Fisheries and Food considers both terms equivalent, yet they are not because the first is a legal model and the second is a conceptual model representing the same reality.

The PS model includes a governing body related with other entities responsible for economic policy. The Production Systems Committee (PSC), comprised of representatives of the links in the production chain, is the legal entity of the PS which manages the planning and organization of production (forms of coordination), integration of links (agreements or strategic alliances) and commercial product protection.

The PS and PSC were created to strengthen national food production of eleven basic strategic commodities such as fish, including tilapia, therefore only the PSC was developed at a national scale with representatives from each state for each product. The state PSC in collaboration with the respective Rural Development District are responsible for developing strategies to strengthen productivity and competitiveness of the chain links located in its territory. The strategies involve materials management to support production, social organization, and development of marketing strategies, technology training, business development, and technological development (Diario Oficial de la Federación, 2001).

The full and effective framework proposed for chains is not recognized in one paragraph; between sections and articles of the Law are the elements that describe their methodology. This legal framework forms the foundation of strategic planning aimed at the articulation of the chain, which shows that the production chain is the responsibility of the PSC.

Based on the logic of global industries, the production chain model is valid, operable, and easy to understand and can incorporate elements according to its perspective, even Law facilitates social organization for productive purposes. Thus, actors can take action to strengthen the PS, the PSC, and the chain itself for the benefit of sectorial and territorial development.

Therefore, this model is capable of organizing agricultural production within a free market economy where actors are homogeneous, are owners of the means of production, where information flows throughout the entire system, and economic benefit is always pursued through the pricing system; a situation contrary to what exists in Mexico. Only large-scale agricultural enterprises competitively incorporate international trade, causing a deficit in the trade balance due to the loss of national markets for agricultural, horticultural, fruit, livestock, fishery and aquaculture products (Ayala Garay et al., 2011; SAT et al., 2016). In particular, national aquaculture production of tilapia is fighting a strong commercial battle against imported products; if a primary product is not competitive, the conservation of its domestic market is at risk.

1.7 Description of the Tilapia Agrifood-Chain

Considering that the aim of aquaculture is growing organisms for human consumption in an aquatic environment, it exists within the collective of *agro-operations* (Engle & Stone, 2005), and its production process (Jensen et al., 2010) can be analyzed with a systemic approach. The Tilapia Agrifood-Chain (TAC) involves an interdependent group of agents working together to bring their products from the aquaculture system or agroecosystem to the consumer (Thorpe & Bennett, 2004). Tilapia is an aquaculture good that is sold in different markets and in different forms (Engle & Stone, 2005), so the analytical model and chain organization vary according to production complexity (See Appendix A).

The chains for high consumption products (commodities) focus on the production, processing and distribution of merchandise for export (Gibbon, 2001), and involve a network of agents and complex relationships with coordinated management and efficiency to meet international market needs and regulations. Its products, whole frozen tilapia, frozen fillets, and fresh fillets, provide a dynamic flow on the international market by increasing demand related to its low sales prices (Young & Muir, 2002).

The chain of high-quality products focuses on increasing the value of prime materials from the aquaculture facility to the consumer's table (Jensen et al., 2010). It involves companies, organizations and related institutions by focusing on agreements to increase the competitiveness of the chain through strategic resource management (Jespersen et al., 2014). Processed tilapia products with export quality and high sales prices target specialized markets (Young & Muir, 2002).

The fresh aquaculture products chain was developed at national, regional, and local levels (GLOBEFISH, 2014). Its production flow, biological input-production-processing-marketing-consumption, characterize its structure (Reta et al., 2007). For these chains, the primary inputs are the hatcheries, and product quality is heterogeneous due to the technological diversity of its production systems (Hernández et al., 2002). The value that is added to whole tilapia, refrigerated or on ice, is limited to filleting or artisanal deboning (Vivanco et al., 2010), and their sales prices vary according to quality and demand (Watanabe et al., 2002).

Given the nature of the aquaculture products offered live in regional markets (Thapa, Dey, & Engle, 2015), the production system is a production chain that behaves like a value chain. Fish must be maintained under optimal conditions, physically and biologically, until they are purchased by consumers, and thus considered as high quality products (Young & Muir, 2002), and this requires a specialized technology base that is reflected in a high sales price for fresh animal produce. The transport system is considered as a link chain, not a service due to its importance in the process (Puduri et al., 2011). In the marketing link it includes the use of aquaculture systems to maintain and display living organisms in supermarkets, markets, and restaurants (Thapa et al., 2015), as well as other points of urban and rural sale (Lango-Reynoso et al., 2015). The production process is based on an organizational model where the building of partnerships among members of the links is intended to coordinate chain resources to meet the needs of the product market.

Aquaculture chains follow the general model established by FAO (Piñones, Acosta, & Tartanac, 2006), where their differences are related to the final product and its productive organization. The value chain is distinguished from the supply chain according to the degree of integration and coordination among the actors, as well as differentiation and competitiveness (Note 20) achieved by managing its resources (Rugman & Verbeke, 2002).

Thus, the developmental stages of the tilapia aquaculture chain are observed in its organizational forms as a production chain, value chain, or supply chain.

2. Methodology

Research conceptual definitions were built within the aquaculture development framework, agrifood chain approach was used to describe Tilapia (*Oreochromis* spp.) production-consumption process.

To build the theoretical structure of the agrifood chain from the sociopoietic territorial approach, which pretends to explain the behavior of the commercial interchange processes between producers and traders of live tilapia, as part of the production consumption links, a deductive critical analysis was developed of the basic and secondary literature (Sautu et al., 2005), related with chain models and the general neo-institutional company theory, including autopoietic social systems, and motivational values, to link to the empirical evidence of corporate, motivational and social systems behaviors. Search sources were: JSTOR, Elsevier, Science Direct Freedom Collection, Taylor and Francis Group, and Google Academic.

From August 2011 to June 2016 a study case (Martínez, 2011) was established on the live tilapia selling points model. This technology was previously transferred by rural innovation processes (Reta et al., 2011) to a market network established at Sotavento region, Veracruz state, Mexico. The Study Case include social, technological, and economic aspects related to the network creation, operation and technical support, to outline performance.

3. Results

3.1 Theoretical Proposal for the Tilapia Agrifood-Chain from a Sociopoietic Territorial Approach

Given that an agrifood-chain is the link for consumption of an agricultural product, and according to the description of the agricultural ecosystem as an autopoietic unit (Martínez-Dávila & Bustillo-García, 2010), we propose the Tilapia Agrifood-Chain using a Sociopoietic Territorial Approach (TAC/STA) as a functional social organization composed of specialized technological, economic, and human elements that distinguish and limit its environment. This organization is a closed structure of self-generated communications and relations which flow from energy, economics, materials, information, and services exchanges needed to go from the production to the consumption of an agricultural good (Figure 3).

The performance of the TAC/STA responds to the effect of interaction between decisional communications and components, rather than the addition of elements (Arnold, 1989). In its environment, situated in an area defined by its culture, is the society that created it, in addition to other specialized functional social systems that recognize and interact with equalities, differences, and hierarchical transversalities in economic, political, scientific, educational, and jurisprudence scope (Figure 2) (Arnold-Cathalifaud, 2008). Inside the environment also are productive actors or individuals controlling production systems, characterized as Psychic Systems (PS) (Note 21) (Casanova-Pérez et al., 2015b), who, through their knowledge of the environment and communicative abilities regarding perceived influence, through its self-reflexive, act to supply or disrupt the mechanisms responsible for recurrent and presupposed operational actions that maintain chain structure.

The cultural system of the territory (Luhmann, 2006), perceived by actors in the chain, influences the development of selective criteria or values used in decision-making process of individual consciousness systems about the characteristics of the information, that will be to communicated within the chain structure (Arnold-Cathalifaud, 2008). Such information refers to that which exists in the system environment regarding resource management of agricultural enterprises (Casanova-Pérez et al., 2015a; Penrose, 2003).

The cultural system of the territory (Luhmann, 2006), perceived by the actors in the chain, influences the development of selective criteria or values used in decision-making systems (Casanova-Pérez et al., 2015a; Penrose, 2003).

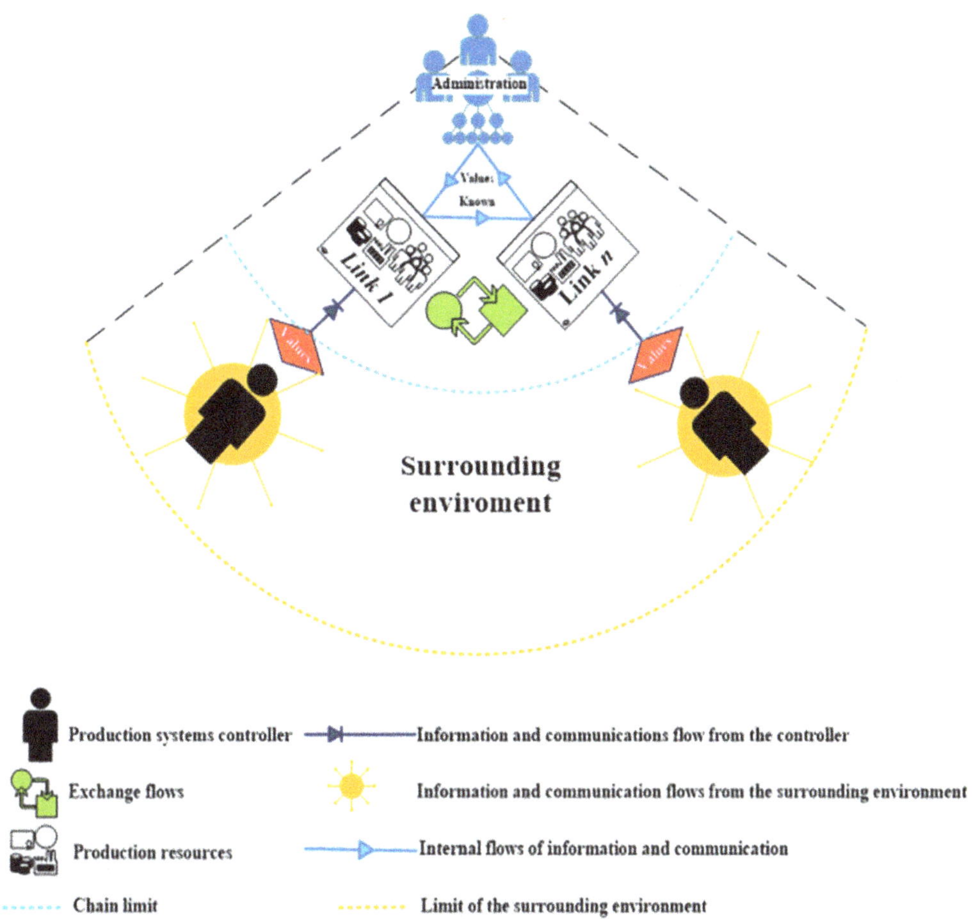

Figure 3. Conceptual diagram of an agrifood-chain using a sociopoietic territorial approach

When information comes to the chain, it is reproduced in a self-referential form, behaving like a recursive series of actions, which determine the cultural capital of the chain (Bourdieu, 1986). It manifests itself in the management practices of the elements or organization resources and exchanges of energy, materials and financial resources, necessary to achieve its social function, and to maintain its status of conservation, growth, or reduction, depending on the existing possibilities. A change in the status of the TAC/STA, in relation to the dynamics of its environment, is only generated by its structure, part of an empirical self-reflection process that accounts for its status and possibilities of accepting an exogenous innovation that maintains its operations and preserves system viability (Arnold-Cathalifaud, 2008).

The viability of the chain, reflected in its capacity of structural coupling to the demands of the market, is its ability to evolve through auto-interventions or innovations that strengthen its capacities to coordinate resources (Arnold-Cathalifaud, 2008; Penrose, 2003). However, such interventions can come from outside when they are discriminated against by rational choices of actors through appropriate interventions by agents of change (Arnold-Cathalifaud, 2008).

Acceptance of exogenous innovations related to collective actions in an organizational social system, such as alliances or partnerships, is conditioned by a payment which goes beyond social benefit (Olson & Zeckhauser, 1966). The search for this benefit causes disturbances in the organizational continuum (Olson, 1965), so it is necessary to identify the nature of incentives to guide the choices of actors, who are subjected to innovation, toward acceptance and adoption (Knoke, 1988).

The formation of alliances among links of the TAC/STA can be seen as a communicative coordination structure with predictive capabilities (Nooteboom, 2009), and is an innovation tool whose members are conditioned by the interests of individuals to join (Knoke, 1988), responding to what they value as members of the society in which they developed (Coase, 1994).

The cultural system lies in the production system environment and is made up of information communicated linguistically through a self-constructed collective memory, containing the values and beliefs that influence the attitudes and behaviors of the actors in relation to their social system (Willke, 1993).

Given that the system is a rational consciousness, which acts as a receiver and generator of information communicated linguistically between TAC and their environment (Casanova-Pérez et al., 2015a), coordination of system resources *via* partnerships should be instituted to achieve its social function and be perceived by actors as restrictive recursive actions.

In conclusion, an agrifood-chain is an autonomous body that reproduces itself through self-generated resources, thus its self-produced performance for its actions is equal to the state of its resources generated by information circulating on the network of human interactions that maintains organization structure. Motivational values of the people who transmit information to the communication structure guide all actions, including intervention, because they follow their natural direction.

4. Discussion

In the state of Veracruz, Mexico, the Tilapia Agrifood-Chain is composed of five links formed by specialized companies (Table 3) (Reta et al., 2007). Fresh tilapia, as its main product (Rangel-Lopez et al., 2014), does not compete with the price and presentation of products nationally, nor with imported tilapia, and thus has lost part of its market. The lack of competitiveness (North, 1994) is related to the low articulation of production among links, without which it is not possible to coordinate the resources and system capabilities (Arrow, 1969).

As a business strategy to increase the competitiveness of the Tilapia Agrifood-Chain based on its own resources and the nature of the territory, a network marketing of live tilapia was established at specified points of sale (Lango-Reynoso et al., 2015). Building partnerships between the links is necessary for quality and quantity, and to maintain product quality during its transit through the chain to meet the regional market (Table 4).

Table 3. Components of the Tilapia Agrifood-Chain in Veracruz, Mexico

Links	Principal actors	Production factors	Exchange flows	Specialized functional systems
Biological inputs	Producer of biological inputs	Natural resources	Economics	Financing
Nursery feeding	Tilapia producer	Economic capital	Materials	Government
Transportation	Transport	Production methods	Services	Input providers
Marketing	Marketer	Human resources	Energy	Education
Consumption	Consumer	Technological resources	Information	Research
		Technical and scientific knowledge		

Table 4. Trade issues between producers and marketers of live tilapia

Actors	Problem	Alternative Solution	Link requirements
Marketers	Supply: inconsistent supply volume and frequency, variation in product quality (size, appearance, health), high prices, lack of specialized transportation service	Wide range of suppliers	Stable production. Improved aquaculture, harvest, and post-harvest practices. Preferred prices without seasonal variation. Product delivery to the home.
Producers	Sales: Inconsistent demand in quantity and frequency	High volume of inventory and seasonal production	Volume and frequency are stable for purchasing

Based on the performance of "Point of Sales of Live Tilapia" as a marketing strategy to increase the competitiveness of the agrifood-chain in the Sotavento region of the state of Veracruz (Lango-Reynoso et al., 2015), the system is not articulated. Thus, the system cannot be promoted and managed by its organization through cooperative agreements among the actors to coordinate their production resources and generate cognitive capacities and learning, which differs from other competing systems; strategies needed to maintain and grow the participating companies (Nooteboom, 2009).

This chain is a functional system located in a defined territory, which is responsible for meeting the social demand for tilapia as a food product through an integrated network of relationships based on communication. This system is coupled to the social environment by a representative governing body of production agents (Arnold-Cathalifaud, 2008), formally established in the territory (Diario Oficial de la Federación, 2001) and called the Systems Committee for Tilapia Products (SCTP).

The actors involved (PS) influence decisions about management practices and organization of resources of the structure. Through interaction with other recognized functional systems in the environment (scientific, educational, financial, political and inputs), the PS receives information to turn it into knowledge transmitted as information on the structure of the chain and, once there, becomes shared knowledge that influences the behavior of the system. This is a recurring process with double contingency, because every productive and organizational action responds to a self-referred decision from the previous action, self-reproduced (Luhmann, 1998).

An innovation focused on the development of social capital (Ostrom & Ahn, 2009) and use of endogenous resources in the chain (Muchnik & Sautier, 1998) alters the traditional management practices and organization of its production resources (Penrose, 2003), and its competitive performance (Soosay, Hyland, & Ferrer, 2008). These interventions are viable only when the functional structure of the chain is aware of its status through self-observation and self-recognition. This change will be achieved by entering new knowledge into the system through the exchange of information on cooperative work (Arnold-Cathalifaud, 2008) and production differentiation (Sankaran & Suchitra, 2006).

To accept alliances prior to intervention, the structure recognizes motivational PS values (Schwartz, 2006) to design motivating stimuli to guide the collective action of innovation.

Recognizing that the social values of the chain actors motivate their decisions on the management of individual and collective resources (Schwartz, 2006), it follows that the PS involved are unknown to or inexperienced in the ways and means of collective work (Soosay et al., 2008). The value of a relational resource such as production structure alliances is that it does not take advantage of developing companies, or the chain, limiting their competitiveness.

Among the cultural (PS) and structural (SCTP) there are controversies that result in inefficiencies and even threats to the survival of aquaculture firms (Allaire & Firsirotu, 1984). Resistance to change is reflected in traditional practices, hence the importance of teamwork is incipient and probably due to mistrust of joint work, the alleged loss of control in their companies due to collective work, and changes in power relations. To know and accept the idea of partnerships, they must satisfy their motivational needs beyond direct benefits and insist that there are strategic actions for their companies (Olson, 1965).

Culture (PS) can change interest in an organizational system (Denison, 1991). To accept an intervention, they must initially develop intervention strategies according to individual values. Afterwards, they can intervene in the proposed participatory change in the chain and its structure by spreading collective values. In an environment of cooperation and coordination of resources to improve competitiveness (Lado & Wilson, 1994), the new chain values (cultural capital) will be part of their intangible resources (Bourdieu, 1986) and the foundation of its social capital (Note 22) (Ostrom & Ahn, 2009) to benefit the production and market activity of the actors (García & Taboada, 2012).

The design of the chain based on its cultural environment and endogenous resources allows for adoption among firms. Thus, the production sector will be politically representative and have management capacity to promote human, social and economic development in the region (IICA, 2010). Under this system (Note 23) the interweaving of necessary relations will develop for firms to grow by following regulatory and organizational standards for production actions and group values, thus laying the foundation for consensus decisions on activities for collective and individual benefit (Ostrom et al., 2003).

Given the agrifood-chain designed as an institution and then translated into an organized business, and that such actions can influence markets, politics and other institutions (Dossi & Lissin, 2011), the system is able to establish formal rules, informal norms and their system application in order to provide the space necessary for economic activity (North, 1994).

5. Conclusion

In conclusion, the self-referencing social systems approach allows us to understand the state of the Tilapia Agrifood-Chain, and to recognize that the nature of the recurring actions taken to solve their business problems come from the shared culture within the territory in which it operates. They can only change their status by introducing new information to the structure formed by the communication network. Partnerships as a strategy for coordination of resources in the chain will only be accepted when people change their interests to benefit the collective.

Chain growth depends on the management of system resources independently of the individual and the achievement of an entrepreneurial culture. Values influence the decisions of the actors on management practices of its resources and capabilities, loss of markets, and the competitiveness of the aquaculture chain.

References

Abric, J. C. (1984). A theoretical and experimental approach to the study of social representations in a situation of interaction. In R. M. Far & S. Moscovici (Eds.), *Social representations* (pp. 169-183). Cambridge: Cambridge University Press.

Alchian, A., & Demsetz, H. (2009). Production, information costs, and economic organization. In R. S. Kroszner & L. Putterman (Eds.), *The Economic Nature of the Firm: A Reader* (pp. 173-196). Cambridge University Press. http://dx.doi.org/10.1017/CBO9780511817410.015

Alexander, J. C., & Colomy, P. (1990). Neofunctionalism today: reconstructing a theoretical tradition. In G. Ritzer (Ed.), *Frontiers of social theory: The new syntheses* (pp. 33-67). New York: Columbia University Press.

Allaire, Y., & Firsirotu, M. E. (1984). Theories of organizational culture. *Organization Studies, 5*(3), 193-226. http://dx.doi.org/10.1177/017084068400500301

Arnold, M. (1989). Teoría de Sistemas. Nuevos Paradigmas: Enfoque de Niklas Luhmann. *Revista Paraguaya de Sociología, 75*(26), 1-56.

Arnold-Cathalifaud, M. (2008). Las organizaciones desde la teoría de los sistemas sociopoiéticos. *Cinta de Moebio, 32*, 90-108. http://dx.doi.org/10.5902/6383

Arroyo, G. (1986). La biotecnologia y el análisis de las cadenas o sistemas agro-alimentarios y agroindustriales. *Revista Centroamericana de Economía, 19*(25), 247-264.

Ayala Garay, A. V., Sangerman-Jarquín, D. M., Schwentesius Rindermann, R., Almaguer Vargas, G., Barrera, J., & Luis, J. (2011). Determinación de la competitividad del sector agropecuario en México, 1980-2009. *Revista Mexicana de Ciencias Agrícolas, 2*(4), 501-514.

Ballestero, E. (2003). La economía agrícola: tendencias y horizontes. *Revista Española de Estudios Agrosociales y Pesqueros, 200*, 503-546.

Bergalli, R. (2005). Relaciones entre control social y globalización: Fordismo y disciplina. Post-fordismo y control punitivo. *Sociologías, Porto Alegre, 7*(13), 180-211. http://dx.doi.org/10.1590/s1517-45222005000100008

Bourdieu, P. (1986). The forms of capital. In J. Richardson (Ed.), *Handbook of Theory and Research for the Sociology of Education* (pp. 241-258). New York: Greenwood Publishing Group. http://dx.doi.org/10.1002/9780470755679.ch15

Bustillo, G. L., Martínez, D. J. P., Osorio, A. F., Salazar, L. S., González, A. I., & Gallardo, L. F. (2009). Grado de sustentabilidad del desarrollo rural en productores de subsistencia, transicionales y empresariales, bajo un enfoque autopoiético. *Revista Científica, 19*(6), 650-658.

Caldentey, P. (2003). Neoinstitucionalismo y Economía Agroalimentaria. *Contribuciones a la economía.* Retrieved June 15, 2016, from http://www.eumed.net/ce/

Casanova-Pérez, L., Martínez-Dávila, J. P., López-Ortiz, S., Landeros-Sánchez, C., López-Romero, G., & Peña-Olvera, B. (2015a). El agroecosistema comprendido desde la teoría de sistemas sociales autopoiéticos. *Revista Mexicana de Ciencias Agrícolas, 6*(4), 855-865.

Casanova-Pérez, L., Martínez-Dávila, J. P., López-Ortiz, S., Landeros-Sánchez, C., López-Romero, G., & Peña-Olvera, B. (2015b). Enfoques del pensamiento complejo en el agroecosistema. *Interciencia, 40*(3), 210-216.

Casar, J. T., & Ros, J. (1983). Problemas estructurales de la industrialización en México. *Investigación Económica, 42*(164), 153-186.

Castro, S. A., & Nader, M. (2006). La evaluación de los valores humanos con el Portrait Values Questionnaire de Schwartz. *Interdisciplinaria, 23*(2), 155-174.

Ceja, P. H. (2009). *Estudio sobre: Evaluación de Medidas de Control a las Importaciones de Tilapia y sus Productos Mediante Medidas Arancelarias y la Aplicación Estricta de Procedimientos de Trazabilidad.* México: SAGARPA/CONAPESCA.

Coase, R. H. (1994). *La Empresa, el Mercado y la Ley*. Madrid: Alianza editorial

Coase, R. H. (1996). La naturaleza de la empresa. In E. L. Suárez (Trans.), In O. Williamson & S. Winter (Eds.), *La Naturaleza de la Empresa: Orígenes, Evolución y Desarrollo* (pp. 29-48). México: Fondo de Cultura Económica.

CONAPESCA. (2011). *Anuario Estadístico de Pesca y Acuicultura.* México: SAGARPA.

CONAPESCA. (2012). *Anuario Estadístico de Pesca y Acuicultura.* México: SAGARPA.

CONAPESCA. (2013). *Anuario Estadístico de Pesca y Acuicultura.* México: SAGARPA.

CONAPESCA. (2014). *Anuario Estadístico de Pesca y Acuicultura.* México: SAGARPA. Retrieved May 14, 2016, from http://www.conapesca.sagarpa.gob.mx/wb/cona/cona_anuario_estadistico_de_pesca

Cook, M. L., & Chaddad, F. R. (2000). Agroindustrialization of the global agrifood economy: bridging development economics and agribusiness research. *Agricultural Economics, 23*(3), 207-218. http://dx.doi.org/10.1016/s0169-5150(00)00093-1

Costas, D. P. (2013). *Agronegocios: Entre la Sociedad del Riesgo y el Neoliberalismo.* Buenos Aires: Instituto de Investigaciones Gino Germani, Facultad de Ciencias Sociales, Universidad de Buenos Aires.

Cuevas, R. V. (2010). Análisis del enfoque de cadenas productivas en México. *Textual, 56*, 83-93.

Da Silva, J. G. (1994). Complejos agroindustriales y otros complejos. *Agricultura y Sociedad, 72*, 205-240.

Davis, J. H., & Goldberg, R. A. (1957). *A Concept of Agribusiness.* Boston: Division of research. Graduate School of Business Administration. Harvard University.

Demsetz, H. (1997). *La Economía de la Empresa.* Madrid: Alianza Editorial.

Denison, D. (1991). *Cultura Corporativa y Productividad Organizacional.* Colombia: Editorial Legis

Diario Oficial de la Federación. (2001). *Ley de Desarrollo Rural Sustentable* (pp. 41-80). Secretaria de Agricultura Ganadería Desarrollo Rural Pesca y Alimentación. México.

DiMaggio, P. J., & Powell, W. W. (1991). *The New Institutionalism in Organizational Analysis* (Vol. 17). United States of America: University of Chicago. Press Chicago, IL.

Dossi, M., & Lissin, L. (2011). La acción empresarial organizada: Propuesta de abordaje para el estudio del empresariado. *Revista Mexicana de Sociología, 73*(3), 415-443.

Engle, C. R., & Stone, N. M. (2005). Aquaculture: Production, Processing, Marketing. In W. G. Pond & A. W. Bell (Eds.), *Encyclopedia of Animal Science (Print)* (pp. 48-51). New York: Marcel Dekker.

Fajnzylber, F. (1995). Latin American Development: From the "Black Box" to the "Empty Box." In B. H. Koo & D. H. Perkins (Eds.), *Social Capability and Long-Term Economic Growth* (pp. 242-265). UK: Palgrave Macmillan. http://dx.doi.org/10.1007/978-1-349-13512-7_12

Flores, N. A. (2013). *Diagnóstico de la Acuicultura de Recursos Limitados (AREL) y de la Acuicultura de la Micro y Pequeña Empresa (AMYPE) en América Latina* (December Vol.). Roma: FAO.

Friedman, T. L. (2006). *La Tierra es Plana.* Madrid: Ediciones Martínez Roca.

Gandlgruber, B. (2010). *Instituciones, Coordinación y Empresas: Análisis Económico más allá de Mercado y Estado* (1st ed.). Barcelona: Anthropos Editorial. Universidad Autónoma Metropolitana.

Gandlgruber, B. B. (2004). Abir la caja negra: teorías de la empresa en la economía institucional. *Análisis Económico, XIX*(41), 19-58.

García, G. A., & Taboada, I. E. L. (2012). Teoría de la empresa: las propuestas de Coase, Alchian y Demsetz, Williamson, Penrose y Nooteboom. *Economía: Teoría y Práctica, 36*, 9-42.

Gibbon, P. (2001). Upgrading primary production: A global commodity chain approach. *World Development, 29*(2), 345-363. http://dx.doi.org/10.1016/S0305-750X(00)00093-0

GLOBEFISH, H. (2014). Tilapia Highlights. *Una Actualización Trimestral Sobre los Mercados Mundiales de Productos Pesqueros* (Vol. 3, pp. 28-30). FAO/GLOBEFISH Highlights.

Goldberg, R. A. (1979). El Enfoque de Sistemas y los Agribusiness. *Cuadernos de Desarrollo Rural, Septiembre*(1), 8-18.

González, M. L. M. (2002). *La Industrialización en México* (1st ed.). México: Universidad Nacional Autónoma de México. Instituto de Investigaciones Económicas. Miguel Ángel Porrúa.

Gutman, G. E., & Gorenstein, S. (2003). Territorio y sistemas agroalimentarios. Enfoques conceptuales y dinámicas recientes en la Argentina. *Desarrollo Económico, 42*(68), 563-587. http://dx.doi.org/10.2307/3455905

Hadis, B. F., Berger, P. L., Luckmann, T., & Hadis, B. F. (1976). La construcción social de la realidad. *Desarrollo Económico, 16*(60), 641. http://dx.doi.org/10.5944/rif.4.2005.5447

Hernández, M. M., Reta, M. J. L., Gallardo, L. F., & Nava, T. M. E. (2002). Tipología de productores de mojarra tilapia (*Oreochromis* spp.): base para la formación de grupos de crecimiento productivo simultaneo (GCPS) en el Estado de Veracruz, México. *Tropical and Subtropical Agroecosystems, 1*(1), 13-19.

IICA. (2010). *Desarrollo de los Agronegocios y la Agroindustria Rural en América Latina y el Caribe: Conceptos, Instrumentos y Casos de Cooperación Técnica.* San José, Costa Rica: Instituto Interamericano de Cooperación para la Agricultura (IICA).

Jensen, T. K., Nielsen, J., Larsen, E. P., & Clausen, J. (2010). The Fish Industry—Toward Supply Chain Modeling. *Journal of Aquatic Food Product Technology, 19*(3-4), 214-226. http://dx.doi.org/10.1080/10498850.2010.508964

Jespersen, S. K., Kelling, I., Ponte, S., & Kruijssen, F. (2014). What shapes food value chains? Lessons from aquaculture in Asia. *Food policy, 49*, 228-240. http://dx.doi.org/10.1016/j.foodpol.2014.08.004

Knoke, D. (1988). Incentives in collective action organizations. *American Sociological Review, 53*(3), 311-329. http://dx.doi.org/10.2307/2095641

Kreps, D. (1990). Corporate Culture and Economic Theory. In J. Alt & K. Shepsle (Eds.), *Perspectives on Positive Political Economy* (pp. 90-143). New York: Cambridge University Press. http://dx.doi.org/10.1017/cbo9780511571657.006

Lado, A. A., & Wilson, M. C. (1994). Human resource systems and sustained competitive advantage: A competency-based perspective. *Academy of Management Review, 19*(4), 699-727. http://dx.doi.org/10.2307/258742

Landreth, H., Colander, D. C., & Rabasco, E. (2006). *Historia del Pensamiento Económico* (4th ed.). España: McGraw Hill.

Lango-Reynoso, V., Reta-Mendiola, J. L., Asiain-Hoyos, A., Figueroa-Rodríguez, K. A., & Lango-Reynoso, F. (2015). "Live Tilapia": Diversifying livelihoods for rural communities in México. *Journal of Agricultural Science, 7*(10), 101-112. http://dx.doi.org/10.5539/jas.v7n10p101

Lawrence, K. F. (2009). Reviewed Work: Absentee Ownership: Business Enterprise in Recent Times: The Case of by Thorstein Veblen. *Political Science Quarterly, 39*(3), 509-512. http://dx.doi.org/10.2307/2142605

Leonardo, W. J., Bijman, J., & Slingerland, M. A. (2015). The Windmill Approach: Combining transaction cost economics and farming systems theory to analyse farmer participation in value chains. *Outlook on Agriculture, 44*(3), 207-214. http://dx.doi.org/10.5367/oa.2015.0212

Liverman, D. M., & Vilas, S. (2006). Neoliberalism and the environment in Latin America. *Annual Review of Environment and Resources, 31*, 327-363. http://dx.doi.org/10.1146/annurev.energy.29.102403.140729

Loría, E. (2009). Sobre el lento crecimiento económico de México: una explicación estructural. *Investigación Económica, 68*(270), 37-68.

Luhmann, N. (1998). *Sistemas Sociales: Lineamientos para una Teoría General* (1st ed., Vol. 15). Anthropos Editorial.

Luhmann, N. (2006). *La Sociedad de la Sociedad.* México: Editoral Herder -Universidad Iberoamericana

Luhmann, N. (2013). La economía de la sociedad como sistema autopoiético. *Revista Mad. Magister en Análisis Sistémico Aplicado a la Sociedad, 29*, 1-25. http://dx.doi.org/10.5354/0718-0527.2013.27342

Maclas, A. A. (1999). Tendencias de la restructuración agroindustrial en la actividad lechera mexicana. In B. E. Martínez, M. A. Álvarez, H. L. A. Gracía & M. de C. Del Valle (Eds.), *Dinámica del sistema lechero mexicano en el marco regional y global* (pp. 183-202). México D.F.: Plaza y Valdez Editores.

Mandelbrot, B. (1997). *La Geometría Fractal de la Naturaleza* (2nd ed.). Barcelona: Tusquets Editores.

Martínez, C. P. C. (2011). El método de estudio de caso: Estrategia metodológica de la investigación científica. *Pensamiento y Gestión, 20*(Julio), 165-193.

Martínez-Dávila, J. P., & Bustillo-García, L. (2010). La autopoiesis social del desarrollo rural sustentable. *Interciencia, 35*(3), 223-229.

Maslow, A. H. (1970). In J. Fadiman & C. McReynolds (Eds.), *Motivation and Personality* (2nd ed.). New York: Harper & Row.

Merton, R. K. (1968). *Social Theory and Social Structure*. New York: The Free Press.

Morales, E. A. (2000). Los principales enfoques teóricos y metodológicos formulados para analizar el sistema agroalimentario. *Revista Agroalimentaria, 6*(10), 75-88.

Morin, E. (1995). *Introducción al Pensamiento Complejo* (2nd ed.). Madrid: Editoral Gedisa.

Muchnik, J., & Sautier, D. (1998). *Systèmes Agro-alimentaire Localisés et Construction de Territoires*. ATP CIRAD.

Muchnik, J., Requier-Desjardins, D., Sautier, D., & Touzard, J. M. (2007). Systèmes agroalimentaires localisés. *Economies et Sociétés AG, 29*, 1465-1484.

Nooteboom, B. (2009). *A Cognitive Theory of The Firm: Learning, Governance and Dynamic Capabilities*. UK: Edward Elgar Publishing. http://dx.doi.org/10.4337/9781848447424

North, D. C. (1994). La evolución de las economías en el transcurso del tiempo. *Revista de Historia Económica/Journal of Iberian and Latin American Economic History (Second Series), 12*(03), 763-778. http://dx.doi.org/10.1017/s021261090000481x

Olson, M. (1965). *The Logic of Collective Action: Public Goods and the Theory of Groups*. United States of America: Harvard University Press.

Olson, M., & Zeckhauser, R. (1966). An economic theory of alliances. *The Review of Economics and Statistics, 48*(3), 266-279. http://dx.doi.org/10.2307/1927082

Ostrom, E., & Ahn, T. K. (2009). The meaning of social capital and its link to collective action. In G. T. Svendsen & G. L. Svendsen (Eds.), *Handbook of Social Capital: The Troika of Sociology, Political Science and Economics* (pp. 17-35). USA: Edward Elgar Publishing. http://dx.doi.org/10.4337/97818484474 86.00008

Ostrom, E., Ahn, T. K., & Olivares, C. (2003). Una perspectiva del capital social desde las ciencias sociales: Capital social y acción colectiva. *Revista Mexicana de Sociología, 65*(1), 155-233. http://dx.doi.org/10.2307/3541518

Parales-Quenza, C. J., & Vizcaíno-Gutiérrez, M. (2007). Las relaciones entre actitudes y representaciones sociales: Elementos para una integración conceptual. *Revista Latinoamericana de Psicología, 39*(2), 351-361.

Parsons, T. (2013). *The Social System*. New York: Routledge.

Penrose, E. T. (2003). *The Theory of the Growth of the Firm*. http://dx.doi.org/10.1093/0198289774.003.0002

Piñones, V. S., Acosta, A. L. A., & Tartanac, F. (2006). *Alianzas Productivas en Agrocadenas* (1 st ed.). Santiago, Chile: Oficina Regional de la FAO para América Latina y el Caribe.

Powell, W. (2003). Neither market nor hierarchy. Network forms of organization. In M. J. Handel (Ed.), *The sociology of organizations: Classic, contemporary, and critical readings* (pp. 315-330). United Kindom: SAGE Publications.

Puduri, V. S., Govindasamy, R., Myers, J. J., & O'Dierno, L. J. (2011). Consumer attitude towards pricing of live aquatic products. *Aquaculture Economics & Management, 15*(2), 118-129. http://dx.doi.org/10.1080/13657305.2011.573522

Rangel-López, L., Lango-Reynoso, F., Asian-Hoyos, A., & Castañeda-Chávez, M. de R. (2014). Diagnóstico de la acuacultura en el municipio de Alvarado, Veracruz, México. *Ra Ximhai, 10*(6), 75-81.

Reta, M. J. L., Luna, F. J., Zetina, C. P., Suarez, S. C., Mena, G. J. M., & Ramos, H. A. (2007). *Programa Maestro Tilapia para el Estado de Veracruz*. México: SAGARPA.

Reta, M. J. L., Mena, G. J. M., Asiain, H. A., & Suarez, S. C. (2011). *Manual de procesos de innovación rural (PIR) en la acuacultura*. México: Colegio de Postgraduados.

Rodríguez, G. M. G. (1999). Las particularidades de la globalización de la leche: una propuesta de análisis. In B. E. Martínez, A. M. A., H. L. A. García & M. de C. Del Valle (Eds.), *Dinámica del sistema lechero mexicano en el marco regional y global* (1st ed., pp. 87-125). México: Plaza y Valdes.

Rugman, A. M., & Verbeke, A. (2002). Edith Penrose's contribution to the Resource-Based View of Strategic Management. *Strategic Management Journal, 23*, 769-780. http://dx.doi.org/10.1002/smj.240

Sankaran, J. K., & Suchitra, M. V. (2006). Value-chain innovation in aquaculture: Insights from a New Zealand case study. *R&D Management, 36*(4), 387-401. http://dx.doi.org/10.1111/j.1467-9310.2006.00441.x

SAT, SE, Banco de México, & INEGI. (2016). *Balanza Comercial de Mercancías de México*. SNIEG, Información de Interés Nacional. Retrieved June 10, 2016, from http://www.banxico.org.mx/SieInternet/consultarDirectorioInternetAction.do?accion=consultarCuadro&idCuadro=CE122§or=1&locale=es

Sautu, R., Boniolo, P., Dalle, P., & Elbert, R. (2005). *Manual de metodología: Construcción del marco teórico, formulación de los objetivos y elección de la metodología*. Buenos Aires: CLACSO.

Schwartz, S. H. (2003). A proposal for measuring value orientations across nations. *Questionnaire Package of the European Social Survey* (pp. 259-290).

Schwartz, S. H. (2006). A theory of cultural value orientations: Explication and applications. *Comparative Sociology, 5*(2), 137-182. http://dx.doi.org/10.1163/ej.9789004170346.i-466.55

SHCP. (2016). *Transparencia presupuestaria*. Retrieved February 25, 2016, from http://transparenciapresupuestaria.gob.mx/es/PTP/SED

Soosay, C. A., Hyland, P. W., & Ferrer, M. (2008). Supply chain collaboration: Capabilities for continuous innovation. *Supply Chain Management: An International Journal, 13*(2), 160-169. http://dx.doi.org/10.1108/13598540810860994

Sykuta, M. E., & Chaddad, F. R. (1999). Putting theories of the firm in their place: A supplemental digest of the new institutional economics. *Journal of Cooperatives, 14*(1), 68-76.

Tallec, F., & Bockel, L. (2005). *Commodity Chain Analysis. Constructing the Commodity Chain: Functional Analysis and Flow Charts*. Rome: FAO.

Teece, D., & Pisano, G. (1994). The dynamic capabilities of firms: An introduction. *Industrial and Corporate Change, 3*(3), 537-556. http://dx.doi.org/10.1093/0198290969.003.0006

Thapa, G., Dey, M. M., & Engle, C. (2015). Consumer Preferences for Live Seafood in the Northeastern Region of USA: Results from Asian Ethnic Fish Market Survey. *Aquaculture Economics & Management, 19*(2), 210-225. http://dx.doi.org/10.1080/13657305.2015.1024346

Thorpe, A., & Bennett, E. (2004). Market-driven international fish supply chains: The case of Nile perch from Africa's Lake Victoria. *International Food and Agribusiness Management Review, 7*(4), 40-57.

Vigorito, R. (1977). *Criterios Metodológicos para el Estudio de Complejos Agroindustriales*. México: ILET

Vivanco, A. M., Martínez, C. F. J., & Taddei, B. I. C. (2010). Análisis de competitividad de cuatro sistema-producto estatales de tilapia en México. *Estudios Sociales, 18*(35), 165-207. http://dx.doi.org/10.1016/j.techfore.2010.05.005

Watanabe, W. O., Losordo, T. M., Fitzsimmons, K., & Hanley, F. (2002). Tilapia production systems in the Americas: technological advances, trends, and challenges. *Reviews in Fisheries Science, 10*(3-4), 465-498. http://dx.doi.org/10.1080/20026491051758

Williamson, O. E. (1994). La Organización del Trabajo: Un Enfoque Institucional Comparativo. In L. Putterman (Ed.), *La Naturaleza Económica de la Empresa* (pp. 361-384). España: Alianza Editorial.

Williamson, O. E. (2009). In E. L. Suárez (Trans.), *Las Instituciones Económicas del Capitalismo*. México: Fondo de Cultura Económica.

Willke, H. (1993). Systemtheorie entwickelter Gesellschaften: Dynamik und Riskanz moderner gesellschaftlicher Selbstorganisation. *Grundlagentexte Soziologie*. München: Juventa Verlag.

Young, J. A., & Muir, J. F. (2002). Tilapia: Both fish and fowl? *Marine Resource Economics, 17*(2), 163-173. http://dx.doi.org/10.1086/mre.17.2.42629359

Notes

Note 1. According to the expense reports Productivity Promotion Program for Fisheries and Aquaculture.

Note 2. Based on national figures of apparent consumption (15 3011 t), production (96 827 t), and imports (54 859 t), it is estimated that in the state 30,474 tons of tilapia have been consumed, of which only 63% are state produced, assuming that 100% of the production is consumed locally, although there is evidence that at least 4% of the production goes to other states.

Note 3. Imports of Chinese tilapia started in 2004 and have grown 245% since that time.

Note 4. Under the Law of Sustainable Rural Development, the Production System Committee is the governing body making decisions and promoting actions to favor the competitiveness of domestic aquaculture activities.

Note 5. Luhmann conceives the collective memory as a set of memories and expectations of the social system, the result of communication system operations evaluated positively or negatively.

Note 6. Technological, economic, and market management have been considered as the basis for the competitiveness of Mexican aquaculture chains.

Note 7. The neoclassical economic paradigm, focused on maximizing economic gains, is based on a free market, perfect competition and zero state participation.

Note 8. The term globalization is a neologism referring to the expansion of an economic development system based on the quantitative theory of money that proclaims interest in the value of money and incentives to develop the market.

Note 9. Considering the scarcity of natural resources, climate change, health risks, economic uncertainty, the use of inappropriate technology, and market problems.

Note 10. Including the theory of transaction costs proposed by R. Coase in 1937 in his book *The Nature of the Firm* and developed by Kenneth Arrow in 1973 in his general equilibrium model.

Note 11. Where forms of client-provider organization are proposed.

Note 12. In the case of consumer links, the elements are consumer families.

Note 13. The management body is a command structure (governance) organized through the conduct of companies, which is formed by their habits and rules.

Note 14. Considering attributes, frequency and intensity.

Note 15. The company, hybrid relationships, or the market.

Note 16. These systems are companies, cooperatives, civil associations, state services, and non-governmental organizations.

Note 17. The values of a collectivist society focus on the interests of the entire community.

Note 18. The values of an individualistic society focus on development and personal success.

Note 19. From breaking traditional commercial agreements.

Note 20. For this work, competitiveness is the ability of an agricultural value chain to maintain itself and grow over time, a market share of its main product, and this is derived from the implementation by its management team of specific mechanisms for efficient management of their heterogeneous internal and external resources, the utilization of their productive opportunities, and the development of their learning abilities and accumulation of knowledge. Thus, they can respond to changes in the environment; which contributes to the permanence and economic growth of its actors through the creation of value between links in the chain.

Note 21. A psychic system is a system of awareness, which from the chain approach, is composed of the actors who control the links in the agrifood-chain.

Note 22. Social capital is seen as an attribute of individuals and their relationships that improves their ability to solve problems needing collective action (Ostrom and Ahn, 2009). This form of shared norms, common knowledge and rules of use regarding an institutional arrangement are used to solve common dilemmas (Ostrom et al., 2003).

Note 23. Understood and organized as an institution.

Appendix

Appendix A. Characterization of Tilapia Agrifood-Chain

Model	Approach	Structure	Function	Management Approach	Product	Prices	Production Systems	Type of Business
Supply chain	Global logistics management	Network services and integrated distribution options for multiple agents	Procurement of materials, processing of these materials into intermediate and finished products, and distribution to customers	Facilitating communication, efficient exchange of information and knowledge, coordinated decisions, cooperative work, integrated activities	Quantity and quality with international standards focused on high consumption items (whole fish and frozen fillets)	Low international prices	Intensive systems	Large scale with high technology development
Value chain	Global business	Joint companies linked by flows of information, materials, and financial and service funtions for special products	Add value to the product or service along the links of economic agents ranging from production to consumption	Establishing business relationships (hierarchical alliances or strategic networks), efficiency in resource use	Quantity and quality with international standards focused on products differentiated by having high added value (live fish, fresh fillets)	High international and domestic prices	Intensive systems	Large and medium scal with high technology development
Production chain	Sectorial agribusiness	Links interconnected by flows of energy, capital, materials, and information, all integrated by agents related to a production process	Process of materials into a final product for delivery to consumers	Coordination of actors and activities	Quantity and quality of variables focused on a fresh whole products	Low domestic and regional prices	Semi-intensive and extensive systems	Medium and small-scale with medium and small-scale technology development

Effects of Phosphate Solubilizing Fungi on Growth and Yield of Haricot Bean (*Phaseolus vulgaris* L.) Plants

Firew Elias[1], Diriba Muleta[2] & Delelegn Woyessa[1]

[1] Department of Biology, Jimma University, Ethiopia

[2] Environmental Biotechnology Unit, Addis Ababa University, Ethiopia

Correspondence: Delelegn Woyessa, Department of Biology, Jimma University, P.O. Box 5140, Ethiopia. E-mail: woyessa@yahoo.com; delelegn.woyessa@ju.edu.et

Abstract

Haricot bean (*Phaseolus vulgaris* L.) is one of the most important cash crops and export commodities besides its use in human food and soil fertility improvement. Phosphorus (P) is one of the major bio-elements that limits agricultural production. However, phosphate-solubilizing fungi play a noteworthy role in increasing the bioavailability of soil phosphates for plants. The purpose of this study was to evaluate the effects phosphate solubilizing fungi on the growth of haricot bean plants. Cultural and morphological features were used to tentatively identify the fungal isolates to genus level. Based *In vitro* phosphate solubilization efficieny conducted in both solid and liquid PVK medium following standard procedures, two best isolates were selected and evaluated under greenhouse for their performance on haricot bean. Under greenhouse experiment, shoot height (47.31 cm plant^{-1}), root length (41.01 cm plant^{-1}), nodule number (65.67 plant^{-1}), nodule dry weight (0.59 g plant^{-1}), shoot fresh weight (62.73 g plant^{-1}), shoot dry weight (14.33 g plant^{-1}), number of pod (12.89 plant^{-1}), 50-seed weight (35.89 g plant^{-1}), P content (0.59%) and N content (1.96%) were significantly increased by co-inoculation of two isolates (PSFAP) in the soil amended with rock phosphate (RP) compared to control. Moreover, the highest number of leaves (59.55 leaves plant^{-1}) and root fresh weight (14.19 g plant^{-1}) were recorded as a result of inoculation with isolate PSFP compared to control. The present study indicated the presence of potential plant associated fungi that possess phytobeneficial traits for extending their use as microbial biofertilizers after testing their suitability for the desired purpose.

Keywords: biofertilizers, phytobeneficial traits, rock phosphate, tricalcium phosphate

1. Introduction

Haricot bean (*Phaseolus vulgaris* L.) is the most important pulse crop predominantly grown in the central rift valley for cash crop and export commodity to generate foreign exchange in Ethiopia (Ferris & Kaganzi, 2008). The crop also plays an important role in improving the soil fertility by fixing atmospheric nitrogen with the association of *Rhizobium* species present in the root nodules (Thalooth et al., 2006).

Phosphorus and Nitrogen are the most critical bio-elements that limit Haricot bean production in Ethiopia (Taye, 2007; Girma, 2009; Gifole et al., 2011). Phosphorus (P) is one of the most indispensable macronutrients next to nitrogen for the growth and development of plants (Hameeda et al., 2008).

Although chemical fertilizers are being added to the soil to increase the availability of phosphorus for plants, a large portion of which is rapidly immobilized and becomes unavailable to plants and can lead to an overall reduction in soil fertility after application (Das et al., 2003). This leads to frequent application of chemical phosphatic fertilizers. However, a regular use of chemical fertilizers can cause severe environmental degradation in addition to their escalating costs. For instance, the repeated and injudicious applications of P fertilizers can lead to the loss of soil fertility by disturbing microbial diversity, and consequently reduces yield of crops (Gyaneshwar et al., 2002). Therefore, the necessity to develop economical and eco-friendly fertilizers is steadily increasing (Reddy et al., 2002; Chuang et al., 2007).

An increase in phosphorus availability to plants through the inoculation of phosphate-solubilizing microorganisms in rock phosphate (RP) amended soil has also been reported under greenhouse and field conditions (Whitelaw, 2000; Hameeda et al., 2008). Several authors reported enhanced growth and yield on wheat (Xiao et al., 2009), soybean (Iman, 2008), maize (Richa et al., 2007; Bojinova et al., 2008) and pea (Kevin & Krista, 2001) through inoculation of P-solubilizing fungi (PSF) in rock phosphate (RP) amended soils. Emphasis is therefore, being placed on the possibility of utilization of rock phosphate, which may be made available to plants by microbiologically mediated processes in order to provide efficient and environmentally desirable approach compared to current technology for industrial P fertilizer production (Bojinova et al., 2008).

Among the rhizosphere microbes, the important genera of P-solubilizing bacteria include *Rhizobium*, *Bacillus* and *Pseudomonas* (Wani et al., 2007; Muleta, 2012). *Penicillium* and *Aspergillus* spp. are the dominant P solubilizing filamentous fungi found in rhizosphere (Wakelin et al., 2004). They are widely used as producers of organic acid. *Aspergillus niger* and some *Penicillium* species have been tested for solubilization of RP and other biotechnological importance such as biocontrol, biodegradation and phosphate mobilization (Chuang et al., 2007; Richa et al., 2007; Pandey et al., 2008).

In Ethiopia, few studies on effects of phosphate solubilizing microorganisms on different crops have been undertaken. Accordingly, effects of plant growth promoting rhizobacteria (PGPR) on growth and yield of tef was evaluated by Woyessa and Assefa (2011) and the effects of phosphate solubilizing rhizobacteria with different doses of poultry manure and RP on the growth and yield of tomato was also undertaken by Gebremeskel and Muleta (2011). Furthermore, the effect of phosphate solubilizing fungus on growth and yield of tef was studied by Hailemariam (1993). However, information on the effects of phosphate solubilizing fungi in RP amended soil on growth of crop plants is scanty in the country.

Hence, this study was to isolate phosphate solubilizing fungi and evaluate their effects on the growth, yield and nutrient content (N and P) of haricot bean plant grown in soil amended with RP under greenhouse condition.

2. Materials and Methods

2.1 Description of the Study Area

The study was carried out in Jimma University (JU), Jimma town, located at 353 km to the south-west of the capital, Addis Ababa. The microbial analysis was conducted in Postgraduate and Research Laboratory, Department of Biology. The greenhouse experiment and soil analysis were conducted at College of Agriculture and Veterinary Medicine, Soil Laboratory, JU. The total area of Jimma zone is 18415 km^2 and located between latitudes 7°18′N and 8°56′N and longitudes 35°52′E and 37°37′E.

2.2 Collection of Rhizosphere Soil Sample

A total of one hundred fifty rhizosphere soil samples were collected from 30 plant samples of cabbage (*Birassica interifolia*), faba bean (*Vicia faba* L.), haricot bean (*Phaseolus vulgaris* L.), sugar cane (*Saccharum officinarum* L.) and tomato (*Lycopersicon esculentum* Mill). The rhizosphere soil samples were collected from selected kebeles of Jimma town (Becho Bore, Ginjo Guduru and Awetu Mendera) and Mana district (Sombo Mana, Hunda Toli, Kemise Waraba, Buture and Gudeta Bula) of different farm land sites. The kebeles were purposively selected based on the preliminary survey made to identify potential growing areas for the crop. The samples were randomly collected from agricultural fields within 1 to 2 km interval between the sampling sites (Woyessa & Assefa, 2011). Roots with adhering soils of healthy plants were collected and transferred to sterile plastic bags and transported to the laboratory and stored at 4 °C for further analysis within 24 hrs.

2.3 Isolation of Phosphate Solubilizing Fungi

Collected rhizosphere soil samples were used for the isolation of phosphate solubilizing fungi on Pikovskaya's agar medium (PVK) as described by Pikovskaya (1948). The medium was autoclaved at 121 °C for 15 minutes. About 20 ml of the sterilized molten agar medium was poured into each petri plate and supplemented with 25 μg ml^{-1} chloramphenicol and allowed to solidify before inoculation.

For each sample, the loosely adherent soils were removed by agitating the roots strongly; the root samples with their adhering soil were cut in to pieces (1-2 cm) by sterile scissors and used for isolation. Ten grams of each plant root fragment with adhered soil was aseptically weighed and transferred to 250 ml Erlenmeyer flask containing 90 ml of 0.85% saline solution. The suspension was shaken on 110 rpm for 25 minutes on a rotary shaker and then allowed to settle for 10 min. Aliquots of 1 mL of the supernatant from the sample was transferred to 9 ml of sterile physiological saline solution in test tubes and serially diluted from 10^{-1}-10^{-6}. From appropriate serially diluted soil suspension, 0.1 ml aliquots were transferred and spread plated on Pikovskaya's agar plates and incubated at 25 °C-28 °C for 5-7 days. Fungal isolates that showed clear zones around the colonies were further purified by transferring into Pikovskaya's agar medium. The pure cultures were then preserved on Potato Dextrose Agar (PDA) slant at 4 °C for further investigation.

2.4 Screening of Fungi for Phosphate Solubilization

The Fungal isolates obtained from rhizospheric soils were evaluated on agar plates and liquid culture containing sparingly soluble phosphates for their activity in mobilizing phosphate from insoluble sources (tricalcium and rock Phosphate media).

2.5 Determination of Solubilization Index on Solid Medium

All the isolates were screened under *In vitro* condition for their phosphate solubilization activity following the method described by Iman (2008) on Pikovskaya agar medium. A spot inoculation of fungal isolates was made on to the plates in triplicate under aseptic condition and incubated at 25-28 °C for 7 days. Uninoculated PVK agar plate served as control. Comparative solubilization index measurement was carried out on day seven of incubation by measuring clear zone and colony diameters in centimeter. Phosphate solubilization index was determined by using: ratio of the total diameter (colony + halo zone) and the colony diameter (Edi-Premono et al., 1996).

$$\text{Solubilization Index (SI)} = \frac{\text{Colony diameter} + \text{Halo zone diameter}}{\text{Colony diameter}} \quad (1)$$

2.6 Identification and Characterization of Phosphate Solubilizing Fungi

The characteristics of fresh cultures with best *In vitro* P solubilization efficiency were further characterized with mycological identification keys and taxonomic description (Cheesbrough, 2000) to identify the isolated fungi to the genus level. Identification was based on colony characteristics and microscopic features. Among the colonial characteristics such as surface appearance, texture and colour of the colonies both from upper and lower sides were considered. In addition, conidia, conidiophores, arrangement of spores and vegetative structures were determined with microscopy. The identified fungi were maintained on Potato Dextrose Agar (PDA) slant at (4 °C) for further investigation.

2.7 Selection of Potential Bioinoculants

Based on qualitative solubilization index (SI) on solid medium and quantitative solubilization efficiency using PVK broth containing TCP & RP, two isolates designated as PSFA and PSFP were selected to evaluate their performance as individual inoculum and as co-inoculants (PSFAP) on haricot bean under greenhouse condition.

2.8 Greenhouse Experiment

A greenhouse experiment was conducted to evaluate the effectiveness of PSF in improving the availability of P levels and the growth of haricot bean plants in rock phosphate amended soil conducted at College of Agriculture and Veterinary Medicine (JUCAVM).

2.8.1 Physicochemical Analysis of Soil and RP

The physico-chemical properties of soil and Rock Phosphate samples used for pot experiment were initially characterized. The soil was obtained from uncultivated land of Jimma University, College of Agricultural and Veterinary Medicine (JUCAVM). Composite samples were taken using sterilized polyethylene bag across the field from a depths of 0-30 cm and bulked for laboratory analysis and immediately transported to Jimma University College of Agriculture and Veterinary Medicine, Soil Testing Laboratory. The soil samples were then air-dried, crushed using a mortar and pestle, and sieved through a 2 mm mesh. Bikilal Rock Phosphate was crashed, grounded with pestle and mortar and then sieved to 2 mm mesh size. The sieved samples were stored in new polythene bags for laboratory chemical and physical analyses.

All laboratory analyses of soil and RP samples were done following the procedures as outlined by Sahlemedhin and Taye (2000) manual. The soil samples were air-dried and ground to pass a 2 mm sieve and 0.5 mm sieve (for total N) before analysis. Soil texture was determined by Bouyoucos hydrometer method (Black et al., 1965). The pH and electrical conductivity of the soils were measured in water (1:2.5 soil:water ratio). Organic carbon content of the soil was determined following the wet combustion method of Walkley and Black (Black et al., 1965). Total nitrogen content of the soil was determined by wet-oxidation (wet digestion) procedure of Kjeldahl method (Sahlemedhin & Taye, 2000). The available phosphorus content of the soil was determined by Bray II method. The available potassium was determined by Morgan's extraction solution (Bray & Kurtz, 1945).

2.9 Preparation of Fungal Inoculum and Seed Inoculation

2.9.1 Fungal Inoculum

The inocula of the two selected fungal isolates were prepared according to Zaidi and Khan (2006). Fungal spores from 10-day old cultures were transferred to 100 ml of sterilized potato dextrose broth medium and incubated on orbital shaker at 120 rpm at 25-28 °C for seven days. The cultures broth were filtered through Whatman No. 42 filter paper into a sterile glass bottle and washed with sterilized distilled water under aseptic condition. The pelleted cells were re-suspended with sterilized distilled water. Then the suspensions were adjusted to approximately 10^6 spore cells ml^{-1} with sterilized distilled water by using a haemocytometer for seed inoculation.

2.9.2 Seed Inoculation

For inoculation, a haricot bean seed, Awash Melka variety was kindly obtained from Institute of Jimma Agricultural Research Center, Ethiopia. The healthy seeds were briefly surface sterilized in 0.1% sodium hypochlorite for 2 minutes then washed repeatedly five times with sterile distilled water (Siddiqui & Akhtar, 2007). A total of 90 seeds out of the 120 haricot bean seeds were inoculated by soaking seeds into each suspension of potato dextrose liquid culture medium of the isolates PSFA, PSFP and mixed suspension of PSFA and PSFP with 10% gum Arabic as adhesive for 2 hrs to deliver inoculants of equal volumes of the suspensions about 10^6 cells ml^{-1} from each fungal isolates per seed. For combined inoculations, the liquid cultures of each isolates were mixed in equal proportion (half from each) per seed according to Zaidi and Khan (2006). The inoculated seeds were coated with 5 g $CaCO_3$ and shaken well till fine coating appeared on seeds in polythene bag, and the spore suspension was drained off and seeds were air dried over night aseptically in laminar air flow. The rest 30 seeds were soaked in distilled water amended with $CaCO_3$ (spore free solution) served as control treatment for comparison (Zaidi & Khan, 2006).

2.9.3 Treatments and the Experimental Design

Experiments were conducted in plastic pots having 17 cm diameter and 25 cm deep that had been sterilized with 20% sodium hypochlorite solution, filled with 4.0 kg of non-sterile soil. Rock phosphate of Bikilal was added as phosphatic P (22.5 mg kg^{-1}) and mixed well with the soil before seeding (Zaidi & Khan, 2006). The inoculated seeds were sown at 2 cm depth soil in each plastic pot (5 seeds pot^{-1}) and thinned down to three plants per pot after 5 days of emergence. The experiments were arranged into 8 treatments: soil only (Uninoculated), soil with isolate PSFA, soil with isolate PSFP, soil with both isolates PSFA and PSFP, soil amended with RP, RP amended soil with isolate PSFA, RP amended soil with isolate PSFP and RP amended soil with both isolate PSFA and isolate PSFP. The pots with different treatments were arranged in a randomized complete block design (RCBD) with three replications of each treatment. The plants were watered using tap water every three days or regularly depending on moisture contents in the pots. The plants' seedlings were allowed to grow for 80 days in the greenhouse under natural conditions (12 hrs photoperiod, temperatures of 16-28 °C, and relative humidity of 65%).

2.9.4 Measurement of Growth Parameters and Yield Components

Growth parameters and yield components of haricot bean plant were measured following the technique of Zaidi and Khan (2006). At 45 days of plant age from each pot, 3 plants per pot (9 plants per treatment) were chosen to measure some morphological characteristics such as length of the shoot, root, and number of leaves. At 80 days of harvest, the whole plants were carefully uprooted from the pots, washed gently under running tap water to remove the adhering soil particles. Plants were pulled carefully without damaging to the shoot and roots. Shoot and root height was measured and the mean was calculated and expressed in centimeter. Shoot and root fresh and dry weights, number of pods/plant, number of nodules and 50-seed weight (g) were recorded. The shoots and roots from each growth unit were placed in paper bags and dried at 70 °C for 48 hours and their dry weight mean was expressed in g $plant^{-1}$. The nodules were collected, counted and their dry weight was determined in the same way as the shoots and roots.

2.9.5 Nitrogen and Phosphorous Analysis

Oven dried leaves were grounded and digested in 15 ml $HClO_4$ and 5 ml HNO_3 then analyzed for the major nutrients using the standard procedures. The total Phosphorus content was determined by triple acid digestion followed by subsequent estimation by Vanadomolybdate phosphoric yellow colour method and total nitrogen content of the shoots was determined by modified "Wet" Kjeldahl method according to Sahlemedhin and Taye (2000).

2.9.6 Nitrogen Analysis

Ground shoot sample (0.3 g) was transferred into digestion tube; 2.5 ml of the digestion mixture was added to each digestion tube, swirled carefully to moisten the ground shoot samples and allowed to stand for 2 hours. The tubes were placed on heating block and heated at 100 °C for 2 hours. After two hours, the tubes were removed from the block and allowed to cool. Three 1ml of 30% H_2O_2 was added successively into each digestion tube and mixed thoroughly. The digestion tubes were again placed on the preheated block and heated at 300 °C until the digest turned to colorless or light yellow. The tubes were removed from the block, cooled to room temperature and 48.3 ml of distilled water was added to each tube, mixed and then allowed to stand overnight. On the next day, the content of each digestion tube was mixed again by shaking, filtered on a 100 ml volumetric flask and brought to the volume with distilled water. Each 100 ml of the acid digest was transferred into a macro - Kjeldahl tube and 20 mL of boric acid solution was measured from a dispenser flask into 250 mL Erlenmeyer flask corresponding to the number of samples.

Two drops of mixed indicator solution were added to each 20 ml of 2% boric acid solution, mixed thoroughly and placed under the condenser. After adding 75 ml of 40% NaOH solution to each digestion tube containing the digest, it was fitted to the corresponding holder and distillation was started. When the distillation was completed, that is, when about 80 ml of the distillate had been collected to boric acid, the flask was removed and distillation process of another sample was continued. Titration was then performed by using 0.1 N H_2SO_4 until the colour of the distillate turned from green to pink at the end point and the utilized H_2SO_4 for titration was recorded volumetrically. Finally the percent of N_2 content of the samples were calculated after correcting for the blank as described by Sahlemedhin and Taye (2000).

2.9.7 Phosphorous Analysis

The total phosphorus content of the shoots was determined by Vanadomolybdate phosphoric yellow colour method of Jackson as outlined by Sahlemedhin and Taye (2000). Five hundred mg of leaf sample were taken in a 250 mL capacity conical flask and were added with 2.5 mL concentrated HNO_3. The flasks were swirled to moisten the entire sample and placed on a hot plate at 180 °C to 200 °C. Five ml of tri-acid mixture (conc. HNO_3, conc. H_2SO_4 and 60% $HClO_4$ in the ratio of 10:1:4) were added to predigested sample and further digestion was carried out at 180 °C to 200 °C on a digestion mantle until the content in the flask became clear white.

The contents of the flasks were cooled and 10 to 15 mL of 6 N HCl added and stirred well. The acid digest was transferred to 50 ml volumetric flask and the volume was made up to 50 mL with distilled water. From this wet oxidized digested sample, five milliliter of aliquot was mixed in 10 mL of Barton reagents and total volume was made as 50 mL. The Barton reagent was prepared by Ammonium molybdate (25 g) was dissolved in 300 mL of distilled water and Ammonium metavenadate (1.25 g) was dissolved in boiling water (300 mL), and then cooled and 250 mL of concentrated HNO_3 was added and cooled at room temperature. Both solutions were mixed and volume was made up to 1 L with distilled water. The samples were kept for half an hour and phosphorus was determined by spectrophotometer using standard curve using various concentrations of standard 10 ppm KH_2PO_4 solution.

2.10 Data Analysis

All data were subjected to one-way analysis of variance (ANOVA) using the (SAS) Statistical Analysis System software Program (Version 9.1). Treatment means were compared using Tukey's test and differences were accepted as significant when $p < 0.05$.

3. Results

3.1 Isolation, Characterization and Selection of Efficient Isolates for Greenhouse Evaluation

Nine fungal isolates that showed larger Solubilization index (SI) and were preliminarily selected as better phosphate solubilizers to evaluate their efficiency on PKV broth using TCP and RP as inorganic phosphate sources. Of the fungal isolates which showed better SI, two of the isolates coded as PSFA and PSFP were found to show the highest (2.85 and 2.39) SI, respectively.

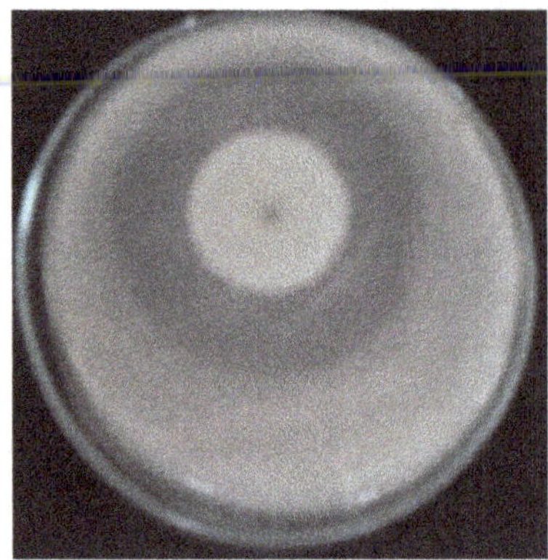

A. PSFA B. PSFP

Figure 1. Insoluble phosphate solubilization studies on Pikovskaya's agar plate (PVK): (A) and (B) two efficient superior phosphate solubilizing isolates (large haloes)

3.2 Physical and Chemical Analysis of Soil Samples and Rock Phosphate (RP)

From the physico-chemical characteristics of the soil and rock phosphate (RP) used for pot experiments, it was confirmed that both the soil and RP were slightly acidic (Table 1).

Table 1. Chemical and physical characteristics of soil and RP used in greenhouse study

Samples	pH-H_2O	EC (dS m)	OC (%)	TN (%)	AVP (ppm)	AVK (ppm)	Sand (%)	Silt (%)	Clay (%)
Soil	6.34	0.097	3.14	0.27	10.27	151	28	44	28
RP	6.52	-	-	-	189.19	100	-	-	-

Note. EC = electrical conductivity, OC = Organic carbon, TN = Total nitrogen, AVP = available Phosphorus, AVK = available Potassium.

3.3 Effects of PSF on the Shoot and Yield Parameters of Haricot Bean Plants

All the test fungi significantly ($p < 0.05$) promoted the plant height compared to the control except PSFA (Table 2 and Figure 2). Among the treatments, the effect of co-inoculation of fungal isolates PSFA (*Aspergillus* sp.) and PSFP (*Penicillium* sp.) produced the highest shoot length (47.31 cm plant^{-1}) followed by *Aspergillus* sp. PSFA inoculation which gave a shoot length of 45.48 cm plant^{-1}in the RP amended soil (Figure 2 and Table 2).

All inoculation with fungal isolates either alone or in co-inoculation had significantly ($p < 0.05$) higher shoot fresh weight than the control (Figure 2 and Table 2).

Soil + PSFA **Soil + RP + PSFA** **Soil + PSFAP**

Soil Only **Soil + RP + PSFAP**

Figure 2. Phosphate solubilizing fungal isolates for improving growth of haricot bean grown in pots amended with rock phosphate after 45 days of sowing in the greenhouse

Note. RP: rock phosphate.

Among the treatments, PSFA (*Aspergillus* sp.) along with RP amended soil and PSFP (*Penicillium* sp.) without RP showed a maximum shoot fresh weight of 58.78 and 56.52 (g/plant), respectively compared to the control. However, the effect was more pronounced (62.73 g/plant) in case of co-inoculation with PSFA (*Aspergillus* sp.) and PSFP (*Penicillium* sp.) in the RP amended soil (Table 2). Furthermore, the treatment showed a significant ($p < 0.05$) difference in shoot dry weight, number of leaves and pods plant^{-1} as well as weight of 50-seeds (Table 2). The plant inoculated with fungal isolates designated as PSFAP which are *Aspergillus* sp. and *Penicillum* sp. in the presence of RP gave the maximum shoot dry weight (14.33 g/plant) followed by inoculation of the same combination (PSFAP) without RP (12.63 g/plant) and single inoculation of *Aspergillus* sp. (PSFA) in the presence of RP (12.58 g/plant) compared to the control (Table 2).

Table 2. Effect of PSF on shoot growth and yield of haricot bean in the rock phosphate amended soil in the greenhouse

Treatments	SH (cm)	SFW (g plant^{-1})	SDW (g plant^{-1})	LN plant^{-1}	PN plant^{-1}	50-seed Weight (g)
Soil only (control)	40.43^{e}	46.34^{d}	7.67^{e}	42.44^{c}	11.36^{c}	34.44^{cd}
Soil + RP	41.53^{d}	46.39^{d}	8.09^{e}	42.44^{c}	11.38b^{c}	34.57^{cd}
Soil + PSFA	40.42^{e}	50.94^{c}	9.54^{d}	43.89^{c}	11.93^{c}	34.79^{bcd}
Soil + PSFP	43.66^{c}	56.52^{b}	11.19^{c}	57.89^{a}	11.36^{c}	34.44^{cd}
Soil + PSFAP	44.96^{b}	51.95^{c}	12.63^{b}	58.00^{a}	11.71^{c}	34.83^{bcd}
Soil +RP + PSFA	45.48^{b}	58.78^{b}	12.58^{b}	52.00^{ab}	12.51^{ab}	35.50^{ab}
Soil + RP + PSFP	43.70^{c}	51.11^{c}	11.67^{c}	49.78^{b}	12.32^{abc}	35.27^{abc}
Soil+ RP+ PSFAP	47.31^{a}	62.73^{a}	14.33^{a}	58.00^{a}	12.89^{a}	35.87^{a}
LSD	0.64	4.13	0.96	3.8	0.6	0.87
CV	0.51	3.13	3.14	2.67	1.76	0.88
P-value	*	*	*	*	*	*

Note. SH = shoot height, SFW = shoot fresh weight, SDW = shoot dry weight, LN = leaves number, PN = pod number and RP = rock phosphate; Mean values followed by the same superscripts within a column are not significantly different at $p < 0.05$. LSD = Least significant difference, CV = Coefficient of variation.

Maximum number (58.00) of leaves plant^{-1} were observed in plants treated with (PSFAP) both in the presence and absence of RP followed by sole inoculation of PSFA without RP (57.89 number of leaves plant^{-1}). The non-inoculated treatments showed the lowest number (42.44) of leaves plant^{-1}.

Increase in number of pods/plant was recorded with the application of single or dual inoculation of the test PSF isolates and accordingly, co-inoculation of (PSFAP), resulted in the largest number (12.89) of pods plant^{-1} when the soil enriched with RP compared to the control. The least number of pods 11.36 plant^{-1} was obtained from uninoculated treatment (Table 2). Similarly, the highest mean weight of 50-seeds 35.87 and 35.50 g plant^{-1} was obtained from co-inoculation of (PSFAP) and single inoculation with PSFA in the rock phosphate amended soil, respectively (Table 2). The least mean value 34.44 g plant^{-1}was obtained from control.

3.4 Effects of PSF Isolates on the Root Growth and Nodulation of Haricot Bean Plants

There were significant variation ($p < 0.05$) among the treatments regarding root length, root fresh and dry weight, nodule number and nodule dry weight as depicted in Table 3 and Figure 3. The highest root length (41.06 cm) plant^{-1} was recorded in case of RP amended soil with dual inoculation of the fungal isolates *Aspergillus* sp. plus *Penicillium* sp. (PSFAP) followed by single inoculation of *Penicillium* sp. (PSFP) and *Aspergillus* sp (PSFA) in the soil amended with RP (36.56 and 36.50 cm plant^{-1}), respectively.

The root fresh weight was significantly greater in treatments with inoculum both with single and co-inoculation than in control except with inoculation of *Penicillium* sp. (PSFP) without RP (Table 3).

Table 3. Effect of phosphate solubilizing fungi inoculation on root growth and nodulation of haricot bean in the rock phosphate amended soil in the greenhouse

Treatments	RL (cm plant^{-1})	RFW (g plant^{-1})	RDW (g plant^{-1})	NN (plant^{-1})	NDW (g^{-1} plant)
Soil only	25.24^{e}	7.60^{d}	2.66^{c}	30.33^{f}	0.20^{d}
Soil + RP	27.39^{d}	9.30^{c}	2.66^{c}	30.00^{f}	0.22^{cd}
Soil + PSFA	27.59^{d}	9.97^{c}	2.67^{c}	38.33^{d}	0.27^{c}
Soil + PSFP	27.59^{d}	7.65^{d}	2.66^{c}	32.33b^{ef}	0.25^{cd}
Soil+ PSFAP	33.06^{c}	11.97^{b}	2.71^{bc}	46.00b^{c}	0.55^{a}
Soil +RP + PSFA	36.50^{b}	14.19^{a}	3.13^{a}	57.67^{b}	0.55^{a}
Soil + RP + PSFP	36.56^{b}	13.78^{a}	2.95^{ab}	37.33^{de}	0.40^{b}
Soil +RP + PSFAP	41.06^{a}	14.09^{a}	2.96^{ab}	65.67^{a}	0.59^{a}
LSD	1.99	0.83	0.26	5.72	0.04
CV	2.17	2.92	3.17	4.74	3.38
P-value	*	*	*	*	*

Note. RL = root length, RFW = root fresh weight, RDW = root dry weight, NN = nodule number, NDW = nodule dry weight and RP = rock phosphate; Mean values are followed by the same superscripts within a column are not significantly different at (Tukey, $p < 0.05$). LSD = Least significant significance difference, CV = Coefficient of variation.

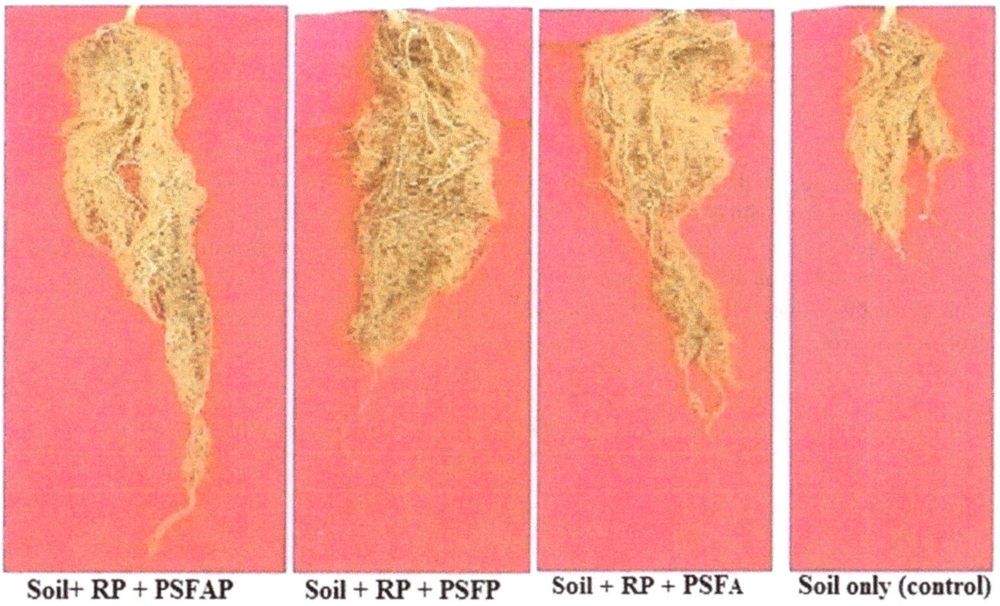

Figure 3. Effect of phosphate solubilizing fungal isolates on root growth of haricot bean grown in pots amended with rock phosphate in the greenhouse

Note. RP: rock phosphate.

The PSF isolates either in single or mixed inoculation promoted root length with a concomitant increase in the root dry weight of haricot bean plant compared to the control (Table 3). The single inoculation of *Aspergillus* sp. along with RP showed the highest increase in root dry weight (3.13 g plant^{-1}) followed by co-inoculation of *Aspergillus* sp. together with *Penicillum* sp. and *Penicillum* sp. along with RP - amended soil, resulting in 2.96 and 2.95 g plant^{-1}, respectively.

On the other hand, the maximum numbers of nodules (65.67) plant^{-1} was counted upon inoculation with *Aspergillus* sp. plus *Penicillium* sp. (PSFAP) followed by single inoculation of PSFA with 57.67 nodules plant^{-1} in the RP amended soil (Table 3 and Figure 4).

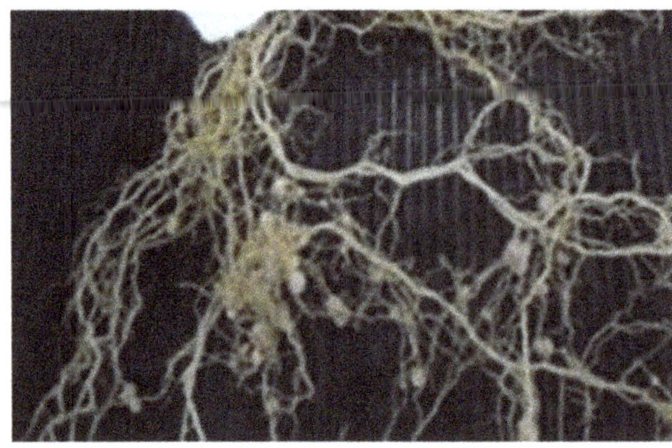

A. Soil +RP + PSFAP

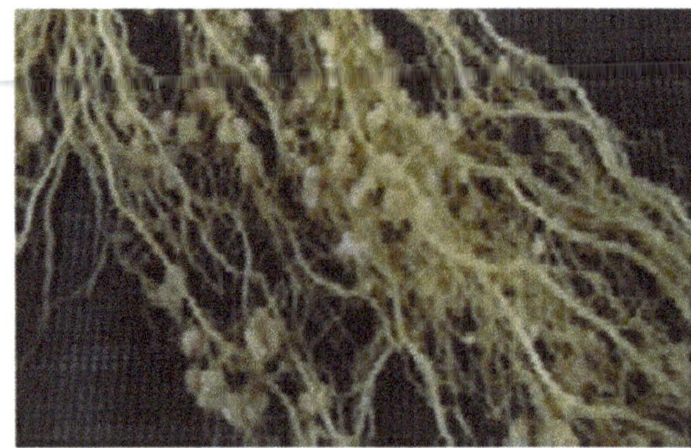

B. Soil only (control)

Figure 4. Effect of phosphate solubilizing fungal isolates on nodulation status of haricot bean grown in soil amended with rock phosphate: A) Inoculated; B). Control Soil only

Note. RP: rock phosphate.

Similarly, the highest nodule dry weight (0.59 g $plant^{-1}$) was recorded in response to co-inoculation (PSFAP) in RP amended soil. Sole inoculation of PSFA in soil amended with rock phosphate and co-inoculation of (PSFAP) without RP also resulted in significantly higher (0.55 g $plant^{-1}$/plant) compared to uninoculated control (Table 3).

3.5 Effects of PSF on Phosphorus and Nitrogen Content of Haricot Bean Plants

The treatments showed significantly variable results regarding the contents phosphorus and nitrogen contents in the leaves of haricot bean (Table 4). Accordingly, maximum content of phosphorus in haricot bean leaves was observed in plants treated with rock phosphate either single or co-inoculation of fungal isolates. The total phosphorus content of plant was significantly ($p < 0.05$) increased when treated with a mixed (PSFAP) fungal isolates in RP amended soil with P content in the leaves 0.58% followed by sole inoculation with PSFA and PSFP with RP enriched soil with P content of 0.54% and 0.52%, respectively. Similarly, the highest (1.92%) percentage of nitrogen content in the haricot bean leaves contained with coinoculation of the fungal isolates (PSFAP) followed by the isolates PSFA and PSFP in the presence of RP enriched soil each with (1.63%) nitrogen content in the leaves (Table 4). On the other hand, the lowest percentage (0.89%) of nitrogen was obtained from the control.

Table 4. Nitrogen and Phosphorus content of haricot bean plant leaves grown in the greenhouse

Treatments	Nitrogen %	Phosphorus %
Soil only	0.89^{d}	0.43^{e}
Soil + RP	0.89^{d}	0.46^{cd}
Soil + PSFA	0.90^{d}	0.44^{de}
Soil + PSFP	0.91^{d}	0.46^{cd}
Soil + PSFAP	1.07^{c}	0.48^{c}
Soil +RP + PSFA	1.63^{b}	0.54^{b}
Soil + RP + PSFP	1.63^{b}	0.52^{b}
Soil +RP+ PSFAP	1.92^{a}	0.58^{a}
LSD	0.07	0.03
CV	2.05	1.88
P-value ($p < 0.05$)	*	*

Note. RP = rock phosphate, Mean values are followed by the same superscripts within a column are not significantly different at (Tukey, $p < 0.05$). LSD = List significant significance difference, CV = Coefficient of variation.

4. Discussion

The present study showed the occurrence of *Aspergillus*, *Penicillium* and *Fusarium* species that were capable of solubilizing sparingly soluble phosphorus from the rhizosphere of different plants collected from farm lands. Similarly, Chuang et al. (2007) isolated P-solubilizing fungi *Aspergillus niger* and *Penicillium* spp. from various subtropical and tropical rhizospheric soil samples of cabbage and maize. In the current study *Aspergillus* spp. and *penicillum* spp. were commonly isolated from the rhizosphere soils of faba beans (*Vicia faba* L.) and haricot bean plants (*Phaseolus vulgais* L.). This result is supported by the findings of Abdul Wahid and Mehana (2002) who had isolated *Aspergillus niger*, *Aspergillus fumigatus* and *Penicillum pinophilum* from the rhizosphere of faba beans (*Vicia faba* L.), kidney bean (*Phaseolus vulgais* L.), and peas (*Pisum sativum* L.).

Phosphate solubilizing fungi belonging to the genera *Aspergillus* and *Penicillium* were also isolated from the rhizospheric soils of sugar cane. Mahamuni et al. (2012) also isolated several phosphate solubilizing fungal groups including *Aspergillus* spp., *Alternaria* spp., *Curvularia* spp., *Penicillium* spp. and *Trichoderma* spp. from the rhizosphere of sugarcane and sugar beet plants. Other studies on the PSF associated with rhizosphere of sugar cane roots showed the occurrence of phosphate solubilizing *Penicillium citrinum* Thom (Yadav, 2010).

The PSF isolates, PSFA and PSFP significantly increased the shoot height of haricot bean plant compared to uninoculated plant. The highest shoot height was observed in case of soils amended with rock phosphate and with dual inoculation of phosphate solubilizing isolates. The increase in plant shoot height due to PSF inoculation along with RP application could be the result of increased transformation of P into available forms as indicated by the increased P in plants by the PSF isolates PSFAand PSFP. Similar results of increase in shoot height due to PSF inoculation has been reported on groundnut plant (Malviya et al., 2011), wheat (Xiao et al., 2009), chickpea (Yadav et al., 2011) and soy bean (Iman, 2008).

The application of *Aspergillus* sp. (PSFA) and *Penicillum* sp. (PSFP) either individually or in combination led to increase in number of leaves compared to control. The effect was more pronounced with dual inoculation of PSFAP in the presence of RP and other treatments. This could be attributed to the highest values of available P and enhanced P uptake compared to other treatments due to the synergistic effect of fungal inoculants. These results are in harmony with those obtained by Patil et al. (2012) who reported that PSF both singly or in combination gave maximum number of leaves on maize plants due to the activity of P solubilization and released growth-promoting substances. In addition, El-Yazeid and Abou-Aly (2011) found that the integrated treatment of P-solubilizers and application of rock-P significantly led to large number and area of leaves as well as the photosynthetic pigments in tomato plant.

The shoot fresh weight was significantly increased by the inoculation of PSF compared to control and treatment receiving only rock phosphate. The highest shoot fresh and dry weight values obtained as a result of applying combined phosphate solubilizing fungal isolates (PSFAP) in rock phosphate amended soil may be due to the role of the two test fungi in solubilization and mineralization efficiency of phosphate pool in plant rhizosphere, which in return increased the level of available nutritional elements required.

Thus, increased transformation of P into available forms as indicated, increased content of P in plants by inoculation of PSFAP which consequently enhanced cell elongation, and multiplication and overall shoot growth of haricot bean plant. These results are in line with the findings of numerous workers (Chuang et al., 2007; Kapri & Tewari, 2010; Panhwar et al., 2011; Patil et al., 2012), who have reported increase in plant growth and fresh matter of different crop plants due to inoculation of phosphate solubilizing fungi along with phosphate sources. The increase in the dry weight of shoot may also be due to greater solubilization of P by the rhizosphere microorganisms that would lead to better symbiotic N_2-fixation by the legumes and the latter one was found to have maximum contribution in increasing dry matter production. These findings are similar to the earlier reports of Saber et al. (2009) where rock phosphate coupled with phosphate solubilizing *A. niger* and *Penicillium* sp. gave more significant dry biomass compared to uninoculated control.

The increase in root parameters such as highest root length, fresh and dry root weights could be attributed to high P-solubilizing ability of the fungal inoculants through which they might have contributed to the enhanced root growth thereby creating more root surface area for uptake of nutrients from the soil. Sharma et al. (2012) reported that one of the advantages of supplying of the plants with phosphorus along with PSF is to create deeper and more abundant roots. In addition, PSF can also release plant growth-promoting substances (Nenwani et al., 2010) that might be another probable means for enhanced root growth in haricot bean plant. Similar evidence on the increased root length with inoculation of PSF (*Aspergillus niger* and *Penicillium* sp.) along with RP has been reported by several workers (Mittala et al., 2008; Yadav et al., 2011; Malviya et al., 2011).

In the present study, application of PSF showed remarkable difference in nodule count and dry weight. The better nodulation in the case of co-inoculation could be due to the favorable synergistic effects of the isolates in mobilizing more P to make available to the plants, which eventually promoted root development that might also have provided rhizobia more infection sites than others. The higher availability of P in the rhizosphere of plants inoculated with PSF probably induced good proliferation of root, providing enough number of sites for rhizobia to form more number of nodules. Rudresh et al. (2005) reported that increase in root growth provides more access to nodulating rhizobia. In another study, Saber et al. (2009) demonstrated that increase in the number and weight of nodule in mung bean with inoculation of *A. niger* and *Penicillium* sp. in RP amended soil.

The highest pod number and seed weight were recorded with inoculation of mixed isolates along with RP application. These highest values in the current study could be due to improved nutrients particularly P both its availability and up take since the availability of nutrients in the soil in sufficient quantities greatly determines the growth and yield of plants. For instance, Agasimani et al. (2002) observed significant increase in number of pods plant^{-1} in groundnut due to inoculation of P-solubilizing fungus (*Aspergillus awamori*). Rudresh et al. (2005) also reported that increase in pod number and seed weight of chickpea grown in phosphate-deficient soil amended with insoluble rock phosphate due to *Trichoderma* inoculation under both glasshouse and field conditions. Mittala et al. (2008) have reported the effect of six phosphate-solubilizing fungi (two strains of *Aspergillus awamoria* and four of *Penicillium citrinum*) isolated from rhizosphere of various crops caused increased growth, number of pods and increased seed production of chickpea plants in pot experiments.

Single and co-inoculation of the isolates (PSFA and PSFP) with or without RP amendment significantly improved the level of phosphate content in haricot bean plant compared to the control. Inoculants (mixed) which solubilized higher amount of phosphate enhanced the content of P over control and other treatments. The highest significant increase in the percentage of phosphate was observed in the co-inoculation of the fungal isolates (PSFAP) along with rock phosphate. This may be due to better utilization of soluble phosphorus from the pool of available phosphorus in the soil due to solubilization of native and added phosphorus by the actions of the PSF isolates. This result is in agreement with previous reports of Xiao et al. (2009) on wheat, Mittala et al. (2008) on chickpea, Richa et al. (2007) on maize and Agasimani et al. (2002) on groundnut plants.

Increased nitrogen content in the leaves of haricot bean plants was recorded by the inoculation of PSF isolates compared to the control. The increase in nitrogen content of haricot bean might be due to increased nodulation as well as the positive interaction between P-solubilizing fungal isolates and root nodulating bacteria. Malviya et al. (2011) demonstrated that the P-solubilizing fungi enhanced the N uptake from soil solution. The significant increase in phosphate and nitrogen uptake by haricot bean plants grown in soil inoculated with PSF proved that these isolates have not only the capability to solubilize RP *in vitro* but also that this phenomenon can be observed *in vivo* with a beneficial effect for plant growth.

Based on the above findings, the use of these PSF as bioinoculants may help to minimize the chemical fertilizer application, reduce environmental pollution and consequently may promote sustainable agriculture. This report suggests further screenings of phosphate-solubilizing microorganisms that requires extensive and consistent research activities to identify and characterize more rhizosphere competent phosphate-solubilizing fungi for their ultimate application as potential solubilizers of fixed soil phosphate to use in natural environment.

5. Conclusions

All the selected isolates were capable of mobilizing TCP and RP in PVK broth. The efficiency of phosphate solubilization is significantly higher in Pikovskaya medium containing TCP than in the medium containing RP. Both single as well as mixed inoculation treatments showed better plant shoot and root growth, number of pods 50-seed weight, and nutrients content compared to uninoculated control. Among the inoculation treatments, a combined inoculation (PSFAP) in the presence of rock phosphate was superior over the rest of the isolates and uninoculated control. Accordingly, from the present result, it can be concluded that the amendment of soil with RP along with the application of P-solubilizing fungi could alleviate soil fertility problem as a result of P fixation and contribute to environmental integrity by reducing artificially synthesized P fertilizers to promote sustainable agriculture.

Acknowledgements

The authors would like to kindly acknowledge Jimma University, College of Natural Sciences for funding this research work. The Department of Biology, Jimma University deserves acknowledgement for allowing laboratory facilities and consumables. College of Agriculture and Veterinary Medicine of Jimma University is thankful for allowing the greenhouse experiment.

References

Agasimani, C. A., Mudalagiriyappa A., & Sreenivasa, M. N. (2002). *Studies on phosphate solubilizers in groundnut*. Nation. Sympo. Mineral Phosphate Solubilization, 14-16 Univ. Agril. Sci., Dharwad, India.

Bojinova, D., Velkova, R., & Ivanova, R. (2008). Solubilization of Morocco phosphorite by *Aspergillus niger*. *Biores. Technol., 99*, 7348-7353. http://dx.doi.org/10.1016/j.biortech.2007.08.047

Cheesbrough, M. (2000). *District Laboratory Practice in Tropical Countries Part 2* (pp. 47-54). Cambridge University Press, Cambridge.

Chuang, C. C., Kuo, Y. L., Chao C. C., & Chao, W. L. (2007). Solubilization of inorganic phosphates and plant growth promotion by *Aspergillus niger*. *Biol. Fert. Soils, 43*, 575-584. http://dx.doi.org/10.1007/s00374-006-0140-3

Das, K., Katiyar, V., & Goel, R. (2003). P solubilization potential of plant growth promoting *Pseudomonas* mutants at low temperature. *Microbiol. Res., 158*, 359-362. http://dx.doi.org/10.1078/0944-5013-00217

Edi-Premono, M., Moawad, M., & Vlek, G. (1996). Effect of phosphate solubilizing *Pseudomonas putida* on the growth of maize and its survival in the rhizosphere. *Indo. J. Crop Sci., 11*, 13-23.

El-Yazeid, & Abou, A. (2011). Enhancing growth, productivity and quality of tomato Plants using phosph ate solubilizing microorganisms. *Aust. J. Basic & Appl. Sci., 5*, 371-379.

Ferris, S., & Kaganzi, E. (2008). Evaluating marketing opportunities for haricot beans in Ethiopia. *IPMS (Improving Productivity and Market Success) of Ethiopian Farmers Project Working Paper 7* (p. 68). ILRI (International Livestock Research Institute), Nairobi, Kenya.

Gebremeskel, A., & Muleta, D. (2011). Evaluating the Effects of phosphate solubilizing rhizobacteria with different doses of poultry manure in Jimma town, Ethiopia. In A. Zaidi (Ed.), *Rhizobacteria and Vegetable Growth* (pp. 311-334). LAP Lambert Academic Publishing, New York, America.

Gifole, G., Sheleme, B., & Walelign, W. (2011). The Response of Haricot Bean (*Phaseolus vulgaris* L.) to Phosphorus Application on Ultisols at Areka, Southern Ethiopia. *J. Biol., Agric. and Healthcare, 1*, 38-49.

Girma, A. (2009). Effect of NP Fertilizer and Moisture Conservation on the Yield and Yield Components of Haricot Bean (*Phaseolus vulgaris* L.) in the Semi Arid Zones of the Central Rift Valley in Ethiopia. *Adv. Environ. Biol., 3*, 302-307.

Gyaneshwar, P., Kumar, G., Parekh, L., & Poole, P. (2002). Role of soil microorganisms in improving P nutrition of plants. *Plant Soil, 245*, 83-93. http://dx.doi.org/10.1023/A:1020663916259

Hailemariam, A. (1993). The effect of phosphate solubilizing fungus on the growth and yield of tef (*Eragrostis tef*) in phosphorous fixing soils. *Proceedings of the workshop on the third cycle local research grant, Eth. Sci. & Technol. Res. Report* (pp. 12-14).

Hameeda, B., Harini, G., Rupela, P., Wani, S. P., & Reddy, G. (2008). Growth promotion of maize by phosphate-solubilizing bacteria isolated from composts and macrofauna. *J. Microbiol. Res., 163*, 234-242. http://dx.doi.org/10.1016/j.micres.2006.05.009

Hinsinger, P. (2001). Bioavailability of soil inorganic P in the rhizosphere as affected by root induced chemical changes: A review. *Plant Soil, 237*, 173-195. http://dx.doi.org/10.1590/S1517-83822010005000001

Vessey, J. K., & Heisinger, K. G. (2001). Effect of *Penicillium bilaii* inoculation and phosphorus fertilization on root and shoot parameters of field-grown pea. *Can. J. Plant Sci., 80*, 361-366. http://dx.doi.org/10.4141/P00-083

Kim, Y. H., Bae, B., & Choung, Y. K. (2005). Optimization of biological phosphorus removal from contaminated sediments with phosphate solubilizing microorganisms. *J. Biosci. and Bioeng., 99*, 23-29. http://dx.doi.org/10.1263/jbb.99.23

Mahamuni, S. V., Wani, P. V., & Patil, A. S. (2012). Isolation of Phosphate Solubilizing Fungi from Rhizosphere of Sugarcane & Sugar Beet Using TCP & RP Solubilization. *Asian J. Bioche. and Pharmace. Rese., 2*, 237-244.

Malviya, J., Singh, K., & Joshi, V. (2011). Effect of Phosphate Solubilizing Fungi on Growth and Nutrient Uptake of Ground nut (*Arachis hypogaea*) Plants. *Adv. Biores., 2*, 110-113.

Mittala, V., Singha, O., Nayyarb, H., Kaura, J., & Tewari, R. (2008). Stimulatory effect of phosphate solubilizing fungal strains (*Aspergillus awamori* and *Penicillium citrinum*) on the yield of chickpea (*Cicer arietinum* L. cv. GPF2). *Soil Biol., and Bioche., 40*, 718-727. http://dx.doi.org/10.1016/j.soilbio.2007.10.008

Muleta, D. (2012). Legume responses to arbuscular mycorrhizal fungi inoculation in sustainable agriculture. In M. Khan (Ed.), *Microbes for Legume Improvement* (pp. 224-266). Springer Verlag Wien publishing house, New York, America.

Nenwani, V., Doshi, P., Saha, T., & Rajkumar, S. (2010). Isolation and characterization of a fungal isolate for phosphate solubilization and plant growth promoting activity. *J. Yeast and Fungal Rese., 1*, 9-14.

Pandey, A., Das, B., Kumar, K., Rinu, K., & Trivedi, P. (2008). Phosphate solubilization by *Penicillium* spp. isolated from soil samples of Indian Himalayan region. *World J. Microbiol. Biotechnol., 24*, 97-102. http://dx.doi.org/10.1007/s11274-007-9444-1

Panhwar, Q., Radziah, O., Zaharah, A., Sariah, M., & Mohd, R. (2011). Role of phosphate solubilizing bacteria on rock phosphate solubility and growth of aerobic rice. *J. Environ. Biol., 32*, 607-612.

Patil, R. M., Kuligod, V. B., Hebsur, N. S., Patil, C. R., & Kulkarni, G. N. (2012). Effect of phosphate solubilizing fungi and phosphorus levels on growth, yield and nutrient content in maize (*Zea mays*). *Karnataka J. Agric. Sci., 25*, 58-62.

Pikovskaya, I. (1948). Mobilization of phosphorus in soil connection with the vital activity of some microbial species. *Microbiologia, 17*, 362-370.

Pradhan, N., & Sukla, B. (2005). Solubilization of inorganic phosphates by fungi isolated from Agricultural soil. *African J. of Biotec., 5*, 850-854.

Reddy, M., Kumar, S., Babita, K., & Reddy, M. S. (2002). Biosolubilization of poorly soluble rock phosphates by *Aspergillus tubingensis* and *Aspergillus niger*. *J. Biores. Technol., 84*, 187-189. http://dx.doi.org/10.1016/S0960-8524(02)00040-8

Richa, G. Khosla, B., & Reddy, M. (2007). Improvement of maize plant growth by phosphate solubilizing fungi in rock phosphate amended soils. *World J. Agric. Scie., 3*, 481-484.

Rudresh, D. L., Shivaprakash, M. K., & Prasad, R. D. (2005). Tricalcium phosphate solubilizing abilities of *Trichoderma* spp. In relation to P- uptake & growth yield parameters of chickpea (*Cicer arietinum* L.). *Canadian J. of Microbiol., 51*, 217-226. http://dx.doi.org/10.1139/w04-127

Saber, W. I. A., Ghanem, K. M., & El-Hersh, M. S. (2009). Rock Phosphate Solubilization by Two Isolates of *Aspergillus niger* and *Penicillium* sp. and Their Promotion to Mung Bean Plants. *Res. J. Microbiol., 4*, 235-250. http://dx.doi.org/10.3923/jm.2009.235.250

Sahlemedhin, S., & Taye, B. (2000). Procedures for Soil and Plant Analysis. National Soil Research Center, EARO, *Technical Paper No. 74*, Addis Ababa, Ethiopia.

Sanjotha, P., Mahantesh, P., & Patil, S. (2011). Isolation and screening of efficiency of Phosphate solubilizing Microbes. *Internati. J. Microbiol. Res., 3*, 56-58. http://dx.doi.org/10.9735/0975-5276.3.1.56-58

Selvi, B., Paul, A., Ravindran, A., & Vijaya, V. (2011). Quantitative estimation of insoluble inorganic phosphate solubilization. *International J. of Scien. and Natur., 2*, 292-295.

Sharma, A., Rawat, S., & Yadav, B. (2012). Influence of Phosphorus Levels and Phosphorus Solubilizing Fungi on Yield and Nutrient Uptake by Wheat under Sub-Humid Region of Rajasthan, India. *ISRN Agronomy, 2012*, 9. http://dx.doi.org/10.5402/2012/234656

Siddiqui, Z. A., & Akhtar, M. S. (2007). Biocontrol of a Chickpea ROOT-ROT Disease Complex with Phosphate-Solubilizing Microorganisms. *J. of Plant Pathol., 89*, 67-77.

Taye, B. (2007). *An overview of acid soils their management in Ethiopia.* Paper presented in the third International Workshop on Water Management (Wterman) Project, Haramaya, Ethiopia.

Thalooth, A. T., Tawfik, M. M., & Muhammad, M. H. (2006). A comparative study on the effect of foliar application of zinc, potassium and magnesium on growth, yield and some chemical constituents of Haricot bean plants growth under water stress conditions. *World J. Agric. Sci., 2*, 37-46.

Wahid, A., & Mehana, T. (2000). Impact of phosphate-solubilizing fungi on the yield and phosphorus-uptake by wheat and faba bean plants. *Microbiol Res., 155*, 221-227. http://dx.doi.org/10.1016/S0944-5013(00)80036-1

Wakelin, A., Warren, A., Harvey, R., & Ryder, H. (2004). Phosphate solubilization by *Penicillium* spp. closely associated with wheat roots. *J. Biol. Fert. Soils., 40*, 36-43. http://dx.doi.org/10.1007/s00374-004-0750-6

Wani, A., Khan, M., & Zaidi, A. (2007). Synergistic effects of the inoculation with nitrogen fi xing and phosphate solubilizing rhizobacteria on the performance of field grown wheat. *J. Plant Nutr. Soil Sci., 170*, 283-287. http://dx.doi.org/10.1002/jpln.200620602

Whitelaw, M. (2000). Growth promotion of plants inoculated with phosphate-solubilizing fungi. *J. Adv. Agron., 69*, 99-151. http://dx.doi.org/10.1016/S0065-2113(08)60948-7

Woyessa, D., & Assefa, F. (2011). Effects of Plant GrowthPromoting Rhizobaceria (PGPR) on Growth and Yield of Tef [*Eragrostis tef* (Zucc.) Trotter] under Greenhouse Condition. *Res. J. Microbiol., 6*, 343-355. http://dx.doi.org/10.3923/jm.2011.343.355

Wu, X., Wei, H., Sun, L., & Wang, Q. (2005). Phosphate availability alters lateral root anatomy and root architecture of *Fraxinus mandshurica* Rupr. seedlings. *J. Integr. Plant Bio., 47*, 292-301. http://dx.doi.org/10.1111/j.1744-7909.2005.00021.x

Xiao, C., Chi, R., He, H., Qiu, G., Wang, D., & Zhang, W. (2009). Isolation of Phosphate-solubilizing fungi from Phosphate mines and their effect on wheat seedling growth. *J. Appl. Biochem. Biotechnol., 159*, 330-342. http://dx.doi.org/10.1007/s12010-009-8590-3

Yadav, J., Verma, P., & Tiwari, N. (2011). Plant Growth Promoting Activities of Fungi and their effect on the chickpea plant growth. *Asian J. Biol. Sci., 4*, 291-302. http://dx.doi.org/10.3923/ajbs.2011.291.299

Yang, H., Fan, B., Gong, M., & Li, Q. (2008). Isolation and identification of a novel phosphate-dissolving strain P21. *Acta. Microbiol. Sin., 48*, 51-56.

Zaidi, A., & Khan, M. (2007). Stimulatory effects of dual inoculation with phosphate solubilizing microorganisms and *arbuscular mycorrhizal fungus* on chickpea. *Australian J. Experim. Agri., 47*, 1016-1022. http://dx.doi.org/10.1071/EA06046

Zaidi, A., & Khan, M. (2006). Co-inoculation Effects of Phosphate Solubilizing Microorganismsand *Glomus fasciculatum* on Green Gram-Bradyrhizobium Symbiosis. *Turk. J. Agric., 30*, 223-230.

9

Ordered Logistic Analysis of Farmers' Market Regulations: Who Finds Them Easy?

Jean Dominique Gumirakiza[1] & Amber Daniels[2]

[1] Department of Agriculture, Western Kentucky University, USA

[2] School of Law, University of Louisville, USA

Correspondence: Jean Dominique Gumirakiza, Department of Agriculture, Western Kentucky University, 1906 College Heights Blvd #41066, Bowling Green, KY 42101, USA.
E-mail: dominique.gumirakiza@wku.edu

Abstract

This study applies an ordered logistic regression to data collected in 2015 using in-person survey, mail, and online surveys from fresh produce vendors at farmers' markets within the south central and western Kentucky regions. The purpose was to explain levels of difficult the vendors face when complying with market regulations. Results indicate that an average fresh produce vendor at farmers' market is 26 percent likely going to comply with market regulations easily, 69 percent moderately, and 4 percent hardly. Participating in CSA and "local" labeling programs, years of farming experience, and being a male vendor are associated with finding relatively easy to comply with farmers markets regulations. Market managers and policy makers will find this study useful in ensuring that those regulations pose no greater difficult to the vendors. Likewise, findings are useful to the vendors for they indicate variables that make easier for them to comply with the regulations.

Keywords: farmers' market regulations, fresh produce vendors

1. Introduction

Farmers' markets are an alternative to supermarkets where farmers sell their produce directly to consumers without middlemen. The markets have a set of market regulations for vendors to follow. This structure varies from state to state and market regulations are different (Public Health Law Center, 2014). According to the Kentucky Department of Agriculture (2014), farmers' markets in Kentucky do not have a general set of rules for all markets to follow. Each market operates independently and decides upon their rules and how they will be enforced. Market rules are decided in a variety of ways. Market manager, a committee or the vendors make the regulations. Even though there are no official regulations across all farmers' markets, there are common rules that are seen throughout markets for vendors (Hamilton, 2002). The difficulties that vendors might face when complying with market regulations include paying participation fees, attending vendors' meetings, adhering to market hours of operation, language barriers for those whose English is not their first language, and meeting quality standards.

The levels of difficult to comply with regulations are different among vendors. Some vendors find them easy to comply while others find them difficulty. Very little is known about farmers' market regulations and the difficult they pose to vendors. This study focuses on vendors of fresh produce and explains the levels of difficult they face in complying with the market regulations. Our research questions were: what are the partial probabilities for fresh produce vendors to face difficult in complying with farmers' market regulations? What are vendor characteristics associated with facing greater difficult? We describe the vendor characteristics and explained the probabilities of facing difficult. We then formulate policy suggestions. Although this study uses data collected from fresh produce vendors at farmers' markets in the south central and western Kentucky regions, the findings are applicable to vendors throughout the state and surrounding regions.

This study is significant because it provides policy makers with suggestions helpful in creating a user-friendly legal framework. Farmers' market managers will find this study significant for its results identify and explain vendor characteristics that are associated with various levels of difficulty in complying with market regulations.

This study is relevant for vendors because the findings serve as an advocacy for those who feel burdened by the regulations.

2. Literature Review

This section presents few previous studies that are closely related to this topic. Ernst (2014) indicated an upward trend in farmers' markets in Kentucky as a result of a considerable support from the Kentucky Department of Agriculture and other state agencies. Futamura (2007) found that over the past century more consumers in Kentucky became concerned about their health and diet as a result of industrialized agriculture production. Farmers' markets in Kentucky have also settled on the idea of locally produced food or "made in Kentucky" foods. This gives the consumers greater sense of "where" their food is coming from. Baharanyi and Boateng (2012) found that there are three levels of farmers markets: developed, developing and underdeveloped. The study showed that most of the markets were underdeveloped. The developed markets tend to locate in urban areas with a full paid staff and well-developed physical structure, which brought more customers into the market. The study recommended better facilities and marketing strategies to attract consumers at the underdeveloped farmers' markets.

There are various products available at farmers' markets (McGarry-Wolf, Spittler, & Ahern, 2005; Brown & Miller, 2008; George, Kraschnewski, & Rovniak, 2011; Alonso & O'Neill, 2011). However, Gumirakiza, Curtis, and Bosworth (2014) indicated that 78 percent of consumers go to these markets to primarily purchase fresh produce. Selling at farmers' markets is generally a part-time agribusiness venture. In fact, Varner and Otto (2008) showed that being a vendor was only a part-time job and most successful vendors travel from one market to the next. Ostrom and Drovan (2013) found that farmers' market vendors have a unique relationship with customers. That relationship brings several customers to the markets as opposed to super markets.

Vendors at farmers' markets must comply with a set of regulations. The most crucial regulations enforce food safety to ensure that the food products sold to the public are safe for consumption. Markets can also make the decisions on what can and cannot be sold at their markets. Markets along with vendors must have the proper permits to participate in a farmers market. Hoffman, Dennis, Gilliam, and Vargas (2007) indicated that if a food is ready to eat it should be labeled as such. The study suggested that vendors must use clean utensils, properly store food products to reduce the risk of food poisoning, and no sick person should handle the food being sold at the market. In this regard, Hofmann et al. (2007) argued that receiving a permit to sell food products at farmers' market signifies that the vendor met all requirements of the food code. They also suggested that the permits should be suspended or revoked when vendors fail to comply with the food code.

According to May (2005), farmers have a strong sense of duty to comply with regulations. They were found with low fear of punishment with the regulations. In fact, less than 10% of the farmers who have had some kind of written warning have had actions taken against them. The amount of farmers with a serious punishment from their offense is even less. Along with farmers' market regulations, farmers also have to deal with the changing policy regarding their farming operations. According to Hazell (2010) the policies create more difficulty to access inputs and the markets and changes in regulations for small farms have caused failures in some cases. The fact is that little is known about farmers' market regulations and the difficulty the cause to the vendors. This study contributes to the existing literature by examining factors that explain levels of difficulty in complying with the market regulations.

3. Methodology

3.1 Data Collection Procedures

This study employs data collected from farmers' market vendors of locally grown fresh produce at famers' markets located within the south central and western regions of Kentucky. We chose these two regions for the closeness and convenience to researchers. The research was conducted between March and September in 2015. We used surveys and distributed them via email, mail, and in-person strategies. The survey was created online using Qualtrics software. The software generated a link accessible to respondents who wish to participate online. Researchers used the farmers' market registry to locate the markets, visited them on days of operations, and asked vendors to complete them. Some vendors were too busy to take the surveys immediately because they were interacting with customers. We asked a convenient way to take the surveys; either online or by mail. We provided an online link to those vendors who preferred the online route. We gave copies of surveys to those who chose to complete the surveys at home and requested them to mail them to us upon completion. Another strategy we used is to get a list of email addresses for all farmers' market managers available at Kentucky Department of Agriculture webpage. We emailed the survey's link to the managers within the two regions of our study and asked them to distribute it to all fresh produce vendors at their markets.

In total, we were able to receive forty-six completed surveys out of seventy-eight surveys we distributed. Thus, we had a fifty-nine percent response rate. Compared to the average response rate of forty-five percent reported by Nulty (2008), our data collection strategy led to a higher response rate. The nature of the outcome (dependent variable) has dictated the choice of the model to use for this analysis. Respondents were presented 5 levels of difficult (Very easy, easy, moderate, difficult, and very difficult) they face when meeting farmers' market regulations. We asked respondents to indicate the level that best reflects their situation.

3.2 Theoretical Framework

The dependent variable for this study consists of polychotomous with an ordinal/ranking nature. Kennedy (2008), Greene (2010), and Wooldridge (2009) indicate that an ordered logistic model is appropriate for such a case. Ordered logistic model generalizes the binary logistic analysis which is used as an explained variable to more than two ordinal outcomes (Train, 2009). This analysis is done in a situational framework where we assume that a respondent i faces j ranking levels/alternatives and chooses the one that is the most appropriate to his/her situation. In this framework, there is an observed ordinal variable, Y; which is a function of another variable Y^*, that is not measured (Williams, 2015). In other words, there is a continuous, unmeasured latent variable Y^*, whose values determine what the observed ordinal variable Y equals.

In practice, the Y^* cannot be observed. We instead observe the response Y whereby

$$Y = \begin{cases} 1 \text{ if } 0 < Y^* \le \mu_1, \\ 2 \text{ if } \mu_1 < Y^* \le \mu_2, \\ \ldots, \\ J \text{ if } \mu_{J-1} < Y^* \le \mu_J \end{cases} \tag{1}$$

where the μ's are the unknown threshold parameters to be estimated. In this analysis, j is represented by the 5 levels of difficult of farmers' market regulations: 1 (Very easy), 2 (Easy), 3 (Moderate), 4 (Difficult), and 5 (Very difficult). Respondents expressed levels of difficult by choosing one alternative out of the five rankings.

The latent variable Y^* is given by:

$$Y_i^* = \sum_{K=1}^{K} \beta_k X_{ki} + \varepsilon_i = Z_i + \varepsilon_i \tag{2}$$

Where, X_{ki} represent explanatory variables. The model included dummy variables and their corresponding estimates are interpreted as the probability difference between X_{ij} values of zero and one. The βs are the ordered log-odds estimates that measure the impact that a corresponding explanatory variable has on the likelihood of difficulty compared to the chances of indicating otherwise. The ε_i is a random disturbance term. This disturbance term represents other factors not included in the model and is assumed to have a standard logistic distribution. In Equation (2), the part that is used to estimate the ordered Logit model parameters is:

$$Z_i = \sum_{K=1}^{K} \beta_k X_{ki} = E(Y_i^*) \tag{3}$$

Equation (3) indicates the expected value of the latent variable Y^*, whose values determine what the observed ordinal outcome Y is. The probability that Y will take on a particular value is given by

$$Prob(Y = j / X_{ki}) = \frac{\exp(\beta_k X_{ki})}{1 + \exp(\beta_k X_{ki})} \tag{4}$$

We hypothesized that there is no relationship between chooser's characteristics and the degree of difficult farmers face (null hypothesis), that is; Ho Ξ $\beta_k = 0$. The alternative hypothesis is that there are significant relationship between chooser's characteristics and the degree of difficult farmers face, that is; Ho Ξ $\beta_k \neq 0$.

4. Results

The main purpose of this study is to assess the levels of difficult faced by farmers' market vendors when complying with the market regulations. We examine whether or not vendor characteristics have significant impact on the probabilities of facing difficulties. This section presents two types of results: descriptive and econometric. The descriptive results show basic statistics about the variables of interest. Table 1 shows a numerical description of the variables we include in the ordered logistic model.

4.1 Descriptive Statistics

An average fresh produce vendor was 37 years old with some college degree (lower than a bachelor's degree) and 14 years of farm experience. The average farm size among participants is 212 acres; ranging from 2 acres to 610 acres. 33 percent and 29 percent of these vendors use farm stands and community supported agriculture programs respectively as other market venues. This suggests that the vendors are diversifying their market options. 66 percent of participants in this survey were males, 52 percent indicated that farming is their primary occupation. When asked whether they label their fresh produce as "local", a vast majority (90 percent) indicated doing so. This study found out that the use of social media and/or having market web-based advertisements among farmers' market vendors in the south central and western Kentucky regions is 66 percent.

Table 1. Descriptive statistics

Variables	Stats
Age	37 (13)
FarmSize (acres of land)	212 (309)
FarmStand	.33 (.48)
CSA	.29 (.46)
LocalLabel	.90 (.30)
WebSocialMedia	.66 (.48)
FarmingPrimaryOccupation	.52 (.51)
FarmExperience (number of years)	14 (9)
EducationLevel	3 (some college) (1.5)
Male	.66 (.48)

Note. Standard errors are in parenthesizes.

We asked the vendors to indicate one level of difficult among five ordinal alternatives (very easy, easy, moderate, difficult, and very difficult) they face when complying with farmers' market regulations. Results indicated that only one percent is in the extreme categories (very easy and very difficulty). Consequently, we decided to reduce the alternatives from five to three. To accomplish this, we merged very easy and easy into "easy" category, very difficult and difficult into "difficult" category, and left "moderate" intact. Hence, the dependent variable consists of three levels of difficult (Easy, Moderate, and Difficult). Responses are presented in Figure 1.

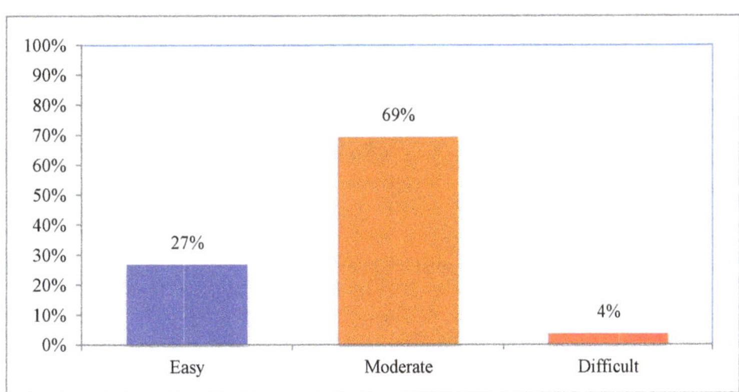

Figure 1. Histogram of levels difficult of farmers' market regulations

This study shows that the majority (69 percent) of farmers' market vendors believe that the market regulations are moderate. There were 27 percent of respondents said they find the farmers' market regulations to be easy to comply. Only 4 percent think that the regulations are difficult. These results are exactly the same as the partial probabilities (in Table 3) that the ordered logistic model produced using mfx2 commend after ologit.

Furthermore, we asked a question to find out the level of satisfaction about how farmers' market regulations are enforced in their market. Results indicate that a majority of farmers' market vendors are either neutral or satisfied. Only six percent of the respondents are either dissatisfied with how farmers' market regulations are enforced. Figure 2 below shows these statistics.

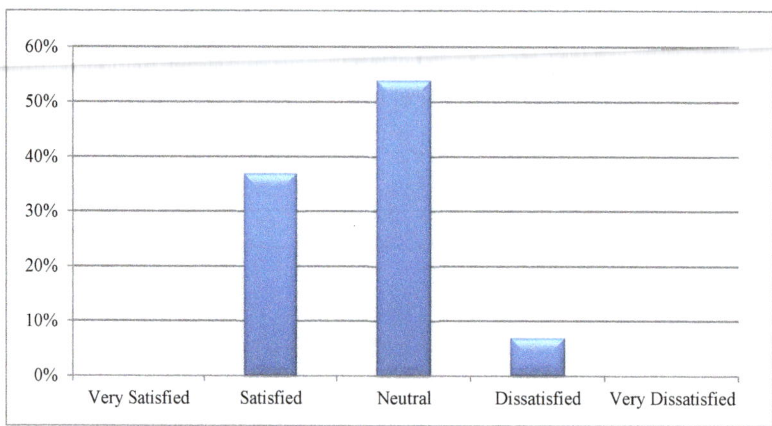

Figure 2. Histogram of levels of satisfaction of how farmers' market regulations are enforced

When asked if they strongly agreed, agreed, strongly disagreed, disagreed or neutral on if the regulations negatively affected their profitability, 13 respondents believe that regulations negatively affected their profitability. This study found out that that the majority of market vendors are either unsure of how regulations affect their profitability or believe that regulations have a positive impact on the profitability of their operation.

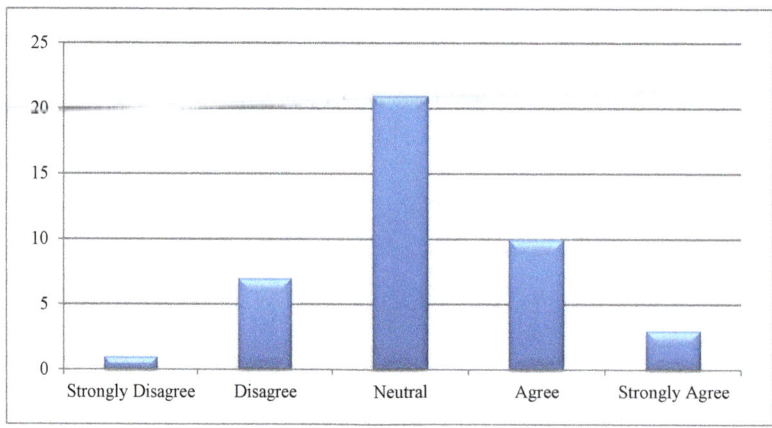

Figure 3. Histogram of if the vendors believed regulation negatively affected their operations

4.2 Model Results

Using stata/MP 13.1 and ologit command, we run the ordered logistic model to estimate the coefficients and odd ratios. Using the mfx2 command after running the ologit model, we computed the probabilities for random farmers' market vendor falling under a specific category of difficulty. Table 2 presents these probabilities.

Table 2. Ordered logistic probabilities using the mfx2 command

Difficult_Level_FM_Regulations	Probabilities	Cumulative probabilities
Easy	0.269	0.269
Moderate	0.694	0.963
Difficult	0.037	1.000

These results suggest that an average fresh produce vendor at farmers' market is 26 percent likely going to comply with market regulations easily, 69 percent moderately, and 4 percent hardly. Overall, this finding implies that the farmers' market regulations pose no great difficult to the vendors.

Coefficients in Table 3 are interpreted as the expected changes in the individual ordered log-odds of falling into the highest level of difficulty versus the lower levels as a result of a one-unit change in the predictor while the other variables in the model are held constant. Because the dependent outcomes are ordered from easy to difficult, a positive coefficient indicates that an increase in the corresponding variable increases the difficulty. The opposite is true for a negative coefficient. The odd ratios indicate the number of times that chances for subjects in the difficult category is multiplied when there a unit change in the specific predictor. The cut1 and cut2 are the estimated cut points on the latent variable used to differentiate the observed levels of difficulty when values of the predictor variables are evaluated at zero.

Table 3. Ordered logistic coefficients and odd ratios

Difficult_Level_FM_Regulations	Coefficients	Odds Ratios
FarmSize	0.003 (0.002)	1.003 (0.002)
FarmStand	-2.910* (1.549)	0.035* (0.101)
CSA	-1.936* (1.419)	0.118* (0.280)
LocalLabel	-3.457* (3.000)	0.128* (0.013)
WebSocialMedia	7.382* (4.006)	1.607* (6.438)
FarmingPrimaryOccupation	-4.008 (3.007)	0.018 (0.055)
FarmExperience	-0.304** (0.082)	.218** (0.100)
EducationLevel	-0.183 (0.498)	0.833 (0.415)
Male	-4.177* (3.070)	0.083* (0.016)
Constant Cut 1	-2.465 (3.129)	0.321 (0.369)
Constant Cut 2	-1.345 (2.201)	0.217 (0.282)

Note. The * and ** represent 10% and 5% respectively. Standard errors are in parenthesizes.

Results in Table 3 indicate that all significant coefficients on all variables (FarmStands, CSA, LocalLabel, FarmExpeience, and Males), but WebSocialMedia are negative. This posits that various market strategies by selling at farm stands, participating in CSA programs, and participating in "local" labeling programs, years of farming experience, and being a male vendor are associated with finding relatively easy to comply with farmers markets regulations. If a farmers' market vendor sells at farm stands, participate in CSA programs, and use "local" labels, his/her ordered log-odds of finding it difficult to participate in farmers' markets would decrease by -2.910, -1.936, and -3.457 respectively. Likewise, males (with a coefficient of -4.177) find it easy to participate in farmers markets than females do. We found that farm size, education, and being a farmer as a primary occupation do not have any significant effect on the levels of difficult of farmers' market regulations.

Furthermore, we found that farm experience is associated with easiness in navigating farmers' market regulations, with a coefficient of -0.304. This means that if a vendor gains one more year of farming experience, his/her ordered log-odds of finding it difficult to participate in farmers' markets would decrease by 0.304. This result is consistent with what one should expect; the longer a farmer participates in farmers markets the less difficult it is to abide by market regulations. Surprisingly, the coefficient on WebSocialMedia is positive; an indication that vendors who have both websites and use social media find farmers' market regulations to be difficult compared to their counterparts. This result proposes that incorporating the farmers' market regulations in these vendors' online tools in simple terms might reduce the difficulty.

The odds ratio in the third column of Table 3 are measures of proportional chances that vendors find it either easy, moderate, or difficult (outcomes) for a unit increase in a specific explanatory variable. Specifically, for a one-unit increase in years of experience with farming, the chances of finding it difficult (versus the combined moderate and easy categories) to comply with market regulations are 0.217 times lower, given that other variables in the model are held unchanged. Similarly, for a one-unit increase in years of experience with farming, the chances of finding it moderate (versus the combined difficult and easy categories) to comply with market regulations are 0.217 times lower. The chances for males to find it difficult (versus the combined moderate and easy categories) to comply with market regulations are 0.0834 times lower. The likelihood of vendors who participate in the CSA program to difficult to comply with market regulations is 0.118 lower compared to those who do not participate. This indicates that encouraging vendors of fresh produce at farmers' market to participile in CSA programs will make it easy for them to comply with market regulations.

5. Conclusion and Policy Implications

This study explains levels of difficult of farmers' market regulations among fresh produce vendors. An ordered logistic model was employed on survey data collected from forty-six fresh produce vendors at farmers' markets in the south central and western Kentucky regions. Data collection was completed in 2015 using an in-person survey, mail, and online strategies. Results indicate that an average fresh produce vendor at farmers' market in the south central and the western Kentucky regions is 26 percent likely going to comply with market regulations easily, 69 percent moderately, and 4 percent hardly. Overall, farmers' market regulations pose no great difficult to fresh produce vendors. This study posits that selling at farm stands, participating in CSA programs, and participating in "local" labeling programs, years of farming experience, and being a male vendor are associated with finding relatively easy to comply with farmers markets regulations. We found no evidence to claim that farm size, education, and being a farmer as a primary occupation affect the levels of difficult of farmers' market regulations among fresh produce vendors.

Based on our findings, we propose to vendors of fresh produce at farmers' markets to diversify their market portfolios by selling at other direct-to-consumer market outlets like farm stands and participating in CSA and "local" labeling programs. Not only this will increase their revenues, but also will enhance their understanding about various regulations; making easy for them to comply with farmers' market regulations. Because we found that vendors who have both websites and use social media find farmers' market regulations to be difficult compared to their counterparts; we encourage these vendors to include farmers' market regulations in their online tools (websites and/or social media accounts) review them frequently. Doing so might increase understanding and easiness to follow the regulations. Findings from this study advocate for any assistance in making easy for vendors to comply with the market regulations.

In terms of policy implications, this study suggests that policies aiming at promoting CSA and farm-stand market arrangements are viable in easing market regulations. We further recommend assistance to facilitate females and those vendors with less farm experience in complying with the farmers' market regulations easily. Vendors with experience should assist less-experienced in the process of complying with the regulations. Because this study noticed that each farmers' market makes its own regulations, we propose a unified set of regulations for all farmers' markets throughout the state with provisions to leave room for individual markets to customize them to meet specific needs. Finally, we propose that future studies examine effects of new rules at the farmers' market on vendors' behavior. For example, if vendors were asked to pay different fees based on specific spots (located at high traffic of consumers, located in the middle, located at less-frequented spots, and the like), it would be interesting to know what their choices and willingness to pay would be and how that might affect their participation. We recognize that the sample size in this study was not large enough. As a result, we suggest further studies with more participants in the regions and/or other states to find similarities and/or differences in findings.

References

Alonso, A. D., & O'Neill, M. A. (2011). A comparative study of farmers' markets visitors' needs and wants: The case of Alabama. *International Journal of Consumer Studies, 35*(3), 290-299. http://dx.doi.org/10.1111/j.1470-6431.2010.00931.x

Baharanyi, N., & Boateng, M. (2012). Assessing the Status of Farmers Markets in the Black Belt Counties of Alabama. *Journal of Food Distribution Research, 43*(3), 74-84.

Dasgupta, S., Eaton, J., & Caporelli, A. (2009). Consumer Perceptions of Kentucky-Grown Chicken Products in Farmers' Markets. *Journal of the Kentucky Academy of Science, 70*(2), 133-40. http://dx.doi.org/10.3101/1098-7096-70.2.133

Ernst, M. (2014). *Selling Farm Products at Farmers Markets*. University of Kentucky Cooperative Extension Service. Retrieved from https://www.uky.edu/Ag/CCD/marketing/farmmarket.pdf

Futamura, T. (2007). Made in Kentucky: The Meaning of "Local" Food Products in Kentucky's Farmers' Markets. *The Japanese Journal of American Studies, 18*, 209-227.

George, R. D., Kraschnewski, L. J., & Rovniak, S. L. (2011). Public health potential of farmers' markets on medical center campuses: A case study from Penn State Milton S. Hershey Medical Center. *American Journal of Public Health, 101*(12), 2226-2232.

Greene, W. H. (2008). *Econometric Analysis* (6th ed.). Upper Saddle River, New Jersey: Pearson Prentice Hall.

Gumirakiza, J. D., Curtis, K. R., & Bosworth, R. (2014). Who Attends Farmers' Markets and Why? Understanding Consumers and their Motivations. *Journal of Food and Agribusiness Management Review 17*(2), 65-82.

Hamilton, N. (2002). *Farmers' Markets Rules, Regulations and Opportunities.* The National Center for Agricultural Law Research and Information of the University of Arkansas.

Hazell, P., Poulton, C., Wiggins, S., & Dorward, A. (2010). The Future of Small Farms: Trajectories and Policy Priorities. *World Development, 38*(10), 1349-361. http://dx.doi.org/10.1016/j.worlddev.2009.06.012

Hofmann, C., Dennis, J., Gilliam, A. S., & Vargas, S. (2007). Food Safety Regulations for Farmers' Markets. *Purdue Extension*, 1-8.

Kennedy, P. (2008). *A Guide to Econometrics* (6th ed.). Malden, Massachusetts: Blackwell Publishing.

Kentucky Department of Agriculture. (2014). *The 2014-2015 Kentucky Farmers' Market Manual and Resource Guide.* Frankfort, KY.

May, P. (2005). Compliance Motivations: Perspectives of Farmers, Homebuilders, and Marine Facilities. *Law and Policy, 27*(2), 317-347. http://dx.doi.org/10.1111/j.1467-9930.2005.00202.x

McGarry-Wolf, M., Spittler, A., & Ahern, J. (2005). A profile of farmers' market consumers and the perceived advantages of produce sold at farmers' markets. *Journal of Food Distribution Research, 36*(1), 192-201.

Nulty, D. D. (2008). The adequacy of response rates to online and paper surveys: what can be done? *Journal of Assessment and Evaluation in Higher Education, 33*(3), 301-314. http://dx.doi.org/10.1080/02602930701293231

Ostrom, M., & Donovan, C. (2013). Summary Report: Farmers Markets and the Experiences of Market Managers in Washington State. *Washington State University Small Farms Program.* Retrieved from http://wafarmersmarkets.org/resource-file/WSU_FMMS_report_Nov_2013.pdf

Public Health Law Center. (2014). Farmers' Market Vendor and Market Rules. Retrieved from http://publichealthlawcenter.org/sites/default/files/resources/08.18.14.Farmers%20Market%20Vendor%20and%20Market%20Rules.pdf

Train, K. E. (2009). *Discrite Choice Methods with Simulation* (2nd ed.). New York City: Cambridge University Press. http://dx.doi.org/10.1017/CBO9780511805271

Varner, T., & Otto, D. (2008). Factors Affecting Sales at Farmers' Markets: An Iowa Study. *Review of Agricultural Economics, 30*(1), 176-89. http://dx.doi.org/10.1111/j.1467-9353.2007.00398.x

Williams, R. (2015). *Ordered Logit Models-Overview*. University of Notre Dame

Wooldridge, J. M. (2009). *Introductory Econometrics: A Modern Approach* (4th ed.). Mason, Iowa: South-Western Cengage Learning.

10

Efficacy of Paper Mill Sludge Along with Organic and Inorganic Nutrients on Growth and Yield of Turmeric (*Curcuma longa* L.)

Bibhuti Bhusan Dalei[1], Bibhuti Bhusan Sahoo[2], Lalatendu Nayak[2], Manoj Kumar Meena[1], Amit Phonglosa[3], Pravamayee Acharya[2] & Niranjan Senapati[2]

[1] AICRP on Niger, Regional Research & Technology Transfer Station (OUAT), Semiliguda, Koraput, Odisha, India

[2] Regional Research & Technology Transfer Station (OUAT), Semiliguda, Koraput, Odisha, India

[3] Regional Research & Technology Transfer Sub-Station (OUAT), Umarkot, Nabarangapur, Odisha, India

Correspondence: Bibhuti Bhusan Sahoo, Regional Research & Technology Transfer Station (OUAT), Semiliguda, Post Box No. 10, Sunabeda, Koraput 763002, Odisha, India. E-mail: bibhutihort@rediffmail.com

Abstract

Red soils are strongly to moderately acidic with low to medium organic matter and poor water retentive capacity. These soils are deficient in macro as well as micronutrients like boron and molybdenum. Being a commercially cultivated crop turmeric production was drastically affected in such type of soil. To defence against the above said crisis an experiment was conducted with seven treatments and replicated thrice, at Regional Research & Technology Transfer Station (OUAT), during *kharif*-2012, under Eastern Ghat High Land zone of Odisha, to assess the efficacy of paper mill sludge (PMS) with a mixture of organic and inorganic fertilizers on turmeric cv. Roma. Results revealed that application of 100% Recommended Dose of Fertilizer with PMS i.e. (T_3) recorded highest fresh rhizome yield of 285.30 q per ha followed by 100% RDF i.e. T_2 with 261.83 q per ha which is at par with T_3. Maximum plant height of 136.97 cm along with highest weight of 73.25 g and 98.27 g of primary and secondary fingers per clump respectively were obtained from T_3.

Keywords: turmeric, paper mill sludge, yield

1. Introduction

Turmeric (*Curcuma longa* L.), one of the Indian customary medicinal plant used in Ayurveda, Unani and Siddha medicine as home remedy for various diseases, botanically it belongs to the family *Zingiberaceae*. It is a perennial plant having a short stem with large oblong leaves and bears ovate, pyriform or oblong rhizomes, which are often branched and brownish-yellow in colour. Turmeric is used as a food additive (spice), preservative and colouring agent in Asian countries, including China and South East Asia. It is also considered as auspicious and is a part of religious rituals. In old Hindu medicine, it is extensively used for the treatment of sprains and swelling caused by injury. In recent times, it is extensively used as digestive aid and treatment for fever, inflammation, wounds, infections, dysentery, arthritis, injuries, trauma, jaundice and other liver problems. It is considered to be safest herb of choice for all blood disorders since it purifies, stimulates and builds blood.

India is projected to have a population of 1.7 billion by 2050 and there is no possibility of increase in cultivable land (Anonymous, 2014). Consecutively India produces 11.9 lakh MT of turmeric from an area of 2.3 lakh ha with an average productivity of 5.1 MT per ha (Saxena et al., 2014). Odisha produces around 0.3 lakh MT of turmeric from an area of 0.025 lakh ha with an average productivity of 12.10 tonne per ha. However, the productivity of turmeric remains constant since 2012 to till date (Anonymous, 2015). On the other hand the consumption pattern of turmeric is increased frequently due to its high therapeutic and nutritional value. To cater the requirement of the increasing population and per capita consumption, there will be required for higher production from a precise unit of land. This demands an increase in average productivity from the same piece of land. But inadequate plant nutrition causes serious disorders in turmeric as well as enormous loss in yield. Higher productivity is possible through quality planting material, balanced nutrition and optimum plant health management. Integrated approach of nutrient management is found beneficial for maintenance of soil fertility and sustaining crop productivity through all possible sources of plant nutrients.

In Odisha turmeric is extensively cultivated in the districts like Kandhamal, Gajapati, Koraput, Rayagada, Nawarangapur and Malkangiri. Red soil covers about 7.14 m ha of lands and being the highest coverage of all soil groups of the state, extend to the above said districts. The soils are strongly to moderately acidic with low to medium organic matter and poor water retentive capacity. These soils are deficient in nitrogen and phosphorus. Micronutrients like boron and molybdenum are highly deficient in these soils. These soils have low cation exchange capacity with high phosphate and sulphur absorption property and deficient in calcium and magnesium. Water soluble phosphates get fixed and become unavailable to crop plants. Applications of in-soluble phosphates two weeks before sowing seeds or mixed application of insoluble rock phosphates and single super phosphate at equal proportion (1:1) makes the best utilization of phosphate. Soil acidity is corrected by application of lime. Application of 1 to 2 ton per ha of paper mill sludge corrects soil acidity as described by Sahu and Mishra (2005).

Keeping these in view, the present experiment was conducted to find out the effect of paper mill sludge (PMS) on turmeric along with various organic and inorganic fertilizers *viz.* Farm Yard Manure (FYM) Vermicompost (VC) and chemical fertilizers as recommended dose of fertilizer (125:100:100 NPK Kg per ha).

2. Materials and Methods

2.1 Experimental Site

The experiment was conducted at Regional Research & Technology Transfer Station (OUAT) Semiliguda under Eastern Ghat High Land zone (18°42′N, 82°30′E, elevation of 884 m.a.s.l.) of Odisha during *kharif* 2012. The soil of experimental field was red and laterite with sandy to clay loam in texture. The pH of the soil was 5.8 with low in organic carbon (0.03-0.05%), available N (150-170Kg per ha), available P (16-18 kg per ha) and available K (152-160 kg per ha).

2.2 Climatic Situation of the Experimental Site

The climatic situation of the experimental site was hot and humid with an annual mean rainfall of 1567 mm, most of which (90%) was received during the month of June to September, mean summer and winter temperature were 34 °C and 12 °C respectively.

2.3 Design of the Experiment

The experiment was laid out in randomized block design (RBD) with seven treatments and replicated thrice. The treatments are T_1-Farmers practice, T_2-100% RDF (125:100:100 NPK Kg per ha), T_3-100%RDF+PMS, T_4-FYM+VC, T_5-FYM+VC+PMS, T_6-50% RDF+FYM+VC, T_7-50% RDF+FYM+PMS.

2.4 Experimental Management

Turmeric *cv.* Roma was planted on 11th June 2012 with a seed rate of 20 q per ha, spacing of 30 cm × 20 cm in a plot size of 10 m × 9 m. The PMS was applied 30 days before planting. FYM, VC and fertilizer was applied as basal dose as per the treatment schedule. Immediate after planting mulching was done by using dry niger stalk and silver oak leaves. All other scheduled cultural operations until the harvest of the crop were followed uniformly to obtain a healthy crop production. At matured stage the above ground dried portion (shoot) was removed carefully before harvesting to obtain good and healthy rhizomes. The rhizomes were harvested subsequently the yield and yield attributes of fresh rhizomes were recorded.

2.5 Statistical Analysis

The data recorded on vegetative growth, yield and yield attributing parameters were subjected to statistical analysis and treatment mean were compared at 5% level of probability as derived by K. A. Gomez and A. A. Gomez (1984).

3. Results and Discussion

3.1 Effect on Growth and Yield Attributes

Effect of paper mill sludge on plant height, number of tillers per plant, number of leaves per tiller, leaf length, leaf breadth, number of primary fingers per clump, number of secondary fingers per clump, weight of primary fingers per clump and weight of secondary fingers per clump are presented in Table 1. All the growth parameters and yield attributes achieved higher values for PMS along with 100% recommended dose of fertilizer. Highest plant height (136.97 cm), number of tillers per plant (4.13), number of leaves per tiller (8.30), leaf length (50.70), and leaf breadth (17.17 cm) were recorded with application of PMS along with 100% RDF(T3). This is due to improvement in soil fertility/nutrient availability to the crop and more plant-available water by application of PMS to the soil.

Highest values for yield attributes like number of primary fingers per clump (4.37), number of secondary fingers per clump (10.93), weight of primary fingers per clump (73.25 g) and weight of secondary fingers per clump (98.27 g) were recorded with application of PMS along with 100% RDF (T3). Lowest plant height (80.57 cm), number of tillers per plant (1.13), number of leaves per tiller (3.60), leaf length (22.67 cm) and leaf breadth (7.53 cm) were recorded with farmers practice. All the growth parameters and yield attributes achieved lower values for farmers practice. It might be due to inadequate supply of nutrients to the plants. Lowest values with respect to yield attributes like number of primary fingers per clump (1.49), number of secondary fingers per clump (4.63), weight of primary fingers per clump (25.36 g) and weight of secondary fingers per clump (41.07 g) were recorded with farmers practice.

Similarly, application of composted paper board mill solid sludge + fly ash + coir pith produced taller and stronger plant growth along with root length and nodules formation in cowpea was observed as described by Prasanthranjan et al. (2004).

Table 1. Effect of paper mill sludge along with organic and inorganic nutrients on growth and yield of turmeric

Treatments	Plant Height at harvest (cm)	Number of tillers per plant	Number of leaves per tiller	Leaf Length (cm)	Leaf Breadth (cm)	Number of primary fingers per clump	Number of Secondary fingers per clump	Weight of primary fingers per clump (g)	Weight of Secondary fingers per clump (g)	Fresh rhizome yield (qper ha)
T1: Farmers Practice	80.57	1.13	3.60	22.67	7.53	1.49	4.63	25.36	41.07	111.63
T2: 100% RDF	125.40	3.73	6.73	45.63	16.33	3.25	10.69	65.47	96.01	261.83
T3: 100% RDF+PMS	136.97	4.13	8.30	50.70	17.17	4.37	10.93	73.25	98.27	285.30
T4: FYM+VC	98.83	1.80	5.23	30.97	9.20	2.77	7.24	47.10	65.23	187.47
T5: FYM+VC+PMS	105.40	2.30	5.50	34.87	12.37	2.89	7.85	48.53	70.56	196.87
T6: 50% RDF+FYM+VC	121.73	2.87	6.37	36.83	13.23	3.19	9.27	54.35	82.91	227.87
T7: 50% RDF+FYM+PMS	122.73	3.50	6.40	43.57	15.30	3.72	9.93	63.27	89.57	253.43
SE m(±)	8.82	0.23	0.38	3.71	0.84	0.11	2.76	1.80	2.84	15.67
CD (P = 0.05)	27.17	0.72	1.17	11.44	2.60	0.32	8.21	5.35	8.42	48.28

Note. Recommended Dose of Fertilizer (RDF), Paper Mill Sludge (PMS), Farm Yard Manure (FYM), Vermicompost (VC).

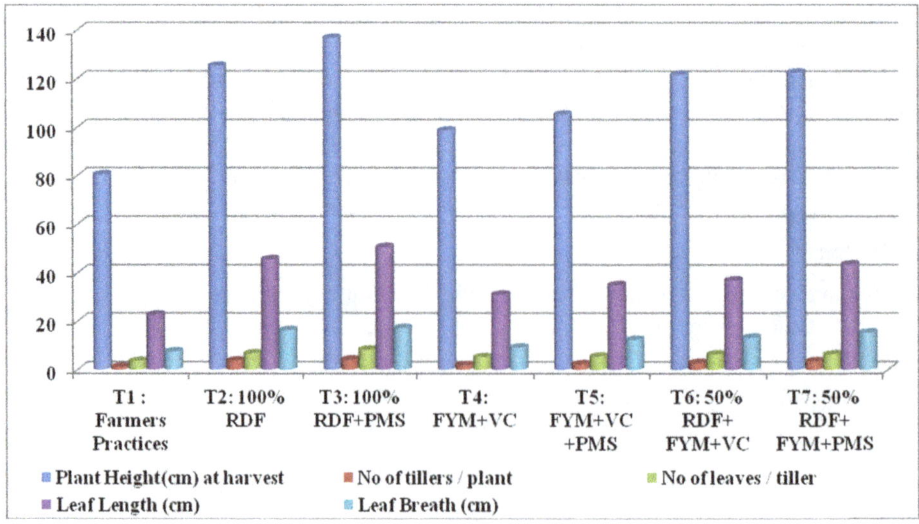

Figure 1. Effect of paper mill sludge along with organic and inorganic nutrients on vegetative growth of turmeric

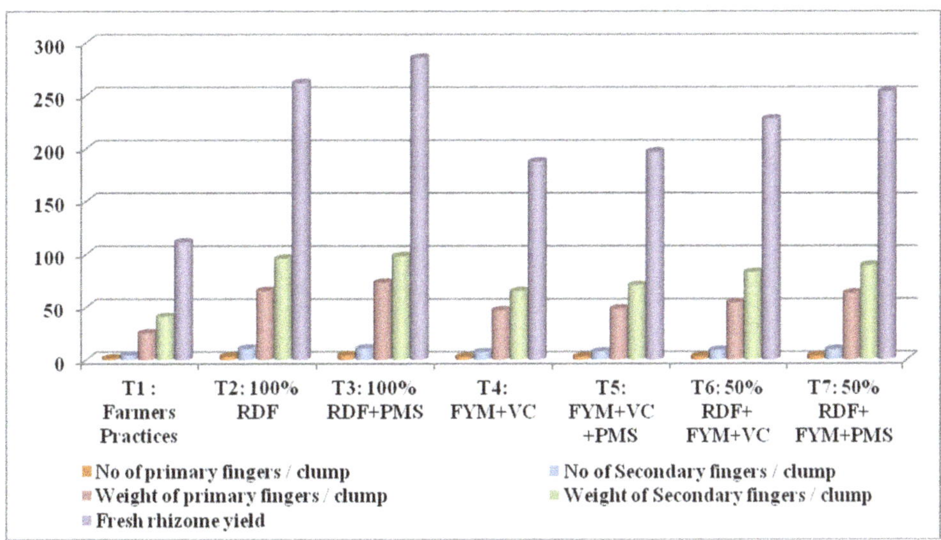

Figure 2. Effect of paper mill sludge along with organic and inorganic nutrients on growth, yield and yield attributes of turmeric

3.2 Effect on Yield

The yield of crop is generally governed by various yield attributing characters. Consequence upon superiority in yield attributes with treatment receiving PMS along with 100% RDF (T3) recorded highest rhizome yield (285.30 q per ha) followed by the treatment having only 100% RDF (261.83 q per ha). There was 9.0% increase in rhizome yield due to application of PMS along with RDF over application of RDF only.

Similarly, application of bio-compost (paper board mill solid sludge + fly ash + coir pith) with RDF recorded 27.45% increased in pod yield over 100 % fertilizer alone in cow pea (Prasanthranjan et al., 2004). Bowen et al. (1995) demonstrated that paper mill sludge amendments in potato could enhance crop yield relative to conventionally fertilized controls. Lowest rhizome yield was recorded with farmers practice.

Hence, Application of 100% RDF along with PMS shows highest extent of response in vegetative growth as well as in yield of turmeric dominating the sole application of 100% RDF and 50% RDF along with FYM, VC or PMS. 100% RDF enhance the better growth and yield of turmeric in comparison to 50% RDF where PMS balances the fertility status by neutralizing the acidity of the soil as described by Sahu and Mishra (2005).

4. Conclusion

Based on the results from the investigation it is concluded that the application of paper mill sludge along with N, P and K of 125:100:100 Kg per ha respectively could be able to produce higher yield in turmeric under Eastern Ghat High Land zone of Odisha.

References

Anonymous. (2014). From Director's Desk. *DOGR News, 18*(1), 1. Retrieved from http://www.dogr.res.in

Bowen, B. D., Wolkowski, R., & Hasen, G. (1995). Comparison of the effect of fresh and composted paper mill sludge on potato growth. *Proceedings of the national pulp and paper society*.

Gomez, K. A., & Gomez, A. A. (1984). *Statistical Procedures in Agricultural Research*. John Wiley, New York.

Prasanthranjan, M., Udayasoorian, C., & Singaram, P. (2004). Impact of paper mill solid sludge, bio compost and treated effluent irrigation on growth and yield attributes of vegetable cowpea. *Madras Agricultural Journal, 91*(7, 12), 483-488.

Sahu, G. C., & Mishra, A. (2005). Soil of Orissa and Its Management. *Orissa Review* (pp. 56-60).

Saxena, M., & Gandhi, C. P. (2015). *NHB database 2014 (*p. 6). National Horticulture Board, Ministry of Agriculture, Government of India 85, Institutional Area, Sector-18, Gurgaon-122 015, India. Retrieved from http://www.nhb.gov.in

Saxena, M., Bhattacharya, S., & Malhotra, S. K. (2016). State wise area and production of turmeric. *Horticultural Statics at a glance 2015* (pp. 225-255). Horticultural Statics Division, Department of Agriculture, Cooperation & Family Welfare, Ministry of Agriculture & Farmers Welfare, Govt. of India.

11

Chemical Composition of Drilled Wells Water for Ruminants

Daniel Bomfim Manera[1], Tadeu Vinhas Voltolini[2], Daniel Ribeiro Menezes[1] & Gherman Garcia Leal de Araújo[2]

[1] Universidade Federal do Vale do São Francisco (Univasf), Petrolina/PE, Brazil

[2] Embrapa Semiárido, Petrolina/PE, Brazil

Correspondence: Tadeu Vinhas Voltolini, Embrapa Semiárido, BR 428, Km 152, Zona rural, CEP 56302-210, Petrolina/PE, Brazil. E-mail: tadeu.voltolini@embrapa.br

Abstract

This study aimed to evaluate the chemical composition of water wells and to discuss the results in relation to nutritional requirements and tolerance limits of domestic ruminants. Ten samples of water wells (three replicates) from Brazilian semi-arid were collected and analyzed for their macro and trace minerals levels. A variation was found in the mineral composition of the waters and the macro minerals presenting highest levels were Cl, Mg, Ca and Na, while the predominant trace minerals were Fe and Mn. The concentration of the examined minerals can provide a small contribution to the animal as in the case of P or supply a considerable amount as Cl. The levels of total dissolved solids found in the majority of the samples can be tolerable for ruminants. In some of the samples the presence of Pb, Cd and Cr was found in concentrations higher than the upper recommended limit for ruminants.

Keywords: heavy metals, trace minerals, water source

1. Introduction

Water is a vital nutrient for all human beings and is considered a scarce resource in several regions in the world. In arid and semi-arid regions, animal production present great social and economic importance, where water availability constitutes one of the greatest challenges (Araújo et al., 2010).

Apart from water volumes, Beede (2012) points out that qualitative water parameters are important for the livestock production systems. Valtorta et al. (2008), Arjomandfar et al. (2010) and Visscher et al. (2013) report that ruminants can tolerate and accept water showing considerable salts concentration.

In addition, Brum and Sousa (1985) indicate that water can be an important source of mineral for ruminants. Minerals are important to grazing and feedlot animals as cattle, sheep and goats, and their deficiency may cause nutritional (Herdt & Hoff, 2011), reproductive (Mendonça Junior et al., 2011) and productive disorders (Tokarnia et al., 1999).

At the NRC (2001), it was also pointed out how important it is to observe the presence of minerals potentially toxic in the water such as Pb and Cd in order to avoid the supply to the animals. Cadmium and Pb are elements considered toxic to animals, which may accumulate in the body (Pareja-Carrera et al., 2014).

In the Brazilian semi-arid, subsurface waters can be an important source for the herds, especially from tube wells, but the physical-chemical composition of total dissolved solids (TDS), macro and trace minerals is scarce. On the other hand, it is important to understand the potential of water to be offered to animals and to propose measures to watering ruminants.

The objective was to examine the mineral composition of well water in the Brazilian semi-arid and to discuss the results based on the nutritional requirements and tolerance levels of ruminants.

2. Materials and Methods

This study lasted from March 2011 to March 2012. Water samples were collected from drilled wells in the Brazilian semi-arid in the municipalities of Jaguarari, Santa Brígida and Pintadas, in Bahia State; Lagoa Grande and Petrolina, in Pernambuco State; and Tauá, in Ceará State, Brazil. We collected ten samples of drilled wells water, with three replicates.

The mineral composition of all samples was determined at the Soil Laboratory of Embrapa Semiárido. The concentrations of potassium (K), phosphorus (P), calcium (Ca), magnesium (Mg), sulfur (S), copper (Cu), iron (Fe), manganese (Mn), zinc (Zn), sodium (Na), chlorine (Cl), nickel (Ni), lead (Pb), cadmium (Cd) and chromium (Cr) were determined according to the methodologies described by Nogueira et al. (2005). The concentrations of the minerals present in the water were obtained as mg/L.

Sodium and potassium levels were determined by flame photometry, whereas the concentrations of Ca and Mg were analyzed by titration, by determining the Ca contents, and subsequently the Ca + Mg contents, and the Mg concentration was defined as the difference. Chlorine and sulfur levels were determined indirectly, first by obtaining the concentrations of chlorines and sulfates and subsequently, by considering the atomic molecular weight, the concentrations of S and Cl were determined.

Phosphorus was determined using a molecular spectrophotometer while the determination of Cu, Fe, Mn, Zn, Ni, Pb, Cd and Cr was performed on an atomic absorption spectrophotometer (model Analyst 100, Perkin Elmer®). The data were presented as descriptive statistics (means and standard deviations).

3. Results and Discussion

The average macro mineral found in greater concentrations were Cl, Mg, Ca and Na with 1.010, 270, 240 and 230 mg/L, respectively. Chlorine was also the element with highest variation: from 430 to 2.580 g/L. The average of S was 100 mg/L, whereas K was 10 mg/L. Phosphorus content was found at very low concentration, at the order of 80 mg/L (Table 1).

The sum of macro and trace minerals, on average, was 1.940 mg/L (1.94 g/L) ranging from 750 to 4.660 mg/L. According to Brazilian legislation lesser than 0.5 g/L (500 mg/L) of total dissolved solids (TDS), in the water, is considered fresh, between 0.5 g/L to 30 g/L are classified as brackish and with more than 30 g/L are saline (CONAMA, 2005). Oliveira and Maia (1998) examined the physical and chemical composition of waters from different aquifers in the state of Rio Grande do Norte, (496 samples from water wells, 129 samples from subsurface waters) and observed 0.8 to 4.0 dS/m of electrical conductivity, equivalent to 510 to 2.560 mg/L of TDS, with 1.0 dS/m = 640 mg/L.

Nine of the ten samples found in the current study showed TDS levels that were within the limits proposed by Beede (2012), which are 3.000 mg/L, in order to avoid problems to the animals. Kattnig et al. (1992), Bahman et al. (1993), Valtorta et al. (2008), and Arjomandfar et al. (2010) examined the productive response of bovines receiving water containing high TDS levels, over 1.400 mg/L and observed no negative effect on productive response. Valtorta et al. (2008) observed an increase in water intake with 10.000 mg/L of TDS, which was also observed by Visscher et al. (2013) when studying the tolerance of bulls receiving saline water with 100, 5.000 or 10.000 mg/L of a NaC1 and KCl mixture.

According to McGregor (2004) bovines can also tolerate waters containing 2.300 to 11.000 mg/L of TDS, however, the author points out the necessity of adapting animals to saline water as an important factor to improve their tolerance. Abou Hussein et al. (1994) examined increased salinity up to 17.000 mg/L of TDS of total water intake of goats and sheep. Increased salt content from 9.500 to 17.000 mg/L of TDS reduced feed intake of both sheep and goats. Water intake with 9.500 mg/L of TDS reduced the feed intake of sheep but not of goats.

Table 1. Macro mineral concentration (mg/L) of the water from drilled wells in the Brazilian semi-arid

Sample	P	K	Ca	Mg	S	Na	Cl	TDS
1	30	10	210	80	80	120	430	960
2	-*	10	200	180	160	220	740	1,510
3	-*	20	220	540	200	180	1,430	2,590
4	250	20	160	290	40	220	1,060	2,040
5	10	20	510	660	120	760	2,580	4,660
6	40	10	180	130	110	260	740	1,470
7	140	20	450	140	70	140	670	1,630
8	-*	20	170	200	100	160	770	1,420
9	90	20	190	420	60	130	1,360	2,270
10	60	10	110	80	80	60	350	750
Mean	80	10	240	270	100	230	1,010	1,930
SD	80	10	130	200	50	200	650	1,110
Guidelines[a]	0.70	20	200	100	-[1]	300	-[1]	3,000
Guidelines[b]	-[1]	-[1]	> 500	> 125	-[1]	> 20	-[1]	> 3,000

Note. P* = phosphorus (mg/L); K = potassium; Ca = calcium; Mg = magnesium; S = sulfur; Na = sodium; Cl = chlorine; SD = standard deviation; TDS = total dissolved solids (macro and trace mineral sum); [a,b] Beede (2012) ([a]: concentrations above which problems likely could occur in livestock; [b]: possible cattle problems); [1]: non presented; * = non detected.

According to NRC (2001), TDS concentrations below 1.000 mg/L are considered safe in the bovine drinking water. Concentrations from 1.000 mg/L to 2.999 mg/L are generally considered safe, but may cause diarrhea in animals that are not adapted to these hydric sources. Water containing 4.600 mg/L of TDS may be rejected by the animals without previous experience, cause diarrhea or reduce productivity compared to the waters with lower TDS. Waters with 5.000 to 6.999 mg/L of TDS should be avoided by pregnant or lactating animals while waters with more than 7.000 mg/L should not be offered to cows (NRC, 2001).

Domestic ruminants have been used and accepted saline waters with considerable TDS levels (Valtorta et al., 2008; Arjomandfar et al., 2010), but there are differences among the species as to their tolerance. Masters et al. (2007) reported the order in which animals are tolerant to salinity as camels > sheep > goats > cattle. In studies conducted in the Brazilian semi-arid with cows and goats that ingest waters with 650 to 8.960 mg/L, it was also observed that ruminants accepted and tolerated waters of up to 8.960 mg/L of TDS (Albuquerque, 2012; Alves, 2012; Costa, 2012).

Costa (2012) did not find differences in productive performance of sheep, nor in intake and digestibility of DM and nutrients (neutral detergent fiber, crude protein, total carbohydrates and total digestible nutrients) when they were exposed to waters with 5.760 mg/L of TDS, superior to their found in the water of drilled wells in the current study.

Tolerance to saline water is also related to the different physiological stages of the animal (Squires, 1993; NRC, 2001), as well as the type of salt or the mineral, and not only the TDS (Squires, 1993; Valtorta et al., 2008). All the examined waters in the present study presented high Cl levels, considering the potential water intake of a lamb weighting 20 and showing 100 g/day of weight gain. The sodium supply through the brackish waters is also elevated compared to the daily requirements of this animal. Beede (2012) does not indicate maximum tolerable levels of Cl in the drinking water, but suggests 300 mg/L of chlorides (an ionic form of Cl). The average Cl levels in the water of drilled wells (1.010 mg/L) ranged from 350 to 2.580 mg/L.

In the case of Na, Beede (2012) suggests 300 mg/L as the maximum level which is higher than the average obtained in nine out of ten water samples. Only one of the samples showed more than 300 mg/L of sodium. For young ruminants (veal calves), Beede (2012) suggests significantly lower Na levels (20 mg/L) because this category can be less tolerant to sodium.

Sodium chloride is considered the main component of TDS, apart from other salts or elements that also contribute to TDS (NRC, 2001). In animal nutrition, sodium and chlorine are considered jointly and, on average, sodium levels found were high, representing 82.80% of mineral requirements of a lamb weighting 20 kg. Chlorine also showed high concentration, with a potential contribution to the animal that exceeds the daily requirements, especially when considering the waters with larger concentrations.

The excess of NaCl in the water can reduce feed intake and increase intake of water (Peirce, 1957; Wilson & Dudzinski, 1973). Increase in milk production (Solomon et al., 1995) and greater efficiency in milk production (Guadalupe et al., 2015) were observed in lactating cows receiving water desalinated compared to saline water. However, saline water has been used in domestic ruminants without health damage reports (Kii & Dryden, 2005; Valtorta et al., 2008; Visscher et al., 2013). According to Peirce (1957), Wilson (1975) and Squires (1993), 1.0 to 1.2% of NaCl in the water (10 to 12 g/1.000 g of the solution) does not cause health damages to the animals.

According to NRC (2005), it is described NaCl toxicity from intake of 4.0% of DM, corresponding to 25.20g/day for an animal eating 630 g/day of DM. Considering a daily water intake of 1.44 L and the average concentration observed in the current study, the NaCl intake through water would be 1.79 g/day, which are below the toxic level.

Considering each analyzed element and considering a growing ovine with 20 kg of body weight, showing 100 g/day of weight gain and presenting macro minerals requirements of (g/day) 1.50 P; 2.90 K; 2.30 Ca; 0.60 Mg; 1.10 S; 0.40 Na and 0.30 Cl (NRC, 2007) and an estimated water intake of 1.44 L/day (NRC, 2007), the daily supply based on the average levels of the analyzed water of drilled wells would be $< 0.01\%$, $< 1.0\%$; 15.03%; 64.80%; 13.00%; 81.80% and $> 100.00\%$, respectively. Phosphorus and K showed a low potential of being supplied by water while Cl, Na and Mg presented a considerable potential to be supplied by water.

Brum and Sousa (1985) examined the mineral composition of drinking water for bovines and found low contribution, below 1% of the daily requirements of cows for Ca; P; Mg; Cu; Co and Zn. On the other hand, the contribution of K and Fe was 9.27% and 16.64%, respectively. In the case of Na, the contribution of saline water, considering a daily intake of 26 L/day would meet 209.73% of Na requirements, which may affect mineral mix intake.

Phosphorus causes environmental worries because, in water, in the form of PO_4, can promote eutrophication (Hubbard et al., 2004). Sinclair and Atkins (2015) identified an excess of phosphorus in the diet as one of the main mineral contaminants on dairy farms. Castillo et al. (2013) verified that high mineral concentration in the water may increase the supply to the animals and consequently in the waste, and according to these authors when the objective is to reduce mineral excretions in waste, it is important to consider the water supply.

Supplementing P through water is one of the strategies to supply this element to ruminants (Karn, 2001), especially when soluble phosphorous are used. The supply of minerals to animals is done through intake of feed, water and soils. According to Duarte et al. (2011) in Brazil, the deficiency of P is the most important among the minerals, resulting in reproductive disorders and reduction in the productive performance. In addition, Prasad et al. (2015) reported that P deficiency can promote bone disorders and weakness in the animals.

Calcium, magnesium, sulphur, sodium and chlorine may be supplied considerably through water wells. In the case of Ca, two of the analyzed water showed 450 and 510 mg/L, which could supply 28.17% and 31.93% of the daily requirements for an ovine of 20 kg with a weight. Considering a diet insufficient to provide mineral to animals, the intake of water can be favorable to complement the supply of minerals. However, when the diet is sufficient to provide minerals as reported by Tebaldi et al. (2000), the excessive supply of a particular element may cause antagonistic or toxic effects.

Beede (2012) proposed 200 mg/L of Ca as the upper limit in the drinking water for livestock farming. Five of the ten water samples showed Ca concentration below 200 mg/L while only one exceeded 500 mg/L. The maximum level of Ca in the diet is 1.50% of DM (NRC, 2005). Considering a lamb with an intake of 630 g DM/day, the daily Ca intake would be 9.45 g/day. The estimated water intake in the current study was 1.44 L/day, so the lamb could receive 0.35 g/day of Ca, which leaves a considerable amount of Ca to be supplied through the diet. Calcium is the main structural component of bones and teeth and its deficiency can cause rickets in young animals, osteomalacia in adults and milk fever in lactating animals. On the other hand, according to McDowell (1989), the excess can cause hypercalcemia or calcification of body tissues.

The highest concentration of Ca compared to the P promoted high Ca:P ratio in the drinking water. According to NRC (2007) the ideal Ca:P is described in the order of 2:1. In the analyzed water samples was 3:1 on average and one sample showed 51:1 of Ca:P. The imbalances in Ca:P can affect mineral absorption.

High levels of Mg were observed in the water of drilled wells. The water samples with greater Mg concentration (420; 540 and 660 mg/L) has potential to provide the requirements of a lamb and those with lower Mg levels (80 mg/L) showed considerable potential to contribute to the animal.

The Australian Water Resources Council (AWRC) (1969) and Ayers and Westcott (1994) suggested 250 to 500 mg/L, depending on the category and physiological state of the animal, as the maximum limits for the concentration of Mg in drinking water for ruminants. The average Mg level (270 mg/L) is higher the limits for lactating animals (250 mg/L). Two water samples showed Mg higher than 500 mg/L, which are above the recommendation. Beede (2012) suggested 100 and < 125 mg/L of Mg in the water as the maximum limit to be allowed for livestock farming and for bovines, respectively. Therefore, only two samples (80 mg/L) would be safe to the animals.

The negative effects of high levels of Mg are the possible restriction in feed intake and diarrhea (NRC, 2001). Fasae and Omolaha (2014) examined Mg levels in the drinking water for ruminants, verified that concentration of this mineral was slightly above the recommended (30 mg/L) and reported possible aesthetic problems. According to NRC (2001), water samples containing 0-60 mg/L of Ca and Mg are considered light, from 61 to 120 mg/L are considered moderately hard, with 121 to 180 mg/L hard and containing more than 180 mg/L very hard, but, on the other hand, hardness of water up to 290 mg/L was not associated with ruminant disorders.

High concentrations of Mg can also affect the absorption of other minerals such as P, as well as an excess of P and K can affect the absorption and metabolism of Mg (NRC, 2001). Magnesium levels above 0.60% of DM in the diet are considered as the maximum level of the daily intake, which may promote intoxication and cause urolithiasis, lethargy, diarrhea, reduction in feed intake and reduce the productive performance (NRC, 2007).

In the case of K, the percentage of the nutritional requirements potentially supplied by water would be low, lower than 1.0% for a growing sheep. Potassium levels varied from 10 to 20 mg/L in the water samples. All the water samples are within the maximum limits suggested by Beede (2012) for use in livestock farming. Potassium deficiency can promote reduction in growth, production and in feed intake (Puls, 1994).

Potassium is also important in the peripartum of lactating animals, especially those showing elevated milk production. In this phase, the animals are more susceptible to metabolic and nutritional disorders such as hypocalcemia and retained placenta (Greghi et al., 2014). A method to reduce hypocalcemia is to supply anionic diets promoting a negative cation-anion balance (Wilkens et al., 2012), which is calculated by the sum of K^+ and Na^+ subtracting the sum of Cl^- and S^{2-} in order to increase Ca absorption.

The potential contribution of water is considerable for S, in order to 26.08% in the sample with the highest concentration (200 mg/L). Beede (2012) does not give any recommendations for maximum limits of S, but suggested 300 mg/L of sulfate (SO_4) for use in livestock farming. Wright (2007) reported 1.000 g/L as a safe sulfate level but above 4.000 mg/L it can be potentially toxic. In all water samples S levels were below 200 mg/L, suggesting low concentration of SO_4, below 300 mg/L, which does not restrict the water intake.

High sulfate concentration may reduce water intake (Beede, 2006). According to Lardner et al. (2013) animals did not choose treated waters and preferred non-treated waters from wells with TDS < 3,000 mg/L and SO_4 < 2,000 mg/L. Sulfate levels of over 2.000 mg/L in the drinking water negatively influenced the animals' intake.

Beke and Hironaka (1991) reported the occurrence of poliencephalomalacia in Hereford calves receiving saline water with predominance of Na and S, suggesting that the elevated intake of S as the causative agent. In the diet, the maximum tolerable S level is 0.30% of DM (NRC, 2005) when the animals are fed with concentrated ingredients.

In the case of trace minerals, Fe showed the highest concentration in the analyzed water samples with 1.36 mg/L, with a large variation, followed by Mn, Pb, Ni, Zn, Cr and Cu, which averaged 0.27, 0.10, 0.08, 0.07, 0.04 and 0.04 mg/L. (Table 2). The potential contribution of water for a lamb with a body weight of 20 kg for Fe, Mn, Cu and Zn would be low, on average, 6.12%, 3.24%, 1.86% and 0.78%, respectively. The water sample with 0.23 mg/L of Cu has a potential to contribute to 10.68% of the daily requirements of a growing sheep.

The AWRC (1969), Ayers and Westcott (1994) and Beede (2012) suggest 0.50 mg/L as the maximum limit of Cu in drinking water for ruminants. The concentrations of Cu in all the water samples were below 0.50 mg/L. In nine of the ten analyzed samples, the concentrations of Cu did not exceed 0.03 mg/L, but in one of them, the concentration of Cu was 0.23 mg/L.

Copper is important to the animal as part of the enzymatic systems and its deficiency can cause anemia. The maximum tolerable level of Cu for a lamb is 15 mg/kg of DM when diets with normal concentrations of Mo (1 to 2 mg/kg of DM) and S (0.15% to 0.25%) are offered. Goats may have greater tolerance than sheep, while for cattle 40 mg/kg of DM is used as the maximum tolerable level. The minerals Cu, Mo, S and Fe should be considered jointly for the animal because the availability of Cu is reduced in the presence of the other three elements (NRC, 2005). According to Mohammed et al. (2014) Cu and P are the main limiting minerals for grazing ruminants in

Asia, Africa and Latin America and copper deficiency can cause growth reduction, bone disorders, diarrhea and infertility. The relationship between trace minerals is important to the absorption of minerals and according to Marques et al. (2013), increased levels of Fe, Mn and Zn limit absorption of Cu.

The water sample with greater Fe may contribute on the level of 50.40%, in the matter of the daily requirements of lamb with a body weight of 20 kg. Iron is the most abundant trace mineral of the body and constituent of hemoglobin. In addition, elevated intake of Fe can promote Cu deficiency in sheep and goats, with the maximum tolerable level of 500 mg/kg of DM (NRC, 2005). In other words, the observed levels in the water are well below the concentrations of Fe considered toxic. In Brazil, Fe has not been identified as a deficient mineral for ruminants, due to the production system based and tropical pastures (Peixoto et al., 2005; Malafaia et al., 2014).

Beede (2006) reported that Fe levels over 0.30 mg/L can reduce water feed intake by the animals and may also promote the proliferation of microorganisms, contributing to mud formation and affect the flux of water in tubing. The predominant form of Fe in the water is important for mineral nutrition because the ferrous form (Fe^{2+}) is soluble in water compared to the ferric form (Fe^{3+}), which is present in the food. The highly soluble iron can interfere in the absorption of Cu and Zn. In the current study, the presented values corresponded to total Fe. Four of the ten water samples showed Fe levels above 0.30 mg/L. One water sample showed 11.20 mg/L of Fe, much higher than the maximum suggested which may cause rejection of water. Al-Khaze'leh et al. (2015) examined water quality for goats in Jordan and also found high levels of Fe.

Genther and Beede (2013) examined the water intake, preference and behavior of dairy cows receiving water with different concentrations, valences and sources of Fe, supplying 0, 4 and 8 mg/L in the form of ferrous lactate. These authors observed that cows spent less time ingesting water with 8 mg/L of Fe in comparison to the other waters. They also observed that the cows' preference was not affected by the sources of P nor by the valences.

Table 2. Trace mineral concentration (mg/L) of the water from drilled wells in the Brazilian semi-arid

Sample	Cu	Fe	Mn	Zn	Ni	Cd	Pb	Cr
1	0.01	0.14	0.04	0.08	0.09	0.01	0.01	0.03
2	0.03	0.92	0.01	0.06	0.08	0.01	0.03	0.06
3	0.23	0.19	0.06	0.25	0.10	0.00	0.22	0.05
4	0.01	11.2	2.18	0.07	0.15	0.00	0.10	0.05
5	0.01	0.11	0.10	0.06	0.16	0.00	0.24	0.06
6	0.01	0.06	0.18	0.06	0.11	0.01	0.09	0.03
7	0.02	0.05	0.03	0.03	0.11	0.00	0.09	0.04
8	0.01	0.06	0.02	0.04	0.05	0.00	0.11	0.04
9	0.02	0.42	0.03	0.02	0.00	0.06	0.01	0.00
10	0.01	0.44	0.04	0.02	0.00	0.08	0.00	0.04
Mean	0.04	1.36	0.27	0.07	0.08	0.02	0.10	0.04
SD	0.07	3.48	0.67	0.07	0.05	0.03	0.08	0.02
Guidelines[a]	0.50	0.40	0.50	25.00	1.00	0.05	0.10	1.00
Guidelines[b]	> 0.60 to 1.00	> 0.30	> 0.05	> 25.00	-[1]	> 0.05	> 0.10	-[1]

Note. Mo* = molybdenum; Co* = cobalt; Cu = copper; Fe = iron; Mn = manganese; Zn = zinc; Ni = nickel; Cd = Cadmium; Pb = Lead; Cr = chromium; SD = standard deviation; [a,b]Beede (2012) ([a]: concentrations above which problems likely could occur in livestock; [b]: possible cattle problems); [1]: non presented.

Water samples with high levels of Mn may contribute on the level of 26.16%, when it comes to of the daily requirements for a growing sheep. According to Zhang et al. (2014), Mn is an essential trace element related to the organic matrix, a cofactor for important enzymes and is also involved in the amino acid metabolism. According to NRC (2007), Mn is described as being slightly toxic to ruminants even at high levels, but can promote antagonism between Mn and Fe. According to NRC (2005), the maximum tolerable level of Mn is 2.000 mg/kg of DM.

In drinking water for ruminants, Ayers and Westcott (1994), Runyan and Bader (1994) and Beede (2012) suggest that 0.05 mg/L of Mn is the maximum limit for domestic animals. Four water samples presented Mn concentrations above this recommendation. According to Beede (2012), concentrations above 0.05 mg/L of Mn affect water intake. Pérez-Carrera et al. (2007) found that 12.50% and 92% of water samples from water table and deep wells, respectively in dairy production systems in Argentina, showed Mn levels above 0.05 mg/L.

The Zn concentrations observed in all water samples were low in relation to the daily requirements of lambs. The water sample containing the highest concentration would supply 2.77% of the daily requirement of a lamb with a body weight of 20 kg. Zinc is involved in the composition of many enzymes that participate in various metabolic pathways and its deficiency can cause reproductive and skeletal disorders, skin and wool anomalies and anorexia (NRC, 2007). Ruminants can tolerate higher levels of Zn than non-ruminants. For sheep, the maximum tolerable level is 300 mg/kg of DM (NRC, 2005). Ten water samples showed Zn levels below 25 mg/L, suggested as the maximum tolerable (Beede, 2012). Smith (1980) examined the addition of zinc sulfate to drinking water for bovines (0.25, 0.50 and 1.0 g/L of Zn) and found a reduction in water intake by 8.0%, 35.0% and 54%, respectively, compared to the animals receiving water without the addition of Zn.

The average levels of Ni, Cd and Cr were below the maximum level suggested by Beede (2012), in the order of 1.00, 0.05 and 1.00 mg/L, respectively, while the concentration of Pb was at the suggested level (0.10 mg/L). Nickel and Pb showed higher average levels (Table 2). Chromium and nickel are elements required by the animals. Chromium is associated with stress responses, glucose metabolism and immune response, while nickel is required by the microorganisms of the rumen (NRC, 2007).

In all analyzed water samples the presence of Ni and Cr was found in concentrations below 1.00 mg/L. Cadmium levels above 0.05 mg/L and Pb above 0.10 mg/L were observed in two and three water samples, respectively. The average concentration of Cd also was below the limits of 0.05 mg/L established by the AWRC (1969) and by Ayers and Westcott (1994).

The NRC (2001) is more restrictive to these elements (heavy metals) and for Ni, Cd, Pb and Cr the suggested upper limits are: 0.25 mg/L, 0.005 mg/L, 0.105 mg/L and 0.10 mg/L, respectively. Cadmium and Pb averages observed in the water exceeded the recommended limits. Five of the ten water samples showed concentrations below the suggested levels for Cd and three for Pb.

Cadmium and Pb are elements considered toxic to the animals, accumulating in the body. Cadmium can cause renal damages (NRC, 2001) and Pb may promote hematologic, reproductive, immune, vascular and renal damages (Pareja-Carrera et al., 2014). They are also antagonists of others minerals. Cadmium interferes in the absorption and metabolism of Zn and Cu, while Pb affects the absorption and function of Zn (NRC, 2001) and the metabolism of Ca (Pareja-Carrera et al., 2014). In addition, Ca, P, Fe and Zn can reduce the absorption of Pb (NRC, 2001).

Valente-Campos et al. (2014) found a great variation in the upper permitted limits of minerals in the water for domestic animals in different countries. Cadmium, for instance, ranged from 0.005 mg/L to 0.14 mg/L. In Brazil, the presented limits varied from 0.01 to 0.05 mg/L. The multiple recommendations of maximum limit for macro and trace minerals makes difficult the understanding of physical-chemical parameters of the water for animals.

The analyzed water presented levels of TDS and other macro minerals that, in accordance with the presented information are not restricted for domestic ruminants as sheep, goats and cattle. However, its use may be dependent of several factors as animal species and physiological stage. In general, water samples presented heterogeneity in mineral composition with different potential contribution to the animal, but minerals levels can be considered in the animal nutrition, avoiding for example antagonistic effects.

The dilution of drinking water with fresh water or rainwater can contribute to use of saline water. The reduction of evaporation rates of water in the reservoirs is also an important strategy to prevent the increase of mineral in the water. Shirley (1985) reports that the bioavailability of minerals in water is similar to those observed in food. However, the mineral bioavailability in each type of water sample is an important information in order to verify the real potential to supply minerals for domestic ruminants.

Trace minerals were found to be more potentially restrictive in the water samples, especially Cd and Pb, while Fe and Mn were found at levels that can promote a rejection or a reduction in water intake by the animal or interfere in the absorption and metabolism of other mineral. Determination of mineral levels in the water is fundamental to promote a safe use of this natural resource. The spatial and temporal variation in the electric conductivity of subterranean waters in the Brazilian semi-arid region (Andrade et al., 2012) and the physical-chemical characterization of water should be more frequent. Additional research is required to establish the limits of the mineral in the drinking water of domestic ruminants.

4. Conclusions

Well waters showed a variation in mineral levels with a predominance of Cl, Mg, Ca and Na as macro minerals and Fe and Mn as trace minerals. The waters showed low levels of P and consequently presented low contribution potential in this mineral, but showed an elevated potential to supply chlorine. Total dissolved solids (TDS) levels in the majority of the samples are within the acceptable range, apart from having the presence of potentially toxic elements such as Cr, Cd and Pb in concentrations above the recommended.

References

AbouHussien, E. R. M., Gihad, E. A., El-Dedawy, T. M., & Abdel Gawad, M. H. (1994). Response of camels, sheep and goats to saline water. 2. Water and mineral metabolism. *Egyptian Journal of Animal Production, 31*, 387-401.

Albuquerque, I. R. R. (2012). *Níveis de salinidade da água de beber para ovinos mestiços Santa Inês* (Dissertação de Mestrado). Universidade Federal da Paraíba, Paraíba.

Al-Khaze'leh, J. M., Reiber, C., Al Baqain, R., & Zárate, A. V. (2015). Drinking water sources, availability, quality, access and utilizationfor goats in the Karak Governorate, Jordan. *Tropical Animal Health and Production, 47*, 163-169. http://dx.doi.org/10.1007/s11250-014-0702-6

Alves, J. N. (2012). *Novilhas Sindi submetidas a ingestão de água com diferentes níveis de salinidade* (Tese de Doutorado). Universidade Federal da Paraíba, Paraíba.

Araújo, G. G. L., Voltolini, T. V., Chizzotti, M. L., Turco, S. H. N., & Carvalho, F. F. R. (2010). Water and small ruminant production. *Revista Brasileira de Zootecnia, 3*, 9326-336. http://dx.doi.org/10.1590/S1516-359 82010001300036

Arjomandfar, M., Zamiri, M. J., Rowghani, E., Khorvash, M., & Ghorbani, Gh. (2010). Effects of water desalination on milk production and several blood constituents of Holstein cows in a hot arid climate. *Iranian Journal of Veterinary Research, 11*, 233-238.

Australian Water Reseources Council (AWRC). (1969). *Quality aspects of farm water supplies*. Canberra, DC: Department of National Development.

Ayers, R. S., & Westcot, D. W. (1994). *Water quality for agricuture.Food and Agriculture*. Rome, Italy, DC: Organization of the United Nations.

Bahman, A. M., Rooker, J. A., & Topps, J. H. (1993). The performance of dairy cows offered drinking water of a low or high salinity in hot arid climate. *Animal Production, 57*, 23-28. http://dx.doi.org/10.1017/S0003 35600006565

Beede, D. K. (2006). *Evaluation of water quality and nutrition for dairy cattle* (p. 24). High Plains Dairy Conference.

Beede, D. K. (2012). What will our ruminants drink? *Animal Frontiers, 2*, 36-43. http://dx.doi.org/10.2527/af.2012--0040

Beke, G. J., & Hironaka, R. (1991). Toxicity to beef cattle of sulfur in saline well water: A case study. *Science of The Total Environment, 101*, 281-290. http://dx.doi.org/10.1016/0048-9697(91)90042-D

Brum, P. A. R. de, & Sousa, J. C. de. (1985). Níveis de nutrientes minerais para gado, em lagoas ("Baías e Salinas") no pantanal Sul Mato-grossense. *Pesquisa agropecuária Tropical, 20*, 1451-1454.

Castillo, A. R., St-Pierre, N. R., Silva del Rio, N., & Weiss, W. P. (2013). Mineral concentrations in diets, water, and milk and their value in estimating on-farm excretion of manure minerals in lactating dairy cows. *Journal of Dairy Science, 96*, 3388-3398. http://dx.doi.org/10.3168/jds.2012-6121

Conselho Nacional do Meio Ambiente (CONAMA). (2005). *Resolução n° 357 de 17 de março de 2005. Dispõe sobre a qualidade dos corpos de água e diretrizes ambientais para o seu enquadramento, bem como estabelece as condições e padrões de lançamentos de efluentes e dá outras providências*. Diário Oficial da República Federativa do Brasil, Brasília, DF. Retrieved June 16, 2015, from http://www.mma.gov.br/conama/res/res05/res35705.pdf

Costa, S. A. P. (2012). *Oferta de água com níveis de salinidade para ovinos Morada Nova* (Dissertação de Mestrado). Universidade Federal do Vale do São Francisco, Petrolina, Pernambunco.

Duarte, A. L. L., Pires, M. L. de S., Barbosa, R. R., Dias, R. B. da C., & Soto-Blanco, B. (2011). Avaliação da deficiência de fósforo em ruminantes por meio de bioquímica sérica. *Acta Veterinaria Brasilica, 5*, 380-384.

Fasae, O. A., & Omolaja, O. E. (2014). Assessment of drinking water quality from different sources in Smallholder ruminant production in Abeokuta, Nigeria. *Food Science and Quality Management, 29*, 39-43.

Genther, O. N., & Beede, D. K. (2013). Preference and drinking behavior of lactating dairy cows offered water with different concentrations, valences, and sources of iron. *Journal of Dairy Science, 96*, 1164-1176. http://dx.doi.org/10.3168/jds.2012-5877

Greghi, G. F., Netto, A. S., Schalch, U. M., Bonato, C. S., Santana, R. S. S., Cunha, J. A., ... Zanetti, M. A. (2014). Suplemento mineral aniônico para vacas no periparto: Parâmetros sanguíneos, urinários e incidência de patologias de importância na bovinocultura leiteira. *Pesquisa Veterinária Brasileira, 34*, 337-342. http://dx.doi.org/10.1590/S0100-736X2014000400007

Guadalupe, G. M. J., Herrera-Monsalvo, C. D., Lara-Bueno, A., López-Ordaz, R., Jaimes-Jaimes, J., & Ramirez-Valverde, R. (2015). Effects of drinking water desalination on several traits of dairy cows in a mexican semiarid environment. *Life Science Journal, 12*, 87-93. http://dx.doi.org/10.7537/j.issn.1097-8135

Herdt, H. H., & Hoff, B. (2011). The use of blood analysis to evaluate trace mineral status in ruminant livestock.Veterinary. *Clinics of North America: Food animal Practice, 27*, 255-283. http://dx.doi.org/10.1016/j.cvfa.2011.02.004

Hubbard, R. K., Newton, G. L., & Hill, G. M. (2004). Water quality and the grazing animal. *Journal of Animal Science, 82*, 255-263.

Karn, J. F. (2001). Phosphorus nutrition of grazing cattle: A review. *Animal Feed Science and Technology, 89*, 133-153. http://dx.doi.org/10.1016/S0377-8401(00)00231-5

Kattnig, R. M., Pordomingo, A. J., Schneberger, A. G., Duff, G. C., & Wallace, J. D. (1992). Influence of saline water on intake, digesta kinetics, and serum profiles of steers. *Journal of Range Management, 45*, 514-518. http://dx.doi.org/10.2307/4002562

Kii, W. Y., & Dryden, G. McL. (2005). Effect of drinking saline water on food and water intake, food digestibility, and nitrogen and mineral balances of rusa deer stags (*Cervus timorensisrussa*). *Animal Science*, 8199-105. http://dx.doi.org/10.1079/ASC41070099

Lardner, H. A., Braul, L., Schwartzkopf-Genswein, K., Schwean-Lardner, K., Damiran, D., & Darambazar, E. (2013). Consumption and drinking behavior of beef cattle offered a choice of several water types. *Livestock Science, 157*, 577-585. http://dx.doi.org/10.1016/j.livsci.2013.08.016

Malafaia, P., Costa, R. M., Brito, M. F., Peixoto, P. V., Barbosa, J. D., Tokarnia, C. H., & Döbereiner, J. (2014). Equívocos arraigados no meio pecuário sobre deficiências e suplementação mineral em bovinos no Brasil. *Pesquisa Veterinária Brasileira, 34*, 244-249. http://dx.doi.org/10.1590/S0100-736X2014000300008

Marques, A. P. L., Botteon, R. C. C. M., Amorim, E. B., & Botteon, P. T. L. (2013). Deficiência de cobre condicionada a altos teores de zinco, manganês e ferro na região do Medio Paraíba, RJ, Brasil. *Semina: Ciências Agrárias, 34*, 1293-1300. http://dx.doi.org/10.5433/1679-0359.2013v34n3p1293

Masters, D. G., Benes, S. E., & Norman, H. C. (2007). Biosaline agriculture for forage and livestock production. *Agriculture, Ecosystems and Environment*, 119-248. http://dx.doi.org/10.1016/j.agee.2006.08.003

Mcdowell, L. R. (1989). *Vitamins in animal nutrition: Comparative aspects to human nutrition* (p. 486). San Diego: Academic Press.

McGregor, B. A. (2004). *Water quality and provision for goats* (p. 19). Australian Government, Rural Industries Research and Development Corporation.

Mendonça Junior, A. F. (2011). Minerais importância de uso na dieta de ruminantes. *Agropecuária Científica no Semi-Árido, 7*(1), 1-13.

Mohammed, A., Campbell, M., & Youssef, F. G. (2014). Serum copper and haematological values of sheep of different physiological stages in the dry and wet seasons of Central Trinidad. *Veterinary Medicine International*, Article ID 972074. http://dx.doi.org/10.1155/2014/972074

National Research Council. (2nd ed.). (2005). *Mineral tolerance of animals*. Washington, D.C.: National Academic Press.

National Research Council. (7th ed.). (2001). *Nutrient requeriments of dairy cattle*. Washington, D.C.: National Academic Press.

National Research Council. (7th ed.). (2007). *Nutrient requeriments of small ruminants*. Washington, D.C.: National Academic Press.

Nogueira, A. R. A., Souza, G. B. (2005). Tecido vegetal. *Manual de laboratórios: Solo, água, nutrição vegetal, nutrição animal e alimentos* (p. 334). São Carlos: Embrapa Pecuária Sudeste.

Oliveira, de M., & Maia, C. E. (1998). Qualidade físico-química da água para irrigação em diferentes aquíferos na área sedimentar do estado do Rio Grande do Norte. *Revista Brasileira de Engenharia Agrícola e Ambiental, 2*, 17-21. http://dx.doi.org/10.1590/1807-1929/agriambi.v02n01p17-21

Pareja-Carrera, J., Mateo, R., & Rodríguez-Estival, J. (2014). Lead (Pb) in sheep exposed to mining pollution: Implications for animal and human helath. *Ecotoxicologyand Environmental Safety, 108*, 210-216. http://dx.doi.org/10.1016/j.ecoenv.2014.07.014

Peirce, A. W. (1957). Studies on salt tolerance of sheep for sodium chloride in the drinking water. *Australian Journal of Agricultural Research, 8*, 711-722. http://dx.doi.org/10.1071/AR9630815

Peirce, A. W. (1963). Studies on salt tolerance of sheep. V. The tolerance of sheep for mixtures of sodium chloride, sodium carbonate, and sodium bicarbonate in the drinking water. *Australian Journal of Agricultural Research, 14*, 815-823. http://dx.doi.org/10.1071/AR9630815

Peixoto, P. V., Malafaia, P., Barbosa, J. D., & Tokarnia, C. B. (2005). Princípios de suplementação mineral em ruminantes. *Pesquisa Veterinária Brasileira, 25*(3), 195-200. http://dx.doi.org/10.1590/S0100-736X 2005000300011

Pérez-Carrera, A., Moscuzza, C., Grassi, D., & Fernández-Cirelli, A. (2007). Composición mineral del agua de bebida en sistemas de producciónlecheraen Córdoba, Argentina. *Veterinaria México, 38*, 153-164.

Prasad, C. S., Mandal, A. B., Gowda, N. K. S., Sharma, K., Pattanaik, A. K., Tyagi, P. K., & Elangovan, A. V. (2015). Enhancing phosphorus utilization for better animal production and environment sustainability. *Current Science, 108*, 1315-1319.

Puls, R. (1994). *Mineral levels in animal health diagnostic data* (2nd ed.). Clearbrook, Canada: Sherpa International.

Runyan, C., & Bader, J. (1994). Water quality for livestock and poultry. *Water quality for agriculture* (p. 186). FAO Irrigation and Drainage Papers No. 29. FAO, Rome.

Shirley, R. L. (1995). Water requirements for grazing ruminants and water as a source of minerals. In L. R. Mcdowell (Ed.), *Nutrition of grazing ruminants in warm climates* (pp. 182-186). Orlando: Academic Press.

Sinclair, L. A., & Atkins, N. E. (2015). Intake of selected minerals on commercial dairy herds in central and northern England in comparison with requirements.*The Journal of Agricultural Science, 153*, 743-752. http://dx.doi.org/10.1017/S0021859614001026

Solomon, R., Miron, J., Ben-Ghedalia, D., & Zomberg, Z. (1995). Performance of high producing dairy cows offered drinking water of high and low salinity in the Arava Desert. *Journal of Dairy Science, 78*, 620-624. http://dx.doi.org/10.3168/jds.S0022-0302(95)76672-3

Squires, V. R. (1993). Australian experiences with high salinity diets for sheep. In H. Lieth, & A. A. Al Masoom (Eds.), *Towards the Rational Use of High Salinity Tolerant Plants* (Vol. 1). *Deliberations about High Salinity Tolerant Plants and Ecosystems* (pp. 449-457). Kluwer Academic Publishers, Dordrecht. http://dx.doi.org/ 10.1007/978-94-011-1858-3_46

Tebaldi, F. L. H., Silva, J. F. C., Vasquez, H. M., & Thibaut, J. T. L. (2000). Composição Mineral das Pastagens das Regiões Norte e Noroeste do Estado do Rio de Janeiro. 1. Cálcio, Fósforo, Magnésio, Potássio, Sódio e Enxofre. *Revista Brasileira de Zootecnia, 29*, 603-615. http://dx.doi.org/10.1590/S1516-359820000002 00038

Tokarnia, C. H., Dobereiner, J., Moraes, S. S., & Peixoto, P. V. (1999). Deficiências e desequilíbrios minerais em bovinos e ovinos: Revisão de estudos realizados no Brasil de 1987 a 1998. *Pesquisa Veterinária Brasileira, 19*, 47-62. http://dx.doi.org/10.1590/S0100-736X1999000200001

Valente-Campos, S., Nascimento, E de S., & Umbuzeiro, G. de A. (2014). Water quality criteria for livestock watering—A comparison among different regulations. *ActaScientiarum.Animal Sciences, 36*, 1-10. http://dx.doi.org/10.4025/actascianimsci.v36i1.21853

Valtorta, S. E., Gallardo, M. R., Sbodio, O. A., Revelli, G. R., Arakaki, C., Leva, P. E., ... Tercero, E. J. (2008). Water salinity effects on performance and rumen parameters of lactating grazing Holstein cows. *International Journal of Biometeorology, 52*, 239-247. http://dx.doi.org/10.1007/s00484-0118-03

Visscher, C. F., Witzmann, S., Beyerbach, M., & Kamphues, J. (2013). Watering cattle (young bulls) with brackish water—A hazard due to its salt content? *Tierärztliche Praxis Großtiere, 41*, 363-370. http://dx.doi.org/10.1111/j.1475-2743.1986.tb00700

Wilkens, M. R., Oberheide, I., Schröder, B., Azem, E., Steinberg, W., & Breves, G. (2012). Influence of the combination on of 25-hydroxyvitamin D_3 and a diet negative in cation-anion difference on peripartal calcium homeostasis of dairy cows. *Journal of Dairy Science, 95*, 151-164. http://dx.doi.org/10.3168/jds.2011-4342

Wilson, A. D. (1975). Influence of water salinity on sheep performance while grazing on natural grassland and saltbush pastures. *Australian Journal of Experimental Agriculture and Animal Husbandry, 15*, 760. http://dx.doi.org/ 10.1071/EA9750760

Wilson, A. D., & Dudzinski, M. L. (1973). Influence of the concentration and volume of saline water on the food intake of sheep, and on their excretion of sodium and water in urine and faeces. *Australian Journal of Agriculture Research, 24*, 245-256. http://dx.doi.org/ 10.1071//AR9730245

Zhang, H. H., Zhou, N., Zhang, T. T., Bao, K., Xu, C., Song, X. C., & Li, G. Y. (2014). Effect of different manganese levels on growth performance and N balance of growing Mink (*Neovision vision*). *Biological Trace Element Research, 160*(2), 206-211. http://dx.doi.org/10.1007/s12011-014-0008-6

12

Influence of Water Management and Silica Application on Rice Growth and Productivity in Central Java, Indonesia

Adha Fatmah Siregar[1,2], Ibrahim Adamy Sipahutar[2], Husnain[2], Heri Wibowo[2], Kuniaki Sato[1], Toshiyuki Wakatsuki[1] & Tsugiyuki Masunaga[1]

[1] Faculty of Life and Environmental Sciences, Shimane University, Matsue, Japan

[2] Indonesian Soil Research Institute, Bogor, Indonesia

Correspondence: Tsugiyuki Masunaga, Faculty of Life and Environmental Science, Shimane University, Matsue, Shimane 690-8504, Japan. E-mail: masunaga@life.shimane-u.ac.jp

Abstract

Rice cultivation in our study site at Central Java, Indonesia, is constrained by water scarcity and blast disease problems. A field experiment was thus conducted to evaluate the effect of water management and silicon (Si) application (with 500 kg ha^{-1} of silica gel) on improving rice growth and productivity and blast disease infection in Jakenan, Central Java. Split plot in randomized complete block design with 4 replications was used. The results showed that two water saving methods, intermittent (IT) and aerobic rice (AR) increased rice yield compared to conventional flooding water management. Further, IT showed better root growth and hence lodging resistance and decreased blast disease infection. IT had higher yield potential compared to AR although the yield of IT and AR were not statistically different. Si application gave significant effect on reducing leaf and neck blast infection and also increased stomata density ($p < 0.01$) in all water treatments. Si application did not result in increased yield but exhibited potential towards improving rice plant growth and production. Since Si fertilizer was never used in rice cultivation in Indonesia, the study reveals that IT combine with Si application was a suitable management for rice production in dry season in water limited Central Java region.

Keywords: blast disease, rice yield, Si application, stomata density, water management

1. Introduction

Indonesia is a country with a diverse tropical environment and plentiful annual precipitation, rice is widely grown and become the most important crop in Indonesia. The current condition of water management in our rice cultivation is still dominated by continuous flooding. This continuous fooding is suitable to apply in Indonesia because there is uncountable natural abundance water in the form of high rainfall in Indonesia. Conversely, certain areas such as Jakenan, Central Java province experience occasional water shortage. Annual rainfall in Indonesia is 2000-3000 mm $year^{-1}$ (Statistics Indonesia, 2016). However certain areas such as in Jakenan, Central Java, annual rainfal is 1100-2000 mm $year^{-1}$.

Rainfed lowland rice in Central Java covers about 83,638 ha (Ministry of Agriculture of Republic of Indonesia, 2016) where farmers practice a high degree of crop intensification. At the onset of the rainy season, a direct seeding crop (locally they call "gogo rancah") is grown with and rainfall is the source for irrigation. Immediately after the harvest of direct seeding crop, the second tranplanted crop (walik jerami) is grown under minimum tillage in submerged water condition. Earlier studies showed that the direct seeding crop season had higher yield than the second transplanted crop season, about 3.5-6.5 Mg ha^{-1} and 1.2-3.0 Mg ha^{-1} respectively (Mamaril et al., 1994; Wihardjaka et al., 1999). It showed that continous flooding as employed in the second transplanted crop season did not improve the yield. On the other hand, the direct seeding method had disadvantages such as poor seedling establishment and plant lodging occurrence which could influence on the yield (Yoshinaga, 2005).

Several cultivation methods have been adopted to improve rice production in Indonesia such as improved varieties, fertilizers, and irrigation. However, appropriate water management and silica (Si) application have not been applied in Indonesia. Related to water management, as mostly Indonesian farmers apply continuous flooding, intermittent as water management is not fully adopted. Nevertheless, previous study stated that continuous flooding can results in lodging due to the degeneration of surface roots that grow within the top 5 cm of the soil (Kar et al., 1974). Rice plants grown in aerated soil condition develop larger root systems than rice grown under continuous flooding conditions, where root die back due to lack of oxygen. Lodging is a major constraint to rice production, especially in high yielding varieties with long stem. It causes direct loss in grain yield and quality and has some indirect effects such as hindering harvesting operations (Fallah, 2000). Lodging problem could be affected by many factors i.e root growth, panicle type, plant height, starch content, silica content as well as cultivation condition (Li et al., 2009; Yang et al., 2000; Ma & Yamaji, 2006).

Silicon (Si) is the second most abundant element after oxygen in the earth's crust and most soils contain considerable quantities of the element (Savant et al., 1997; Singer & Munns, 2006). However, certain soils are low in plant-available Si which occurred in soil with highly weathered, leached, acidic and low in base saturation. Si has been shown to be a beneficial element for rice which contributes to improve resistance of rice to blast disease, lodging problem, absorption of elements such as N, P, and K. Si is absorbed by plants as monosilicic acid (H_4SiO_4) (Jones & Handreck, 1967). Once absorbed, silicic acid condenses into a hard polymerized silica gel known as plant opal on epidermal surfaces (Yoshida et al., 1962).

Related to lodging resistance, as Si deposited on epidermal surface, it is supposed to stiffen stems and leaves of rice plants to improve rice plant resistance to lodging. Previous study reported that Si treatment serves to impart more strength to the stem to resist breaking than those plants in non Si treatments by increasing the number of silicated cells and Si content in stalks even at higher levels of nitrogen (Sadanandan & Varghese, 1968). Si contributes to increase the mechanical strength as the culm wall and a vascular bundle become thicker and larger (Shimoyama, 1958).

Application of Si fertilizer is routine for rice cultivation in Japan, China, Brazil and other countries (Ma & Takahashi, 2002; Korndorfer & Lepsch, 2001). Meanwhile in Indonesia, the farmers have never used it in rice cultivation. There are some studies on soil available Si on paddy field of Indonesia. Darmawan et al. (2006) reported that over the past three decades, soil Si availability has decreased by 11-20%. Husnain et al. (2008) reported that dissolved Si concentration in irrigation water in Indonesia has also decreased by 10-20% in the same period. Husnain et al. (2011) stated that paddy soils contained available Si less than 300 mg SiO_2 kg^{-1}, a deficiency criterion proposed by Sumida (1992), in 76% out of total 92 paddy soils examined in West Sumatra, and 22.5% out of total 59 paddy soils in West Java, while in Central Java and East Java, it was less than 3% out of total 43 paddy soils in both provinces. These studies stated increasing risks on rice cultivation such disease and pest attacks, lodging and so on that read in reduction and unstabilization of rice production, and also stated necessity of Si application for rice cultivation in Indonesia. However none of the study examining the effect of Si application on rice cultivation in Indonesia.

Blast disease caused by fungus *Pyricularia grisea* (Cooke) Sacc. [= *Magnaporthe grisea* (Hebert) Barr] is one of the most devastating diseases of rice plant. This disease has become increasingly important, as reflected by the most recent data indicating that 10,604 ha and 11,929 ha of rice field throughout the country were damaged by blast disease in 2010 and 2011, respectively (Wibowo, 2011). Up to the present, fungicides have been used effectively to control blast disease but not with Si application. Our study site has faced water scarcity and blast disease problem in rice cultivation, however up to present the farmers have been applying only continuous flooding as their water management and using fungicide for blast disease control. Therefore in the present study, we conducted a field experiment to evaluate the effect of two water saving methods and Si application on improving rice plant growth and blast disease infection in Central Java.

2. Materials and Methods

2.1 Sites and Soils

Field experiment was conducted at experimental site of *Balai Penelitian Lingkungan Pertanian* (Indonesian Agricultural Environment Research Institute-IAERI), Jakenan, Central Java province, Indonesia during the dry season. This location lies on 06°46′66.7″ S-111°11′91.4″ E.

A field experiment was carried in 2014 to comparing three water management consist of continuous flooding (CF), Intermittent (IT) and Aerobic rice (AR) as main plots (Figure 1). Aerobic rice is a water saving technique for rice cultivation regions where rice is grown without ponded water because of low water availability (Bouman et al., 2007). The plots were in aerobic condition due to water scarcity before we started the water management. Then three weeks after sowing when the rain started, we started to employ three water managements. In CF management, the field was maintained with 5 cm depth of ponded water until flowering stage then at ripening stage of 105 days after sowing (DAS), about 15 days before harvest the field was dried and the outlet was opened. On IT management, the field was flooded about 5 cm water layer for 3 consecutive days then start to interrupt the water supply for 7 consecutive days with closed outlet. This pattern was conducted until panicle initiation stage. Then during flowering stage, the field was in flooding condition about 5 cm water layer and 15 days prior to harvest, the field was dried with opened outlet. In AR management, the field was in flooding condition for 28 days (tillering stage) with 5 cm water layer, after that we started the aerobic condition with closed inlet in following condition until harvest, *i.e.* when the water level drop to 15 cm below the soil surface, we irrigate the field until it reaches 15 cm. 15 days prior to harvest, the field was dried with opened outlet. Field water tube was installed in AR treatment to monitor the water level.

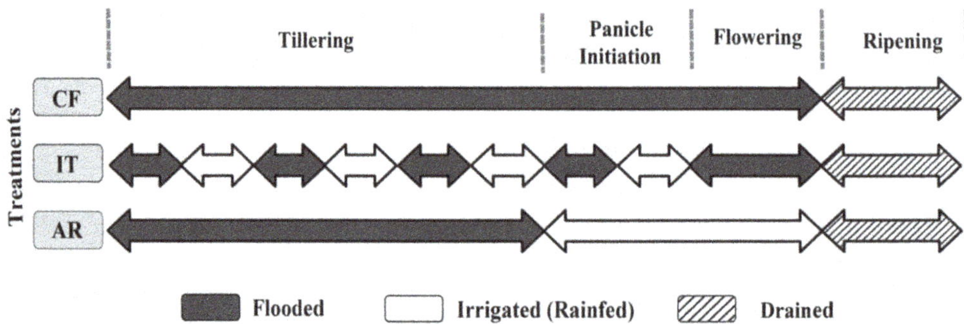

Figure 1. Diagram of water management

Note. CF: 5 cm depth of water until 105 DAS then dried for 15 days before harvest. IT: the field was flooded with 5 cm depth for 3 days then interrupted water supply for 7 days (closed outlet), with this pattern employed until panicle initiation. At flowering stage, IT was in flooding condition with 5 cm depth then the field was dried 15 days before harvest (opened outlet). AR, the field was flooded with 5 cm depth for 28 days (tillering stage) then the field was set in aerobic condition until flowering (keeping water level higher than the soil depth of 15 cm), then the field was dried 15 days before harvest.

Direct seeding was employed, therefore the plots were dried for three weeks before water managements were started.

The sub plot was characterized by two treatments including Si+ and Si- (with and without Si fertilizer). We used local silica gel "Silica gel White" sold as desiccant by IMCO Co. as Si fertilizer. This local silica gel has the spherical shape with the diameter of 2-4 mm and has lower water solubility, 0.1 gg^{-1} 24 h^{-1} compare to Japanese Si fertilizer, Super Inergia, 0.3 gg^{-1} 24 h^{-1}. A rice variety "Ciherang" was used for this study as it is a very common variety recommended by Ministry of Agriculture of Republic of Indonesia. Ciherang rice variety which released in 2000 is an indica rice categorized as short-duration variety (116-120 days) with average yield of 6 ton ha^{-1} and is suitable for planting in rainy and dry season. Split plot in randomized complete block design with 4 replications was used. The plot size was 4 m × 5 m for each treatments. During plotting, we installed plastic sheet about 30 cm into the soil between treatments at the border sides to avoid contamination. Each plots had an inlet and outlet for irrigation.

Initial soil properties (Table 1) showed that the soil in experimental site had low soil available silica below the critical level proposed by Sumida (1992) and Dobermann and Fairhurst (2000): 300 and 86 mg SiO_2 kg^{-1} respectively. The parent material of the experimental site is alluvial (Kadar & Sudijono, 1993).

Table 1. Initial soil properties

Soil Properties	Values	Criteria[a]
pH (H_2O)	4.90	Acid
Total C (g kg^{-1})	7.6	Very low
Total N (g kg^{-1})	0.3	Very low
Exchangeable cations ($cmol_c$ kg^{-1})		
Ca	2.14	Low
K	0.04	Very low
Mg	0.25	Very low
Na	0.15	Low
Available Si (mg SiO_2 kg^{-1})	31.3	Low[b]

Note. [a] Refered to Indonesian Soil Research Institute (2005).

[b] Refered to Sumida (1992).

2.2 Plant Cultivation

Rice cultivation was conducted with direct seeding method due to water scarcity with row spacing 20 cm × 20 cm. Land preparation was done by conventional tillage. Silica gel was applied before sowing the seed as 500 kg ha^{-1}. The rainfall collected in the pond is used for irrigation. The fertilizer dosage was 350 kg ha^{-1} of Urea, 100 kg ha^{-1} of SP-36 and 50 kg ha^{-1} of KCl. Urea and KCl fertilizer were applied two times, at 24 DAS (days after sowing) and 50 DAS. Meanwhile for SP-36 was applied one time at 24 DAS. During this cultivation we did not apply any fungicide for blast disease.

2.3 Sampling and Analysis

The available Si for initial soil analysis was determined using the acetate buffer method (Imaizumi & Yoshida, 1958). Soil samples were extracted in 1 mol L^{-1} acetate buffer (pH 4.0) at a ratio of 1:10 for 5 h at 40 °C with occasional shaking. Although Sumida (1991) reported that the acetate buffer method was not suitable for soils previously amended with silicate fertilizer, this was not a problem because no silicate fertilizer had been applied previously in this experimental site. The extracted Si content in the soil samples was determined using atomic absorption spectrophotometer (Z-5000; Hitachi, Tokyo, Japan). The soil pH was measured using the glass electrode method with a soil:water ratio of 1:2.5 (IITA, 1979; McLean, 1982). For determining soil exchangeable cation, soil samples were extracted with 1M NH_4OAc at pH 7 (Thomas, 1982) and measured by Inductively Coupled Plasma (ICPE-9000 Shimadzu Co, Kyoto, Japan).

Rice leaf samples, the Y-leaf, were collected at 50 DAS, 90 DAS and harvest then analyzed for total Si content. Samples were digested with HNO_3 in a high pressure Tefflon Vessel (Quaker et al., 1970; Koyama & Sutoh, 1987). After heating and digest in 160 °C for 5 hours and cooling overnight, then adding HF 10% and H_3BO_3 4%. The extracted Si content in the plant samples was determined using atomic absorption spectrophotometer (Z-5000; Hitachi, Tokyo, Japan).

Lodging resistance was measured using Force Gauge at 75 and 110 DAS. 10 plant samples were selected from each treatment for lodging resistance measurement. To measured lodging resistance, the stem was bent at 15 cm from the surface of the soil to establish an angle 45° (Yoshinaga, 2005).

Stomata samples were collected with clear nail polish method (Radoglou & Jarvis, 1990) at 50, 80 and 95 DAS. Epidermal impression was prepared by coating the rice leaf surface with nail polish which was peeled off, once nail polish was dried, it was mounted onto a slide by a cello tape. The impression approach was used to determine the number of stomata. The sample was collected only from abaxial leaf surface since the abaxial leaf surface has greater stomata density than the adaxial surface exposed to sun light (Martin & Glover, 2007). Less stomata density on adaxial surface could decrease leaf water transpiration rate (Wang & Clarke, 1993b). These impressions were observed by light microscopy (Olympus BX51) and number of stomata were investigated in a field of 0.04 mm^2 then we calculated the number of stomata per mm^2 leaf area.

Blast disease infection was observed at 50 and 95 DAS for leaf blast and 95 DAS and harvest for neck blast. 10 plant samples were observed from each treatment for blast disease infection. We observed leaf blast disease infection using score value which employed by IRRI System (1996). Score value for each symptom category of blast disease are 0: no lessions; 1: small brown speaks of pin point size or large brown speak without speculating centre; 2: small round dish to slightly elongated necrotic grey spots about 1-2 mm in diameter with distinct brown margin lessions are mostly found on upper leaves; 3: same as score 2, but significant number of lesions are on upper leaves; 4: typical susceptible blast lesion, 3 mm or longer infecting lesions than 2% of leaf area; 5: typical blast lesion infecting 2-10% of leaf area; 6: typical blast lesion infecting 11-25% of the leaf area; 7: typical blast lesion infecting 26-50% of the leaf area; 8: typical blast lesion infecting 51-75% of the leaf area and 9: more than 75% leaf are affected.

Normality test was conducted before analyze the effect of treatments and the outlier data was excluded. To determine the influence of water managements and Si application on parameters, data were statistically analyzed by two way analysis of variance (ANOVA). Significances among the treatments were determined by Tukey's honestly significant difference test at $p < 0.05$. All statistical analysis was performed using IBM SPSS Statistic version 20.0 (IBM SPSS, 2011. Chicago IL, USA).

3. Results

3.1 Plant Growth

Table 2 shows the effects of treatments on plant growth as root weight, shoot weight and number of tillers. On root weight, IT was higher by 25 and 43% for Si+ and 15 and 16% for Si- comparing to CF and AR respectively. IT management also showed higher shoot weight than CF and 5% significant with AR (Table 2).

On Si application, there was no significant difference on root weight, shoot weight and number of tillers.

Table 2. Effect of treatments on root and shoot weight, and number of tillers at harvest

Si application	Water management		
	CF	IT	AR
Root (g m^{-2})			
Si+	360 ± 26.2ab	449 ± 96.5b	314 ± 46.8a
Si-	349 ± 66.1ab	401 ± 72.6b	344 ± 37.2a
Shoot (kg m^{-2})			
Si+	2 ± 0.5ab	2.2 ± 0.4b	1.8 ± 0.2a
Si-	2 ± 0.2ab	2.1 ± 0.3b	1.6 ± 0.1a
Number of tillers (per m^2)			
Si+	349 ± 15.8b	304 ± 41.3b	274 ± 36.5a
Si-	358 ± 23.5b	356 ± 44.8b	279 ± 14.9a

Note. Means followed by the same latter do not differ significantly at 5%; No statistical difference was observed between Si treatments; ± denotes standard deviation.

3.2 Leaf and Neck Blast Infection

On the effect of Si application, the leaf blast infection at 50 and 95 DAS was the lowest in IT (Figure 2). Si application gave significant effect ($p < 0.01$) on reducing leaf blast infection throughout the observation periods in all water management (Figure 2). Si application decreased the leaf blast infection by 62, 45 and 45% at 50 DAS and 62, 29 and 48% at 95 DAS for CF, IT and AR management, respectively. Si application gave significant effect ($p < 0.01$) on reducing neck blast infection throughout the observation periods in all water management as well as it did for the leaf blast (Figure 3). Si application could decrease significantly ($p < 0.01$) the neck blast infection by 72, 86 and 80% at 95 DAS and 75, 69, and 80 % at harvest for CF, IT and AR management comparing to Si- treatments, respectively. Describe, neck blast infection was severer and effect of Si was clearer. Si application clearly showed significant effect on decreasing both leaf and neck blast in the present study.

On the effect of water management, IT had effect on reducing leaf and neck blast infection.

3.3 Rice Yield

The result showed that water management showed significant effect on yield. The rice yield in IT increased 29 and 4% comparing to CF and AR in Si+ treatments, and 60 and 5% to CF and AR in Si- treatment, respectively. We also observed yield component that probably contributed the yield difference among treatments. The 1000-grains weight of IT and AR were not significantly different but were higher than that of CF (Table 3).

Meanwhile on Si application, there was no significant difference on rice yield.

3.4 Lodging Resistance

For The result of lodging resistance at 95 and 110 DAS is shown in Figure 4. The lodging resistance tended to decrease from 95 to 110 DAS in all water management because the shoot weight increase with grain filling and leaf senescence occurs (Yoshida, 1981). IT showed slightly higher lodging resistance than AR did at 95 DAS. Meanwhile, Si application showed no significant effect in all the water management.

3.5 Stomata Density and Length

In all water management, Si application increased stomata density by 8-44% for all the observation period. Meanwhile on stomata length, Si application had no significant effect. Generally in IT, the increase rate was higher than CF and AR. IT and AR had higher stomata density than CF in both Si treatment condition at 80 DAS (Table 4).

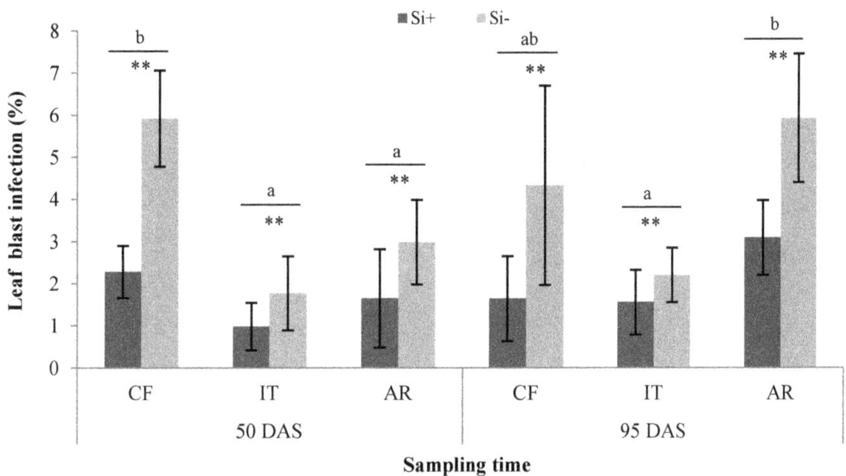

Figure 2. Effect of treatments on leaf blast infection

Note. The different alphabet letters indicates significant difference among water managements each sampling time; ** Significant difference at $p < 0.01$ between Si+ and Si- at each sampling time; Error bars indicate standard deviation among the mean values.

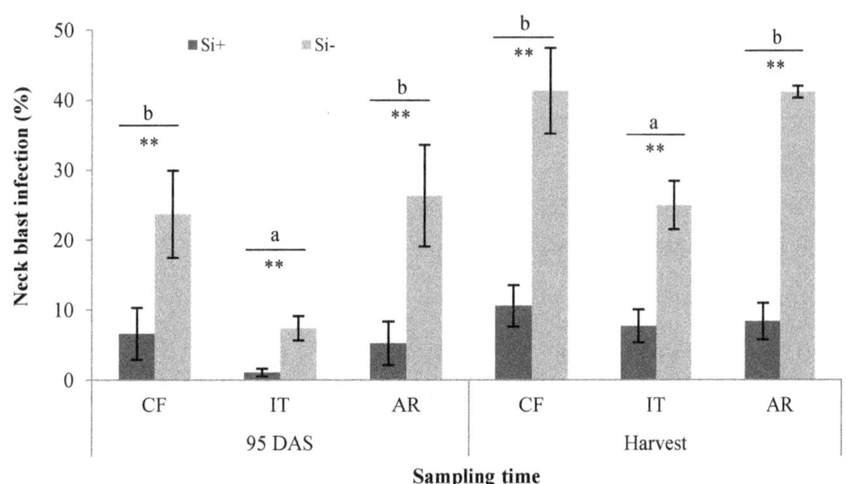

Figure 3. Effect of treatments on neck blast infection

Note. The different alphabet letters indicates significant difference among water managements each sampling time; ** Significant difference at $p < 0.01$ between Si+ and Si- at each sampling time; Error bars indicate standard deviation among the mean values.

Table 3. Effect of treatments on rice yield and yield component

Si application	Water management		
	CF	IT	AR
Yield (g m^{-2})			
Si±	423 ± 41.5a	547 ± 39.7b	524 ± 28.4b
Si-	332 ± 44.7a	530 ± 18.0b	507 ± 42.3b
The 1000-grains weight (g)			
Si±	30 ± 1.5a	32 ± 0.8b	31 ± 0.9b
Si-	29 ± 1.9a	31 ± 0.6b	30 ± 0.8b

Note. Means followed by the same latter do not differ significantly at 5%; No statistical difference was observed between Si treatments; ± denotes standard deviation.

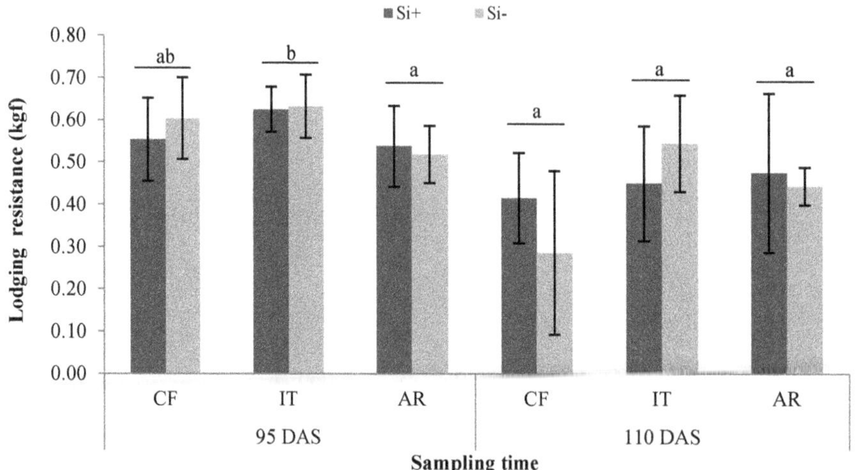

Figure 4. Effect of treatments on lodging resistance

Note. The different alphabet letters indicates significant difference among water managements each sampling time; There was no significant difference between Si+ and Si- at each sampling time; Error bars indicate standard deviation among the mean values.

Table 4. Effect of treatments on stomata density and length

Si Application	Water management								
	50 DAS			80 DAS			95 DAS		
	CF	IT	AR	CF	IT	AR	CF	IT	AR
Stomata density (per mm^2)									
Si+	591±76a	610±33a	534±62a	668±49a	795±45c	724±5b	630±21a	721±19a	730±14a
Si-	464±35a	435±53a	588±25a	519±45a	608±29c	588±25b	580±35a	536±26a	508±43a
	**	**	**	**	**	**	**	**	**
Stomata length (× 10^{-3} mm)									
Si+	11±0.2a	12±0.6a	15±3.2b	14±0.7a	14±0.4b	14±0.4ab	11±0.4a	13±0.7b	13±0.5b
Si-	12±0.7a	13±0.6a	14±0.7b	13±0.3a	14±0.2b	13±0.6ab	13±0.1a	12±0.4b	13±0.5b
	ns	ns	ns	ns	ns	ns	ns	ns	ns

Note. Means followed by the same latter do not differ significantly at 5%; No statistical difference was observed between Si treatments; ± denotes standard deviation.

4. Discussion

The result showed that IT had higher root weight that possibly due to better soil aeration which could increase root growth. Xu et al. (2007) stated that root biomass in intermittent irrigation was higher than in continuous flooding. Moreover, Mishra (2012) found that intermittent irrigation positively affected root length, density and total root mass in rice growth. These were consistent with the results of the present study. On the other hand, several works reported possible negative effects of CF and AR water condition on plant growth. Continuous flooding could degenerate and has been proved to be detrimental to the rice root growth (Kar et al., 1974; Sahrawat, 2000). Low soil water availability and high soil impedance of paddy field could inhibit root growth (Taylor & Gardner, 1963; Cornish et al., 1984). Therefore as the shoot drives water uptake through a plant, root system, properties and distribution include the weight, ultimately determine plant access to water and thus set limit on shoot weight and functioning (Nardini et al., 2002).

Furthermore, higher shoot weigh in IT might be due to better root growth which could enhance water and nutrient uptakes for its higher shoot growth in IT. On number of tillers, CF and IT had higher number of tiller than did AR. Lower number of tiller in AR could be related to the low soil moisture which could induced impaired and reduced tillering numbers (Yoshida, 1981).

Related to blast disease infection, IT had lesser leaf and neck blast infection. This might be because IT had a less favorable soil moisture condition for blast disease life-cycles (Chapagain et al., 2011). In IT, the soil moisture condition repeteadly changed from submerged to non-saturated, rather dry. When the soil became unsaturated, the soil temperature could increase in the day time and could lower the relative humidity in the fields comparing to that in CF. The lower relative humidity were less favorable for rice disease and insect pest (Bin, 2008). Moreover, Xuan and Gergon (2016) stated that to reduce blast development, intermittent irrigation during seedling stage was also effective.

In the aspect of soil moisture condition, AR must have been the driest and should have an advantage in blast infection. However, AR tended to have higher leaf and neck blast infection than IT did. This could be attributed in the difference of rice Si uptake in each water management. Aerobic soil condition observed in AR could decrease the solubility of Si (Winslow, 1995) which could reduce rice Si uptake. Meanwhile the soil in IT treatment repeatedly experienced submerged condition which increases soil Si availability (Fageria et al., 2011).

Regarding with rice Si uptake, IT had higher leaf Si content ($p < 0.05$) than CF and AR at 95 DAS and than CF at harvest (Figure 5). IT water condition could enhance plant Si uptake through both better root growth and higher soil Si availability. Then, higher plant Si content probably reduced blast disease infection as Ma and Takahashi (2002) reviewed several literatures.

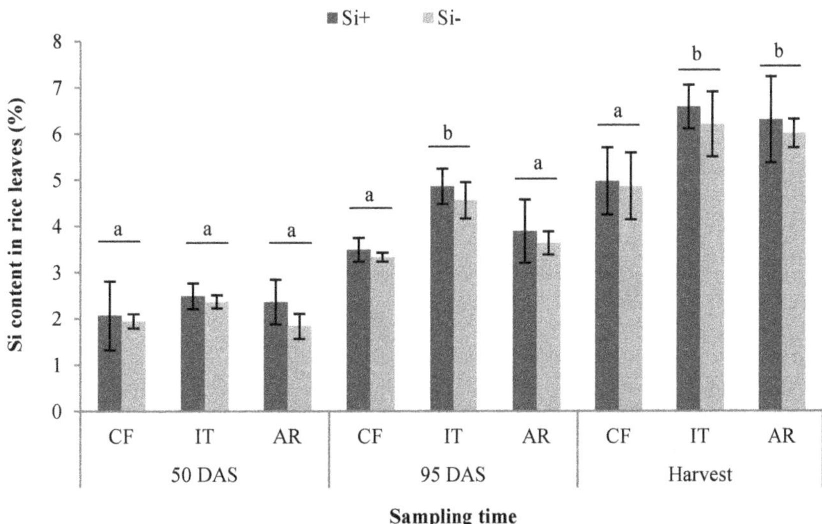

Figure 5. Effect of treatments on Si content in rice leaves

Note. The different alphabet letters indicates significant difference among water managements each sampling time; There was no significant difference between Si+ and Si- at each sampling time; Error bars indicate standard deviation among the mean values.

Si application clearly gave the positive effect on reducing leaf blast infection on Ciherang variety, which agreed with the research results found in West Java (Siregar et al., 2016) where soil Si available was 426.54 mg SiO_2 kg^{-1} and for the other rice varieties in different countries such as in Japan, Brazil and Thailand (Seebold, 1988; Prabhu et al., 2001; Hayasaka et al., 2005; Wattanapayapkul et al., 2011). The present result showed that Si application showed clearer effect on reducing blast disease on rice plant with soil Si available is 31.27 mg SiO_2 kg^{-1}, lower than critical level proposed by Dobermann and Fairhust (2000) 86 mg SiO_2 kg^{-1}.

The Si content in rice leaves was not significant different between Si treatment but tended to increase on Si+ treatment (Figure 5). The mechanism of Si-induced blast resistance has been hypothesized that silicic acid uptake by plant form hard glass-like coating of polymerized SiO_2, so called plant opal, on the epidermal surfaces and this coating will acts as a physical barrier which could block the fungi penetration (Yoshida, 1965; Winslow et al., 1997; Datnoff & Rodrigues, 2005).

In rice cultivation, the yield could be affected by water and nutrients availability (Dobermann & Fairhust, 2000) and to achieve sufficient amounts of these factors, rice plant requires a good rooting ability. Related to root condition, where IT had higher root growth (Table 2) as well as higher yield compare to CF and AR did (Table 3). Higher yield achieved in IT treatment could due to better root growth which could enhance water and nutrient uptake contributing to higher photosynthetic rate (Osaki et al., 1997).

Beside the fact that IT had better root growth, overall IT management also had the lowest leaf and neck blast infection compared to AR and CF management throughout observation period. As water condition in IT with better root growth, it could enhance higher Si uptake which followed by improving the blast resistance of rice plant in IT. The present study showed significant negative correlation between neck blast infection and the rice yield ($r = 0.64$ and $r = 0.65$ at Si+ and Si- treatment respectively; $p < 0.05$) (Figure 6). In general, it is known that blast disease is one of the most destructive for rice production. And neck blast is considered the most important symptom of rice blast because it is more closely related to yield loss (Zhu et al., 2005). Bastiaans (1993) reported that leaf blast could reduce the photosynthetic rate. This meant leaf blast also possibly reduce rice growth and yield. However, blast infection could not fully explain the difference on rice yield shown in Table 3.

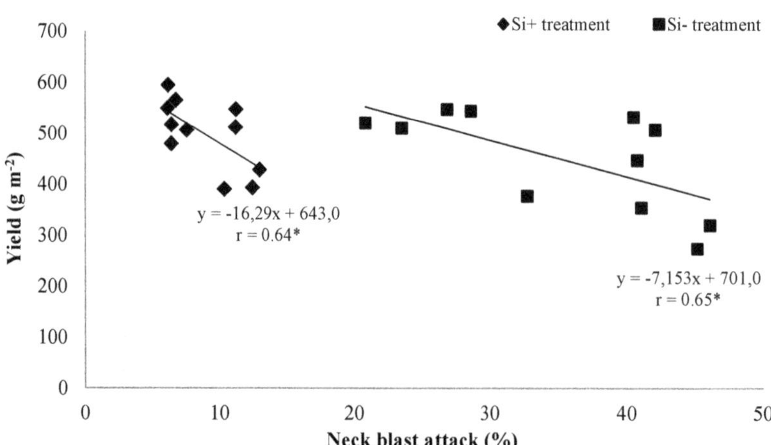

Figure 6. Correlation between neck blast attack with the yield ($p < 0.05$)

Moreover, IT showed higher stomata density compare to CF and AR (Table 4). This result could be took part on increasing the rice yield in IT. Some previous studies (Jones, 1992; Ishimaru et al., 2001) stated that the improvement on morphological characteristics of stomata such as stomata density could improve the yield.

In this present study, although IT had higher yield but it was not significant different with AR. This result might be related with the Si content in rice leaf and the transpiration rate. As shown in Figure 5, the Si content in rice leaves at harvest was not significant different between the IT and AR and higher compare to CF. Transpiration plays a certain role in translocation and accumulation of Si to the tops of rice, i.e leaves and husk, where the transpiration rate is higher at those plant organs. Along with higher Si content in leaves, it will stimulate the translocation of photoassimilated CO_2 to the panicle in rice (Ma & Takahasi, 2002) which could influence on the yield.

IT management showed possibility to improve lodging resistance, which might be due to better root growth. The higher lodging resistance in IT was attributed to higher root weight (Table 2). Previous studies stated that root system was responsible for lodging in rice plant. Higher lodging resistance would require heavier roots and deeper root system (Terashima et al., 1994; Feng-zhuan at al., 2010). Therefore with better root growth in IT management it could improve lodging resistant of rice plant. Meanwhile, Si application showed no significant relationship with the lodging resistance in the present study.

Stomata density showed that generally the increase rate of stomata in IT was higher. However at 80 DAS, showed that IT and AR had higher stomata density than CF in both Si treatment condition (Table 4). On stomata length, the result showed that IT and AR tend to have higher stomata length than CF. It probably indicated the adaptation of rice plant to water limited condition as reported by Spence et al. (1986) and Kramer (1988). The present study showed that Si application clearly gave the positive effect on increasing stomata density on Ciherang variety throughout observation, which agreed with the previous results found in West Java (Siregar et al., 2016). These are in line with the result from Dias et al. (2014), stated that there is indication of Si addition promoted the development of higher stomata density. Si application combined with water saving condition had the highest effect on stomata density increment.

Some of previous studies presumed that Si plays a role in decreasing the transpiration rate by changing the stomata movement rather than affecting its morphology and density (Gao et al., 2006; Zargar & Agnihotri, 2013). According to Marin (2003) benefits of Si application to plants includes direct effect such as structural development and indirect effect like in increasing the photosynthetic rate by improving stomata density. Moreover, apart from the present result that showed Si could improve stomata density, Si also could keep the leaf erect as it is deposited in the leaf therefore Si could stimulate canopy photosynthesis by improving light interception (Ma & Takahasi, 2002).

5. Conclusions

The present study demonstrated that two water saving management increased rice yield comparing with conventional flooding water management. This probable attributed to better grain filling status shown of the 1000-grains weight. Besides this result, IT had better root growth that possibly leaded to improve lodging resistance and shoot growth and also decreased blast disease infection. These results suggested that IT had higher yield potential comparing to AR although the rice yield of IT and AR were not statistically different in this time. This result might be due to Si uptake which IT and AR had higher Si content in leaves, and could promoted on photosynthetic rate.

On Si application, it clearly improved plant resistance to both leaf and neck blast infection and increased stomata density in all water treatments. In this time, these phenomena did not result in the higher yield but exhibited potential improving rice plant growth and production in Central Java region. In conclusion, IT combine with Si application was a suitable management for rice production in dry season in water limited Central Java region.

Acknowledgements

This work was supported by JSPS KAKENHI Grant Number 24405047, 25257405.

References

Bastiaans, L. (1993). Effect of leaf blast on photosynthesis of rice. 1. Leaf photosynthesis. *Netherlands Journal of Plant Pathology, 99*, 197-203. http://dx.doi.org/10.1007/BF01974664

Bin, D. (2008). *Study on environmental implication of water saving irrigation in Zhanghe Irrigation System.* Wuhan University. The project report submitted to Regional Office for Asia and the Pacific, FAO. Retrieved March 28, 2016, from http://www.fao.org/nr/water/espim/reference/study_environment_water_saving.pdf

Bouman, B. A. M., Lampayan, R. M., & Tuong, T. P. (2007). *Water management in irrigated rice: Coping with water scarcity*. Philiphine: International Rice Research Institute (IRRI).

Chapagain, T., Riseman, A., & Yamaji, E. (2011). Achieving more with less water: Alternate wet and dry irrigation (AWDI) as an alternative to the conventional water management practices in rice farming. *Journal of Agricultural Science, 3*(3), 3-13. http://dx.doi.org/10.5539.jas.v3n3p3

Cornish, P. S., So, H. B., & McWilliam, J. R. (1984). Effects of soil bulk-density and water regime on root-growth and uptake of phosphorus by ryegrass. *Aust J Agric Res, 35*, 631-644. http://dx.doi.org/10.1071/AR9840631

Darmawan, Kyuma, K., Saleh, A., Subagjo, H., Masunaga, T., & Wakatsuki, T. (2006). Effect of long-term intensive rice cultivation on the available silica content of sawah soils: Java Island, Indonesia. *Soil Sci Plant Nutr, 52*, 745-753. http://dx.doi.org/10.1111/j.1747-0765.2006.00089.x

Datnoff, L. E., & Rodrigues, F. A. (2005). The Role of silicon in suppressing rice diseases. *APSnet Features*. http://dx.doi.org/10.1094/APSnetFeature-2005-0205

Dias, G. M. G., Soares, J. D. R., Pasqual, M., Silva, R. A. L., Rodrigues, L. C. A., Pereire, F. J., & Castro, E. M. (2014). Photosynthesis and leaf anatomy of *Anthurium* cv. Rubi plantlets cultured *in vitro* under different silicon (Si) concentrations. *AJCS, 8*(8), 1160-1167.

Dobermann, A., & Fairhurst, T. (2000). *Rice: Nutrient disorders and nutrient management*. Singapore and Los Banõs: Potash & Phosphate Institute (PPI), Potash & Phosphate Inst. of Canada (PPIC), and Int. Rice Res. Inst. (IRRI), Los Banõs, Philippine.

Fageria, N. K., Carvalho, G. D., Santos, A. B., Ferreire, E. P. B., & Knupp, A. M. (2011). Chemistry of lowland rice soils and nutrient availability. *Communications in Soil Science and Plant Analysis, 42*, 1913-1933. http://dx.doi.org/10.1080/00103624.2011.591467

Fallah, A. (2000). *Effects of silicon and nitrogen on growth, lodging and spikelet filling in rice (Oryza sativa L.)* (Unpublished doctoral dissertation). University of the Philippines, Los Baños, Philippines.

Feng-zhuan, Z., Zheng-xun, J., Guo-hui, M., Wen-nan, S., Hai-ying, L., Mei-lan, X., & Yan, L. (2010). Dynamics between lodging resistance and chemical contents in japonica rice during grain filling. *Rice Science, 17*(4), 311-318. http://dx.doi.org/10.1016/S1672-6308(09)60032-9

Gao, X., Zou, C., Wang, L., & Zhang, F. (2006). Silicon decreases transpiration rate and conductance from stomata of maize plants. *Journal of Plant Nutrition, 29*, 1637-1647. http://dx.doi.org/10.1080/01904160600851494

Hayasaka, T., Fujii, H., & Namai, T. (2005). Silicon content in rice seedlings to protect rice blast fungus at the nursery stage. *Journal of General Plant Pathology, 71*(3), 169-173. http://dx.doi.org/10.1007/s10327-005-0182-7

Husnain, Wakatsuki, T., Setyorini, D., Hermansah, Sato, K., & Masunaga, T. (2008). Silica availability in soils and river water in two watersheds on Java Island, Indonesia. *Soil Sci Plant Nutr, 54*, 916-927. http://dx.doi.org/10.1111/j.1747-0765.2008.00313.x

Husnain, Aflizar, Darmawan, & Masunaga, T. (2011). Study on soil silicon status in Indonesia. *Proceeding of the 5th International Conference on Silicon in Agriculture*, September 13-18, 2011, Beijing.

IITA. (1979). Selections methods for soil and plant analyses. *Manual Series No. 1*. Ibadan, Nigeria.

Imaizumi, K., & Yoshida, S. (1958). Edaphological studies on silicon supplying power of paddy field. *Bull Natl Inst Agric Sci, B8*, 261-304

Indonesian Soil Research Institute. (2005). *Petunjuk Teknis Analisis Kimia Tanah, Tanama, Air dan Pupuk*. Balai Penelitian Tanah (Indonesian Soil Research Institute), Bogor, Indonesia.

International Rice Research Institute (IRRI). (1996). *Standard evaluation system for rice* (4th ed.). Los Banõs, Philippines: IRRI.

Ishimaru, K., Shirota, K., Higa, M., & Kawamitsu, Y. (2001). Identification of quantitative loci for adaxial and abaxial frequencies in *Oryza sativa. Plant Physiol Biochem, 39*, 173-177. http://dx.doi.org/10.1016/S0981-9428(00)01232-8

Jones, L. H. P., & Handreck, K. A. (1967). Silica in soils, plants and animals. *AdvAgron, 19*, 107-149. http://dx.doi.org/10.1016/S0065-2113(08)60734-8

Jones, H. G. (1992). *Plants and microclimate* (2nd ed.). Cambridge: Cambridge University press.

Kadar, D., & Sudijono. (1993). *Geological map of the Rembang quadrangle, Jawa*. Bandung, Indonesia: Geological Research and Development Centre.

Kar, S., Varade, S. B., Subramanyam, T. K., & Ghildyal, B. P. (1974). Nature and growth pattern of rice root system under submerged and unsaturated conditions. *Il Riso, 23*, 173-179.

Korndorfer, G. H., & Lepsch, I. (2001). Effect of silicon on plant growth and crop yield. In L. E. Datnoff, & G. H. Korndorfer (Eds.), *Studies in plant science* (pp. 133-147). Amsterdam: Elsevier.

Koyama, T., & Sutoh, M. (1987). Simultaneous multi element detemination of soil, plant and animal samples by inductively coupled plasma emission spectrophotometry. *Jpn. J. Soil Sci. Plant Nutr, 58*, 578-585.

Kramer, P. J. (1988). Changing concepts regarding plant water relations. *Plant Cell Environ, 11*, 565-568. http://dx.doi.org/10.1111/j.1365-3040.1988.tb01796.x

Li, H. J., Zhang, X. J., Li, W. J., Xu, Z. J., & Xu, H. (2009). Lodging resistance in japonica rice varieties with different panicle types. *Chinese Journal of Rice Science, 23*(2), 191-196.

Ma, J. F., & Takahashi, E. (2002). *Soil, fertilizer, and plant silicon research in Japan.* Amsterdam: Elsevier.

Ma, J. F., & Yamaji, N. (2006). Silicon uptake and accumulation in higher plants. *Trends Plant Sci, 11*, 392-397. http://dx.doi.org/10.1016/j.tplants.2006.06.007

Mamaril, C. P., Wihardjaka, A., Wurjandari, D., & Suprapto. (1994). Potassium fertilizer management for rainfed lowland rice in Central Java, Indonesia. *Philipp. J. CropSci, 19*, 101-109.

Marin, J. A. (2003). High survival rates during acclimatization of micropropagated fruit tree rootstocks by increasing exposures to low relative humidity. *Acta Horticulturae, 616*(1), 139-142. http://dx.doi.org/10.17660/ActaHortic.2003.616.13

Martin, C., & Glover, B. J. (2007). Functional aspects of cell patterning in aerial epidermis. *Current Opinion in Plant Biology, 10*, 70-82. http://dx.doi.org/10.1016/j.pbi.2006.11.004

McLean, E. O. (1982). Soils pH and lime requirement. In A. L. Page., E. Baker, E. J. Ellis, et al. (Eds.), *Methods of Soil Analysis* (No. 9, Part 2, pp. 199-209). Madison, Wisconsin.

Mishra, A. (2012). Intermittent irrigation enhances morphological and physiological efficiency of rice plants. *Agriculture (Pol'nohospodárstvo), 58*(4), 121-130. http://dx.doi.org/10.2478/v10207-012-0013-8

Ministry of Agriculture of Republic Indonesia. (2016). Harvested area, yield and productivity of rice. *Basis data of agriculture.* Retrieved January 5, 2016, from https://aplikasi.pertanian.go.id/bdsp/hasil_kom.asp

Nardini, A., Salleo, S., & Tyree, M. T. (2002). Ecological aspects of water permeability of roots. In Y. Waisel, A. Eshel, & U. Kafkafi (Eds.), *The Hidden Half* (3rd ed., pp. 683-698). New York: Marcel Dekker.

Osaki, M., Shinano, T., Matsumoto, M., Zheng, T., & Tadano, T. (1997). A root-shoot interaction hypothesis for high productivity of field crops. *Soil Science and Plant Nutrition, 43*, 1079-1084. http://dx.doi.org/10.1007/978-94-009-0047-9_215

Prabhu, A. S., Filho, M. P. B., Datnoff, L. E., & Snyder, G. H. (2001). Silicon from rice disease control perspective in Brazil. In L. E. Datnoff., G. H. Snyder, & G. H. Korndorfer (Eds.), *Silicon in agriculture* (pp. 293-311). Amsterdam: Elsevier, Amsterdam. http://dx.doi.org/10.1016/S0928-3420(01)80022-7

Quaker, N. R., Klucker, P. D., & Chang, G. N. (1970). Calibration of inductively coupled plasma emission spectrophotometry for analysis of the environmental materials. *Anal. Chem, 51*, 885-895.

Radoglou, K. M., & Jarvis, P. G. (1990). Effects of CO_2 enrichment on four polar clones. Leaf surface properties. *Ann Bot, 65*, 627-632.

Sadandan, A. K., & Varghes, E. J. (1968). Studies on the silicate nutrition of rice in the laterite soil of Keral. I. Effect on growth and yield. *MadrasAgri J, 11*, 261-264.

Sahrawat, K. L. (2000). Elemental composition of the rice plant as affected by iron toxicity under field conditions. *Commun Soil Sci Plant Anal, 132*, 2819-2827. http://dx.doi.org/10.1080/0010362000937063

Savant, N. K., Snyder, G. H., & Datnoff, L. E. (1997). Silicon management and sustainable rice production. *AdvAgron, 56*, 151-199. http://dx.doi.org/10.1016/S0065-2113(08)60255-2

Seebold, K. W. (1988). *The influence of silicon fertilization on the development and control of blast caused by Magnoporthe grisea (Hebert) Barr. in upland rice* (Unpublished doctoral dissertation). University of Florida, Gainesville.

Shimoyama, S. (1958). Effects of silicic acid on the lodging tolerance and the alleviation of wind damage to rice plants in the growth stage toward heading. In A. Okuda (Ed.), *Studies on the advancement of yield potentials of crop plants with the adoption of silisic acid inputs* (Vol. 48, pp. 57-59). Rept Res by Min Educ.

Singer, M. J., & Munns, D. N. (2006). *Soils: An introduction* (6th ed.). New Jersey: Pearson Prentice Hall.

Siregar, A. F., Husnain, Sato, K., Wakatsuki, T., & Masunaga, T. (2016). Empirical study on effect of Silicon application on rice blast disease and plant morphology in Indonesia. *Journal of Agricultural Science, 8*(16), 137-148. http://dx.doi.org/10.5539/jas.v8n6p137

Spence, R. D., Wu, H., Sharpe, P. J. H., & Clark, K. G. (1986). Water stress effects on guard cell anatomy and the mechanical advantage of the epidermal cells. *Plant, Cell and Environment, 9*, 197-202. http://dx.doi.org/10.1111/1365-3040.ep11611639

Sumida, H. (1991). Characteristics of silica dissolution and adsorption in paddy soils: Application to soil test for available silica. *Jap J Soil Sci Plant Nutr, 62*, 378-385.

Sumida, H. (1992). Silicon supplying capacity of paddy soils and characteristics of silicon uptake by rice plants in cool regions in Japan. *Bull Tohoku Agric Exp Stn, 85*, 1-46.

Statistic Indonesia. (2016). Retrieved January 6, 2016, from http://www.bps.go.id/linkTabelStatis/view/id/1349

Taylor, H. M., & Gardner, H. R. (1963). Penetration of cotton seedling taproots as influenced by bulk density, moisture content and strength of the soil. *Soil Sci, 96*, 153-156. http://dx.doi.org/10.1097/00010694-1963 09000-00001

Terashima, K., Ogata, T., & Akita, S. (1994). Eco-physiological characteristics related with lodging tolerance of rice in direct sowing cultivation. II. Root growth characteristics of tolerant cultivars to root lodging. *Japanese Journal of Crop Science, 63*, 34-41. http://dx.doi.org/10.1626/jcs.63.34

Thomas, G. W. (1982). Exchangeable cations. In A. L. Page (Ed.), *Methods of soil analysis, Part 2 Chemical and microbiological properties*. *Agronomy* (Vol. 9, pp. 159-165).

Wang, H., & Clarke, J. M. (1993b). Genotypic, intra plant and environmental variating in stomatal frequency and size in wheat. *Can J Plant Sci, 73*, 671-678. http://dx.doi.org/10.4141/cjps93-088

Wattanapayapkul, W., Polthanee, A., Siri, B., Bhadalung, N. N., & Promkhambut, A. (2011). Effects of silicon in suppressing blast disease and increasing grain yield of organic rice in Northeast Thailand. *Asian Journal of Plant Pathology, 5*, 134-145. http://dx.doi.org/10.3923/ajppaj.2011.134.145

Wibowo, B. S. (2011). Sebaran dan perkembangan OPT padi. *Prosiding Seminar Nasional Penyakit Tungro*. Makassar, Indonesia.

Wihardjaka, A., Kirk, G. J. D., Abdulrachman, S., & Mamaril, C. P. (1999). Potassium balances in rainfed lowland rice on light textured soil. *Field Crop Res, 64*, 237-247. http://dx.doi.org/10.1016/S0378-4290 (99)00045-3

Winslow, M. D. (1995). *Silicon: A new macronutrient deficiency in upland rice*. CIAT Working Document No. 149. Cali, Colombia.

Winslow, M. D., Okada, K., & Correa-Victoria, F. (1997). Silicon deficiency and the adaptation of tropical rice ecotypes. *Plant Soil, 188*, 239-248. http://dx.doi.org/10.1023/A:1004298817861

Xu, F. F., Zeng, X. C., Shi, Q. H., & Ye, L. M. (2007). Effects of different irrigation patterns on the growth of rice root. *Agric Res Arid Areas, 25*(1), 102-104.

Xuan, T. H., & Gergon, E. B. (2016). Strategies for the management of rice pathogenic fungi. In S. K. Desmukh, J. K. Misra., J. P. Tewari, & T. Papp (Eds.), *Fungi: Application and management strategies* (pp. 396-423). Boca Taron: CRC Press.

Yang, H. J., Yang, R. C., Li, Y. Z., Jiang, Z. W., & Li, J. S. (2000). Relationship between culms traits and lodging resistance of rice cultivars. *Fujian J AgricSci, 15*(2), 1-7.

Yoshida, S., Ohnishi, Y., & Kitagishi, K. (1962). Chemical forms, mobility, and deposition of silicon in the rice plant. *Soil Sci Plant Nutr, 8*, 107-111. http://dx.doi.org/10.1080/00380768.1962.10430992

Yoshida, S. (1965). Chemical aspects of the role of silicon in physiology of the rice plant. *Bull Natl Inst Agric Sci Series, B 15*, 1-58.

Yoshida, S. (1981). *Fundamentals of rice crop science* (p. 269). Los Banos Laguna, Philippines: International Rice Research Institute.

Yoshinaga, S. (2005). Improved lodging resistance in rice (*Oryza sativa* L.) cultivated by submerged direct seeding using a newly developed hill seeder. *JARQ, 39*(3), 147-152. http://dx.doi.org/10.6090/jarq.39.147

Zargar, S. M., & Agnihotri, A. (2013). Impact of silicon on various agro-morphological and physiological parameters in maize and revealing its role in enhancing water stress tolerance. *J Food Agric, 25*(2), 138-141. http://dx.doi.org/10.9755/ejfa.v25i2.10581

Zhu, Y. Y., Fang, H., Wang, Y. Y., Fan, J. X., Yang, S. S., Mew, T. W., & Mundt, C. C. (2005). Panicle blast and canopy moisture in rice cultivar mixtures. *Phytopathology, 95*, 433-438. http://dx.doi.org/10.1094/PHYTO-95-0433

13

Farmer Perception and Adaptation Strategies on Climate Change in Lower Eastern Kenya: A Case of Finger Millet (*Eleusine coracana* (L.) Gaertn) Production

Madegwa Yvonne[1], Onwonga Richard[1], Shibairo Solomon[2] & Karuku George[1]

[1] Department of Land Resource Management and Agricultural Technology, University of Nairobi, Nairobi, Kenya

[2] Department of Plant Science and Crop Protection, University of Nairobi, Nairobi, Kenya

Correspondence: Onwonga Richard, Department of Land Resource Management and Agricultural Technology, University of Nairobi, P.O. Box 29503-00625 Nairobi, Kenya. E-mail: richard.onwonga@uonbi.ac

The research was financed by The Regional Universities Forum for Capacity Building in Agriculture (RUFORUM).

Abstract

Eastern Kenya, a semi-arid region, is characterized by low and erratic rainfall, high temperatures, and low soil fertility. Climate change has further worsened the situation leading to frequent droughts and hence increased food insecurity. Traditional crops like finger millet are possible solutions to combating changing climate due to their drought resistance nature, ability to produce high yields with little inputs and high nutritional content. It is against this backdrop that a survey was carried out in Mwala and Katangi divisions of Machakos and Kitui counties, respectively, to assess farmer's perception on climate change, coping and adaptation mechanisms in finger millet production systems in smallholder farming systems of lower eastern Kenya. Data was collected, using semi-structured questionnaire, from 120 farmers i.e. 60 in each division. A stratified random sampling procedure, with location as a stratum was used to select respondent's households. A computer random number generator was used to select number of households in each stratum. Maize and beans were the most popular crops grown by over 98% of the farmers in both sub-counties. Farmers also grew drought tolerant legumes; cow peas, green grams pigeon peas and cereals; sorghum and finger millet. Temperature rise was ranked highest with 88% and 98%, followed by prolonged drought with 70% and 72%, irregular rainfall at 69% and 81% and increased wind intensity at 22% and 28% at Machakos and Kitui, respectively, as aspects of climate change perceived by farmers. Farmers had taken up early planting at 88.6% and 93.7%, use of organic inputs at 89% and 92%, introduced new tillage practices, by applying ridges and furrows and tied ridges at 45% and 54%, and by adopting irrigation at 13%, and 9%, as coping strategies to climate change in Machakos and Kitui, respectively.

It can be concluded that farmers in Machakos and Kitui are aware of climate change and its negative effects on crop production. In a bid to minimize crop loss and food insecurity, they have taken up various soil moisture conservation and soil fertility enhancement technologies.

Keywords: arid and semi-arid areas, climate change, coping strategies, finger millet, food security

1. Introduction

The Agricultural sector globally is one of the most vulnerable to effects of climate change (Reilly, 1995). Eastern Kenya being a semi-Arid LandSAL is most vulnerable to effects of climate change due to unstable nature of the environment. The region is prone to low, highly erratic rainfall with long dry seasons, unpredictable rainfall patterns, high evapotranspiration and low soil fertility (Bishaw et al., 2013) leading to poor crop production and food insecurity.

The government of Kenya and efforts through the research community has in recent times promoted adoption of traditional crops in the SALs as a measure to mitigate effects of climate change (Milcah et al., 2013). Such crops include drought tolerant legumes such as cowpeas and green grams and cereals; sorghum and millets. These crops are high in nutrients, drought tolerant and high yielding, disease resistant and have ability to produce yields with little inputs (Holt, 2000; Fetene et al., 2011). However, for their successful adoption understanding and buy in by farmers is needed. In addition, traditional farmers have had their own ways of coping with challenges of crop and food production.

In the past, farmers choose drought tolerant crops to overcome drought challenges, grew a wide range of crops to minimize risks, and cultivated crops that were pest and disease tolerant with high nutritional value (Macharia, 2004). The idea or phenomenon of 'climate change' is new to farmers -they may or may not understand what it is. There is lack of information on how the farmers understand and are facing the challenges of climate change and especially in the SALs of Kenya. Information on crops grown and reasons for their production, farmers' perception on climate change and farmers coping and adaptation mechanisms are therefore, needed. The current study was thus aimed at determining farmer's perceptions and coping mechanisms to climate change in relation to finger millet production.

2. Materials and Methods

2.1 Site Description

The survey was carried out in Mwala and Katangi divisions of Machakos and Kitui counties, respectively, in lower Eastern Kenya.

Machakos county is located between latitudes 0°45′S to 1°31′S from North to South and longitudes 36°45′E and 37°45′E from East to West with a mean altitude of 1714 meters above sea level (Muhammed et al., 2010). Long rains fall between March and May and short rains between October and December. Maximum and minimum temperatures experienced are 24.3 °C and 11.1 °C, respectively (Ellenkamp, 2004). Main agricultural activities include livestock keeping and small scale farming of maize, beans, millet, sorghum, cassava, peas, sweet potatoes and Irish potatoes.

Kitui County is located between latitude of 0°3.7′ and 3°0′ South and longitude 37°45′ and 39°0′ East with an altitude of 1151 meters above mean sea level (Pauw et al., 2008). The region receives average rainfall of about 900mm (Droogers & Loon, 2006). Long rains fall between April and May, while short rain between October and December (Pauw et al., 2008). Maximum and minimum temperatures experienced are 30 °C and 25 °C, respectively. Main agricultural activities included small scale livestock keeping and farming of tobacco, maize, beans, green grams, cowpea pigeon pea and mangoes.

2.2 Farmer Selection and Data Collection

The survey was carried out on a total of 120 farmers in Mwala (60) and Katangi (60). A stratified random sampling procedure, with location as a stratum was used to select respondent's households. A computer random number generator was used to select number of households in each stratum.

Data was collected on; (i) types of crops grown to mitigate against effects of climate change and reasons thereof, (ii) Farmer's perception on climate change parameters which was determined by evaluating the respondents understanding of aspects of climate change experienced and effects of climate change on finger millet production, and (iii) Coping and adaptation measures of respondents to climate change.

2.3 Data Analysis

The data was analyzed for descriptive statistics using Statistical Packages for Social Sciences (SPSS) version 16.

3. Results and Discussions

3.1 Crops Grown in Eastern Kenya and Reasons for Their Production

Farming was the main occupation of respondents with mixed farming being the main agricultural activity at both sites. Maize was the most popular crop grown in the two regions with cultivation being carried out by over 100% and 99% of respondents in Machakos and Kitui, respectively (Figure 1).

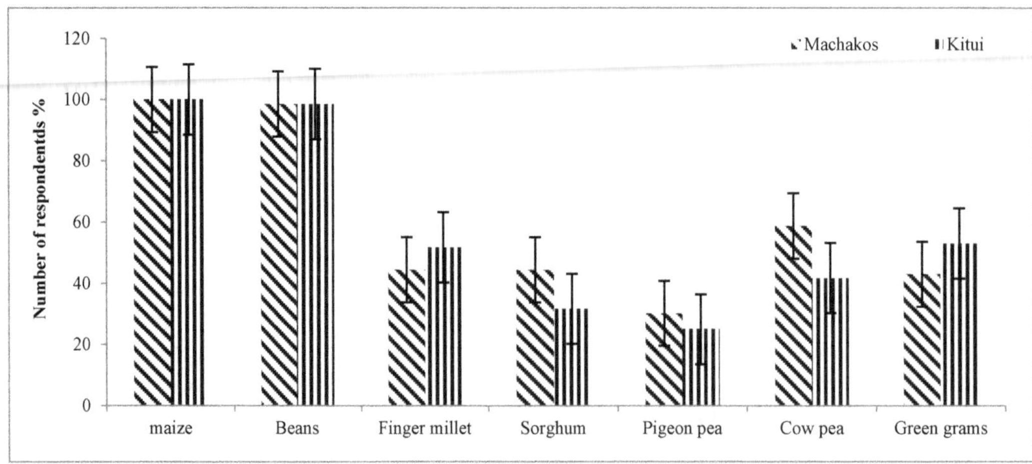

Figure 1. Crops grown in Machakos and Kitui

The farmers identified home consumption and income as the main reasons for cultivation of the cereal. This was due to its importance as a staple food crop consumed daily as ugali (a dish of maize flour cooked with water to a hard consistency) or porridge (a dish of maize and water cooked to a thick paste consistency). The findings are similar to a report by Kibaara (2005) who documented that maize was the main staple food in Kenya accounting for 40% of calories consumed daily. However, other authors (*e.g.* Ngotho, 2013) have reported that over time farmers are opting to invest less area for production of maize and more towards drought resistant crops like cowpea, green grams, sorghum and pigeon pea.

Common beans were the second most important crop in the two regions, with cultivation being carried out by 98.5% of respondents in Machakos, and 98.3% in Kitui (Figure 1). Main reasons for cultivation were reported as income and home consumption (Broughton et al., 2003). High consumption of beans was mainly due to its use in the preparation of the popular food githeri (mixture of maize and beans) playing a major role in high market availability of the crop.

The other crops grown were cowpeas cultivated by 53.7% and 41.7%, green grams cultivated by 43%, and 53% and pigeon peas cultivated by 30.2% and 25% of respondents in Machakos and Kitui, respectively. These crops were grown in small areas ranging from 0.3 ha to 0.6 ha for home consumption, as food and a little sold for income (Figure 1). This can be attributed to the legumes being highly nutritious, having the ability to be produced with little inputs, being able to obtain nitrogen through nitrogen fixation and their drought tolerant nature (Swaminathan et al., 2012).

Green grams, cowpea and pigeon pea crops were cultivated at a slightly higher acreage in Kitui than Machakos. This may be attributed to latter being a higher potential region than Kitui. Resultantly, the climate in Machakos is more suitable for a variety of crops than Kitui. Farmers in Kitui, therefore, focused their production on other drought resistant crops other than maize in order to insure against food insecurity. Ngotho (2013) suggested that with increasing effects of changing climate leading to unreliable rainfall, changing weather patterns and increased disease attacks, production of maize has been decreasing in Kitui. To insure against food insecurity farmers in the region had started to shift away from maize production to planting drought tolerant crops like sorghum, cow pea and pigeon pea among others.

The other crops grown were sorghum cultivated by 44.4% and 31.7% and finger millet cultivated by 44% and 51.7% of the respondents in Kitui and Machakos, respectively (Figure 1). Sorghum was grown at an average area of 0.3 ha, while finger millet was grown on an average land area of 0.05 ha and 0.04 ha in Machakos and Kitui counties, respectively. Both crops were grown for home consumption and income. This was due to the crops drought resistant nature, ability to produce with few inputs and popular consumption as porridge. Mgonja et al. (2007) suggested and advocated for the crops as being suitable for cultivation in dry areas due to their drought resistant nature and ability to produce with low inputs of water and fertilizer.

3.2 Farmers' Perception on Climate Change

Nearly all farmers in Machakos (100%) and Kitui (99%) were aware of climate change. This may have been as result of increased focus on the phenomenon by extension service providers. With increasing effects of climate change and the threat it poses to food security especially in arid and semi-arid regions, extension service providers had been drawing farmers' attention to climate change, its effects, and adaptation and coping mechanisms. This agrees with Baethgen et al. (2003) and Jones (2003) who found that access to extension increased farmer's perception and awareness to climate change.

3.2.1 Aspects of Climate Change Experienced

Most respondents indicated that temperature rise at 88% and 98%, followed by prolonged drought at 70% and 72%, irregular rainfall patterns at 69% and 81%, reduced rainfall and 80% and 84% and increased wind intensity at 22% and 28% in Machakos and Kitui, respectively, were the main aspects of climate change mentioned (Figure 2).

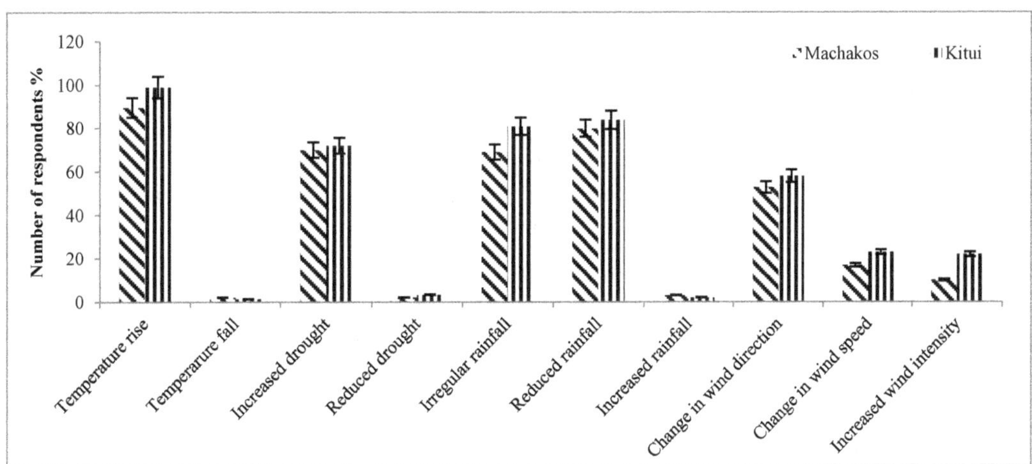

Figure 2. Climate change effects

One reason why the famers at both locations tend to suffer the same effects of climate change of temperature increase, reduced and irregular rainfall and hence increased drought is because, they depend on rain fed agriculture. Consequently, this has made the farmers more sensitive to changes in climate (rainfall and temperature) that may lead to negative effects in yields. Agatsive et al. (2010) noted that communities that depended on rain fed agriculture especially in ASAL regions noticed reduced rainfall, temperature increase and drought faster than other effects of climate change. The is in agreement with the suggestions of Macharia et al. (2012) who in his study in semi-arid regions of Nyeri north and Laikipia east districts indicated that the main indicators of climate change were changes in rainfall patterns, high temperatures and increase of drought leading to low crop yields.

3.2.2 Effects of Climate Change on Finger Millet Production

Of the farmers interviewed, 98%, in both regions, had observed that finger millet production had adversely reduced with increasing effects of climate change (Figure 3).

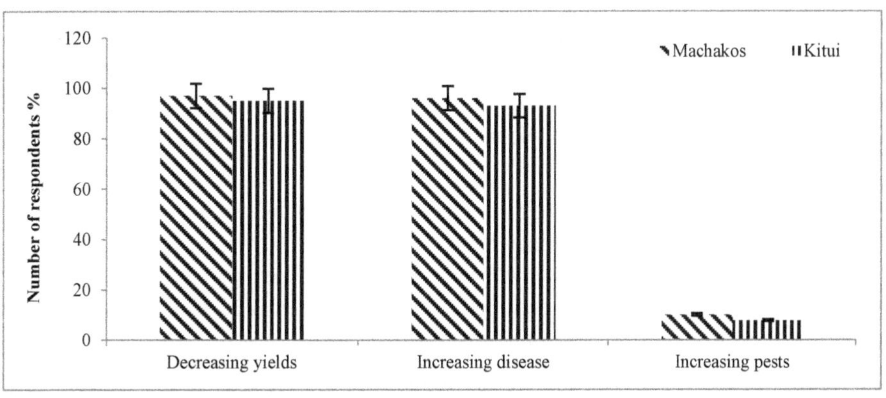

Figure 3. Effect of climate change on finger millet production

The farmers suggested that there was decreased finger millet production because of overdependence on rain fed agriculture which was now being challenged by increased occurrence of drought, reduced rainfall, increased pests and disease attacks as a result of climate change. Increased occurrence of drought and reduced rainfall had led to reduced soil moisture especially during critical plant growth phases leading to low finger millet yields. This is in agreement with findings of Tadesse (2010) who suggested that overdependence on rain fed agriculture by majority of people living in ASALs increased vulnerabilities to climate change parameters such as rising temperatures, rainfall variability, increased occurrence of drought and dependence on rain fed agriculture resulting in low crop yields.

Increase in diseases was also identified as one of the effects of climate change by 96% and 93% of repondents in Machakos and Kitui, respectively. Increased variability to climate change, especially increased temperature and reduced rainfall may have improved environmental conditions for establishment of disease pathogens. Goodman and Newton (2005) found that climate change would negatively affect plant disease resistance resulting in increased disease infestations.

Of the respondents 10.2% in Machakos and 7.7% in Kitui had noticed a slight increase in finger millet insect pests. The increase may be as a result of climate change that may have improved habitat for pest infestation. Harrington and Stork (1995) and Patterson et al. (1999) suggested that higher temperatures were expected to increase rates of development and the number of crop pests.

3.3 Coping and Adaptation Measures to Climate Change

Many respondents had taken up early planting followed by, use of organic inputs, tillage practices and irrigation and use of organic fertilizers as adaptation measures to climate change (Figure 4).

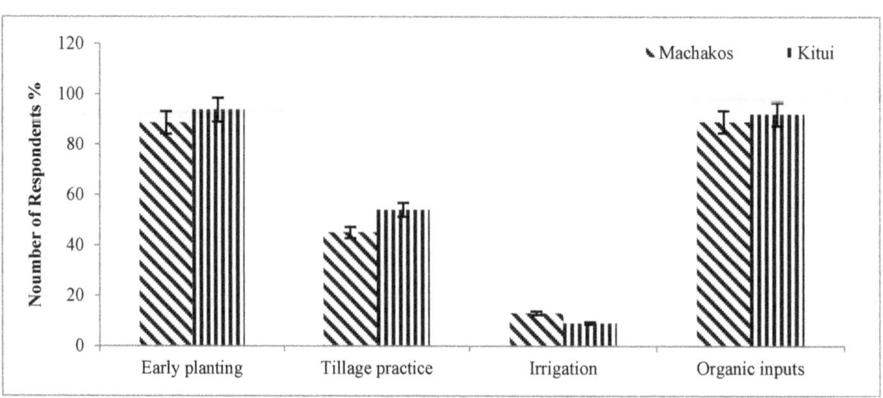

Figure 4. Coping and adaptation mechanisms to climate change

Farmers, 88.6% in Machakos and 93.7% in Kitui, considered early planting as a reliable method to mitigate against the effects of reduced and change in rainfall patterns. As rains had become more unpredictable and unreliable, crop failure was on the rise. Early planting was seen as a viable climate change adaptation mechanism as it allows for optimization of unpredictable, unreliable rainfall by ensuring crops were already established on the farm when the rains commenced. This agrees with work done by Pandey (2009) which showed that farmers commonly practiced early planting as drought coping mechanism so as to maximize on any rainfall events, for improvement of soil water management and consequently improved yields. Similarly, Larson (2010) showed that early planting had a positive effect on crop yields due to better water-use efficiency despite receiving less rainfall.

Respondents, 89% of Machakos and 92% of Kitui indicated using organic inputs as another coping mechanism for dealing with climate change. Through extension service providers farmers were aware of the importance of organic inputs, especially manure for soil water conservation. It has been reported that increase in soil organic matter improves water holding capacity and conductivity through improved soil aggregation and thus helping plants withstand vagaries of drought (Adeleye et al., 2010).

Use of water conserving tillage practices was carried out by 45% of respondents in Machakos and 54% in Kitui to cope with water stress which is a resultant of climate change (Figure 4). Establishment of different tillage practices was seen as a way to optimize on the limited rainfall events, through reduced loss of soil water by erosion, runoff and evapotranspiration. Practices in common use were tied ridges, ridges and furrows and building trenches around the farm. These tillage practices if appropriately established increased soil water conservation (Gichangi, 2012). Methods like ridges and furrows increased the area of soil water infiltration, consequently increasing the amount of plant available soil moisture. Increased soil moisture will lead to higher yields, and as such improve food security. This is in congruent with findings of Tian et al. (2003) and Wang et al. (2008) who reported that conservation tillage was one of the most efficient ways to conserving available soil water through reduced water and nutrient losses through erosion, leaching and runoff.

Irrigation was not a very common practice used to cope with climate change in the two regions. Only 13% and 9% of respondents in Machakos and Kitui, respectively, practiced irrigation (Figure 4). Through irrigation, farmers were able to provide crops with required soil moisture especially during periods of drought. This mirrors views by Kibet (2011) to the effect that irrigation could be used as a reliable supplement for natural rainfall, especially during periods of drought resulting in improved yields.

4. Conclusion

Apart from the popular maize which is used as food and income, farmers in the drought prone regions of eastern Kenya grew drought tolerant legumes including cowpeas, green grams and pigeon peas to mitigate the risks of climate change. Main aspects of climate change experienced were temperature rise, prolonged drought and irregular rainfall patterns. Early planting, tillage practices, use of organic inputs, and irrigation were identified as main adaptation measures to mitigate against the effects of climate change. Farmers have put emphasis on soil moisture conservation and soil fertility in adapting to effects of climate change. However, attention needs to be brought on traditional crops, which are naturally better suited to the harsh arid and semi arid environment, as a viable method to combat climate change. Although some farmers cultivated these crops, they did not consider them as a climate change adaptation mechanisms and focused on their subsistence production. Extension service providers need to educate farmers on traditional crops and their role as a feasible climate change adaptation mechanism.

References

Adeleye, E. O., Ayeni, L. S., & Ojeniyi, S. O. (2010). Effect of poultry manure on soil physic Chemical properties, leaf nutrient contents and yield of yam (*Dioscore arotundata*) on Alfisol in southwestern Nigeria. *Journal of American Science, 6*(10), 509-516.

Agatsive, J., Situma, C., & Ojwang, G. O. (2010). Analysis of Climate Change Vulneability: Risks in the small holde sector. *Case studies in Laikipia and Narok district representing major agro ecological zones in Kenya.* Rome: FAO.

Baethgen, W. E., Meinke, H., & Gimene, A. (2003). *Adaptation of agricultural production systems to climate variability and climate change: Lessons learned and proposed research approach.* Insights and Tools for Adaptation: Learning from Climate Variability, November 18-20, 2003, Washington, DC.

Bishaw, B., Neufeidt, H., Mowo, J., Abdelkair, A., Muruiki, J., & Dalle, G. (2013). *Famers' stategies for adopting to and mitigating climate vaiability and change through agroforestry in Ethiopia and Kenya.* Orlando: Oregon State University.

Broughton, W. J., Hernandez, G., Blair, M., Beebe, S., Gepts, P., & Beebe, S. (2003). *Beans (Phaseolus spp.) – Model food legumes*. Dordrecht: Kluwer Academic Publishers.

Droogers, P., & Loon, A. V. (2006). *Water Evaluation and Planning System, Kitui-Kenya.* Wageningen, WatManSup Research.

Ellenkamp, G. R. (2004). *Soil variability and landscape in the Machakos district, Kenya. A detailed soil survey as part of the study on the influence of soil variability on tradeoffs between agricultural productivity and soil fertility* (MSc thesis). Wageningen University.

Fetene, M., Okori, P., Gudu, S., Mneney, E., & Tesfaye, K. (2011). *Delivering New Sorghum and Finger Millet Innovations for Food Security and Improving Livelihoods in Eastern Africa.* Nairobi: ILRI.

Gichangi, E. M., Wambua, J. M., Kamau, G., Muhammad, L., & Nguluu, S. N. (2012). *Harvesting and conserving rain water using planting furrows and tied ridges in dry areas.* Katumani: KARI.

Goodman, B. A., & Newton, A. C. (2005). Effects of drought stress and its sudden relief on free radical processes in barley. *Journal of the Science of Food and Agriculture, 85*, 47-53. http://dx.doi.org/10.1002/jsfa.1938

Harrington, R., & Stork, N. E. (1995). *Insects in a Changing Environment*. San Diego: Academic Press.

Holt, J. (2000). *Investigation into the biology, epidemiology and management of finger millet blast in low-input farming systems in East Africa*. Retrieved June 8, 2016, from http://www.wisard.org/wisard/shared/asp/projectsummary.asp/Kennumer

Jones, J. W. (2003). *Agricultural responses to climate variability and climate change*. Paper presented at Climate Adaptation.net Conference, Insights and Tools for Adaptation: Learning from Climate Variability, November 18-20, 2003, Washington, DC.

Kibaara, W. B. (2005). *Technical Efficiency in Kenyan's Maize Production: An Application of the Stochastic Frontier Approach* (Unpublished MSc. Thesis). Department of Agricultural and Resource Economics, Colorado State University.

Kibet, C. (2011). *Major Challenges facing Kenyan Agricultural sector Innovations in extension and advisory services*. International Conference, November 15-18, 2011, Nairobi, Kenya.

Larson, E. (2010). *How to Plant Corn for Higher Yields* (Unpubished Thesis). Mississipi State University.

Macharia, P. (2004). *Kenya. Gateway to land and water information: Kenya National Report*. Rome: FAO.

Macharia, P. N., Thuranira, E. G., Ng'ang'a, L. W., Lugadiru, J., & Wakori, S. (2012). Perceptions and adaptation to climate change and variabilty bi immigrant communities in semi-arid regions of Kenya. *African Crop Science Journal, 20*(2), 287-296.

Mgonja, M. A., Lenné, J. M., Manyasa, E., & Sreenivasaprasad, S. (2007). Finger Millet Blast Management in East Africa. Creating opportunities for improving production and utilization of finger millet. *Proceedings of the First International Finger Millet Stakeholder*. Workshop, Projects R8030 and R8445 UK Department for International Development, Crop Protection Programme held September 13-14, 2005 at Nairobi.

Milcah, W., Mulu-Mutuku, L., Dolphine A., Odero-Wanga, L., Adijah, M., Ali-Olubandwa, L., ... Amos, N. (2013). Commercialisation of Traditional Crops: Are Cassava Production and Utilisation Promotion Efforts Bearing Fruit in Kenya. *Journal of Sustainable Development, 6*(7), 48-58.

Muhammad, L. D. M., Mulwa, R., Mwangi, W., Lamgyintuo, A., & Rovere, R. L. (2010). *Characterization of Maize producing households in Machakos and Makueni districts in Kenya*. DTMA-County Report-Kenya. Nairobi, KARI-CIMMYT.

Ngotho, A. (2013). Nairobi Star. *Smallholder farmers in Makueni turn to drought-resistant crops*. Retrieved June 6, 2013, from http://www.agra.org/news-events/smallholder-farmers-in-makueni-turn-to-droughtresistant-crops

Pandey, S. (2009). *Drought coping mechanisms and poverty insights from ain fed faming in Asia*. Rome: International Fund for Agricultural Development (IFAD).

Patterson, D. T., Westbrook, J. K., Joyce, R. J. V., Lingren, P. D., & Rogasik, J. (1999). Weeds, Insects and Disease. *Climatic Change, 43*, 711-727. http://dx.doi.org/10.1023/A:1005549400875

Pauw, W. P., Mutiso, S., Mutiso, G., Manzi, H. K., Lasage, R., & Aerts, J. C. J. H. (2008). *An Assessment of the Social and Economic Effects of the Kitui Sand Dams Community based Adaptation to Climate Change*. Nairobi: SASOL & Institute for Environmental Studies.

Reilly, J. (1995). Climate Change and Global Agriculture: Recent Findings and Issues. *American Journal of Agricultural Economics, 77*, 243-250. http://dx.doi.org/10.2307/1243242

Swaminathan, R., Singh, K., & Nepalia, V. (2012). Insect Pests of Green Gram Vigna radiata (L.) Wilczek and Their Management. *Agricultural Science*. Retrieved July 6, 2014, from http://www.intechopen.com/books/agricultural-science/insect-pests-of-green-gram-vigna-radiata-l-wilczek-and-their-management

Tadesse, D. (2010). *The impact of climate change in Africa*. Pretoria: Institute for Security Studies.

Tian, Y., Su, D. R., Li, F. M., & Li, X. L. (2003). Effect of rainwater harvesting with ridge and furrow on yield of potato in semiarid areas. *Field Crops Research, 84*, 385-391. http://dx.doi.org/10.1016/S0378-4290(03)00118-7

Wang, Q., Zhang, E. H., & Li, F. M. (2008). Runoff efficiency and the technique of micro-water harvesting with ridges and furrows, for potato production in semi-arid areas. *Water Resources Research, 22*, 1431-1443. http://dx.doi.org/10.1007/s11269-007-9235-3

14

Seed Germination and Growth of Cucumber (*Cucumis sativus*): Effect of Nano-Crystalline Sulfur

Luma S. Albanna[1], Nidá M. Salem[1] & Akl M. Awwad[2]

[1] Department of Plant Protection, Faculty of Agriculture, the University of Jordan, Amman, Jordan

[2] Nanotechnology Laboratory, Royal Scientific Society, Amman, Jordan

Correspondence: Akl M. Awwad, Nanotechnology Laboratory, Royal Scientific Society, Amman, Jordan. E-mail: akl.awwad@yahoo.com; akl.awwad@rss.jo

Abstract

The present paper is focused on green synthesis of high purity sulfur nanoparticles (SNPs) and its effect on seed germination and seedling growth of cucumber (*Cucumis sativus*). Synthesized SNPs were characterized by X-ray diffraction (XRD), Fourier transform infrared spectroscopy (FTIR), and scanning electron microscopy equipped with energy-dispersive X-ray spectroscopy (SEM-EDS). The crystalline size of synthesized SNPs as calculated by Scherer equation was 40 nm. SEM analysis of the SNPs is in spherical shape and with a diameter size between 5-80 nm. In the present study, different concentrations of SNPs were used for the treatment of cucumber seeds to study the effect on bioavailability of seed germination and seedling growth of cucumber. The results of this experiment showed that an increase in concentrations of SNPs had significantly increased seed germination and seedling growth of cucumber.

Keywords: sulfur nanoparticles, cucumber, seed germination, bioavailability, rosemary leaves extract

1. Introduction

Nanoparticles have distinctly different size diameter, surface area, chemical and biological activities compared to both individual molecules and bulk materials with the same chemical composition. Sulfur is an essential element for plants. It works as nitrogen-fixing nodules on legumes, in the formation of chlorophyll, proteins, amino acids, vitamins, and enzymes, the plant's resistance to diseases. In fact soils get sulfur from airborne particles, the weathering of minerals in soil, and decomposition of organic materials by microbial activity. Different nanomaterials were used to study their effect on seed germination and seedling growth such as titanium oxide TiO_2 (Silva et al., 2016; Nithiya et al., 2015; Gao et al., 2013; Samadi et al., 2014), zinc oxide ZnO_2 nanoparticles (Lin & Xing, 2007; Jayarambabu et al., 2014; Raskar & Laware, 2014), copper oxide CuO nanoparticles (Moon et al., 2014), iron oxide Fe_2O_3 nanoparticles (Kumar et al., 2015; Canivet et al., 2015), silver nanoparticles (Parveen & Rio, 2015; Razzaq et al., 2016; Hojjat, 2015), nano-crystalline powders of Fe, Co, and Cu (Ngo et al., 2014), Nano-SiO_2 (Siddiqui & Al-Whaibi, 2014), and aluminum oxide nanoparticles (Juhel et al., 2011), However, some of these nanomaterials have many disadvantages due to the difficulty of scale up the process of synthesis, and toxic materials. Developing facile and green methods for synthesizing sulfur nanoparticles are of importance and still a challenge for materials researchers.

The importance of sulfur nanoparticles in different applications, such as antimicrobial agents, fertilizers, and insecticides, the development of green methods is highly essential for Nano-sized sulfur particles. In previous work, we studied the effect of sulfur nanoparticles on plant's growth (Salem et al., 2016). As continuation of our previous work, the main objective of this study is to investigate the effect of green synthesized sulfur nanoparticles by rosemary (*Rosmarinus officinalis*) leaves aqueous extract on seed germination and seedling growth of cucumber.

2. Materials and Methods

2.1 Materials

Sodium thiosulfate pentahydrate ($Na_2S_2O_3{\cdot}5H_2O$, 99.5%), hydrochloric acid (37%, HCl), and ethanol (99.8%) were obtained from E Merck, Darmstadt, Germany. Fresh rosemary (*Rosmarinus officinalis*) leaves were obtained from in and out of the Royal Scientific Society, Jordan. Double distilled and deionized water was utilized for the preparation of leaves extract.

2.2 Preparation of Rosemary (Rosmarinus officinalis) Leaves Aqueous Extract

20 g of dried powder of *rosemary* leaves were mixed with 500 mL deionized water and heated at 80 °C for 10 min. Afterwards the mixture was then cooled at room temperature. The aqueous extract was obtained by filtration on filter paper Whatman No. 1 to remove solid particles. Then the extract centrifuged at 1200 rpm for 5 min to remove heavy biomaterials. The filtrate was stored at room temperature for further experimental work.

2.3 Synthesis of Sulfur Nanoparticles (SNPs)

In this experiment 1.2 g of sodium thiosulfate pentahydrate was dissolved in 100 ml of *rosemary* leaves aqueous extract under stirring on magnetic stirrer at room temperature. Afterwards 10% hydrochloric acid was added drop wise to the sodium thiosulfate solution under stirring for allowing the sulfur precipitations uniformly. The suspended sulfur particles obtained were then centrifuged at 1000 rpm for 5 min at ambient temperature. The supernatant was discarded and the precipitate was repeatedly washed with distilled water and absolute ethanol to get rid any biological materials. The product was finally dried in a vacuum at 60 °C for 4 h for characterization.

2.4 Seeds

The cucumber seeds were purchased from National Seeds, Jordan and prior to starting the experiments; cucumber seeds were stored in dry conditions in the dark to avoid any potential loss of their viability.

2.5 Seeds Germination and Exposure

The seeds were checked for their bioavailability by suspending them in deionized water. The seeds settled to the bottom were selected for further study. The seeds were immersed in a 5% dimethyl sulfoxide (DMSO, $C_2H_6SO_2$, E Merck, Germany) solution 10 min for sterilization and consistency of all experiments. Afterwards cucumber seeds were rinsed three times in deionized water and then soaked in a SNPs suspension at concentrations 100 ppm, 200 ppm, 300 ppm, 400 ppm, and 600 ppm, for 4 h in an incubator at 27 °C. Healthy and uniformly sized seeds were selected and then snow at equal distance in prepared soil pots. Seed germination experiments were carried out with 6 sets. First set considered as control (0 ppm SNPs) for comparison with the treated ones. Each treatment was carried out with three replicates and the results were presented as a mean standard deviation (±SD).

2.6 Seed Germination Application

The seed germination percentage (S_g%), was calculated from the following formula (Jayarambabu et al., 2014):

$$S_g\% = \frac{S_c}{S_s} \times 100 \quad (1)$$

Where, S_s is the number of seed germinated in sample and S_c is the number seed germinated in control.

2.7 Fresh and Dry Mass

The fresh and dry mass of root and stem was quantified through weighing in precision scale. The dry mass of root and shoot was determined after placed in an oven at 60 °C for 24 h giving constant weight.

2.8 Statistical Analysis

Each treatment was conducted with three replicates and the results were presented in mean standard deviation (±SD). All treatments were compared to those controls using t-test paired two samples for means determined at a 5% confidence level ($p < 0.05$).

3. Results and Discussion

The XRD pattern of green synthesized sulfur nanoparticles by rosemary leaves aqueous extract is illustrated in Figure 1. The 2θ peaks at 15.26°, 21.68°, 22.86°, 25.64°, 27.52°, 31.21°, 33.44°, 36.84°, 42.54°, 47.52°, and 51.04° are attributed to the crystal planes of sulfur at (113), (131), (222), (040), (313), (044), (400), (422), (319), (515), and (226), respectively. The sulfur nanoparticles are well-crystalline and the position and the relative intensity of the diffraction peaks match well with the standard monoclinic phase sulfur diffraction pattern (JCPDS N-34-094). The average particle size of the synthesized sulfur nanoparticles was about 20 nm as calculated using Debye-Scherrer formula (Klug & Alexander, 1954).

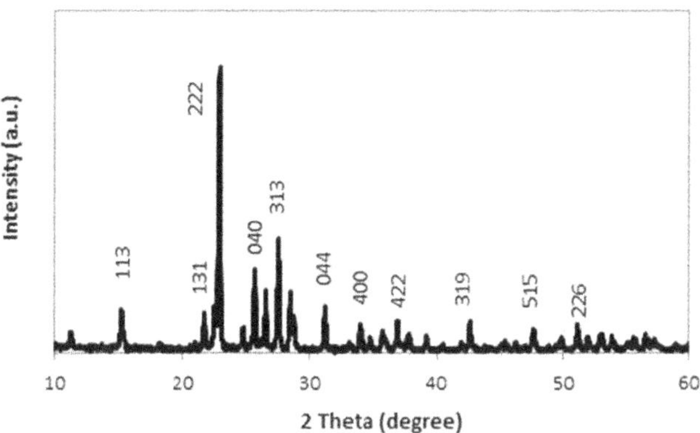

Figure 1. XRD pattern of synthesized sulfur nanoparticles using *Rosemary leaves* aqueous extract

FT-IR spectra of *rosemary* leaves aqueous extract is illustrated in Figure 2. A strong and abroad absorption bands at 3424 cm^{-1} could be ascribed to the stretching absorption band of amino (-NH), hydroxyl (-OH) stretching H-bonded alcohols and phenols. The absorption peaks at 2916 cm^{-1} and 2846 cm^{-1} could be assigned to the asymmetric and symmetric stretching of $-CH_2$ and $-CH_3$ functional groups of aliphatic. The shoulder peak at 1701 cm^{-1} corresponds to stretching carboxyl groups. The band at 1620 cm^{-1} is characteristic of amide carbonyl group in amide I and amide II. The band 1415 cm^{-1} is assigned to the methylene scissoring vibrations from the proteins. C-N stretch of aromatic amines and carboxylic acids gives rise to band at 1373 cm^{-1}. The band at 1022 cm^{-1} assigned to the C-O stretching vibrations of alcohols. The broad peak at 523 cm^{-1} can be assigned to aromatic compounds. These functional groups act as dispersing, capping and stabilizing agents for SNPs during the process of synthesis.

FT-IR spectra of the synthesized SNPs, Figure 3 indicated a new chemistry linkage on the surface of sulfur nanoparticles. This suggests that *rosemary* leaves extract can bind to sulfur nanoparticles through carbonyl of the amino acid residues in the protein of the extracts, therefore acting as stabilizer and dispersing agent prevent agglomeration of sulfur nanoparticles. The main characteristic peaks of rosemary leaves extract were observed in FT-IR spectra of sulphur nanoparticles. The FT-IR spectrum of the sulfur nanoparticles shows a strong and sharp peak at 462 cm^{-1}.

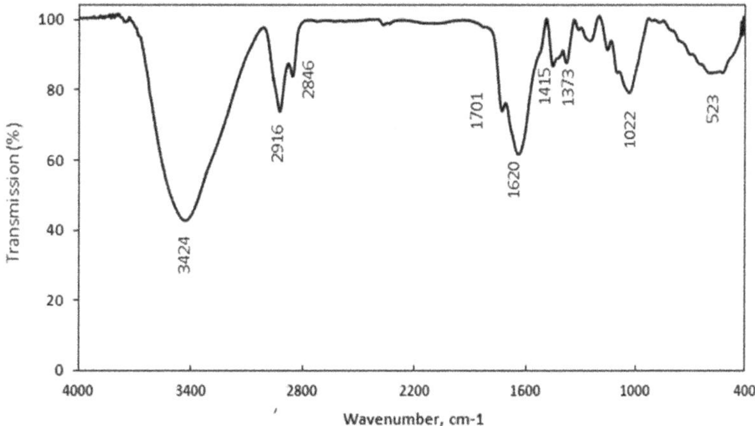

Figure 2. Fourier infrared spectrum of *Rosemary* leaves extract

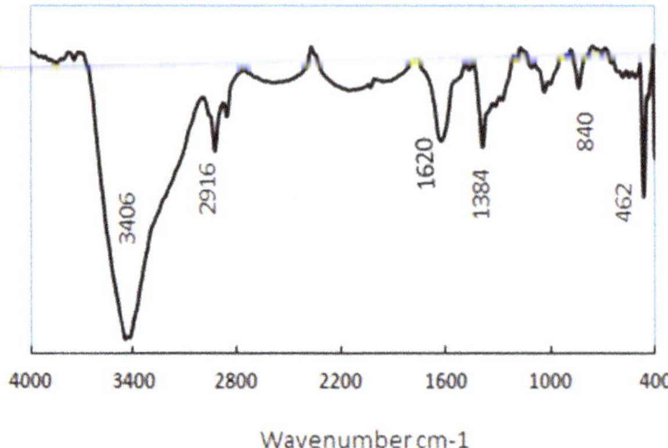

Figure 3. Fourier infrared spectrum of synthesized sulfur nanoparticles

Scanning electron microscopy (SEM) images of synthesized sulfur nanoparticles are illustrated in Figure 4. The crystals of sulfur nanoparticles are spherical in shape. The average diameter particles size is approximately in the range of 20-80 nm.

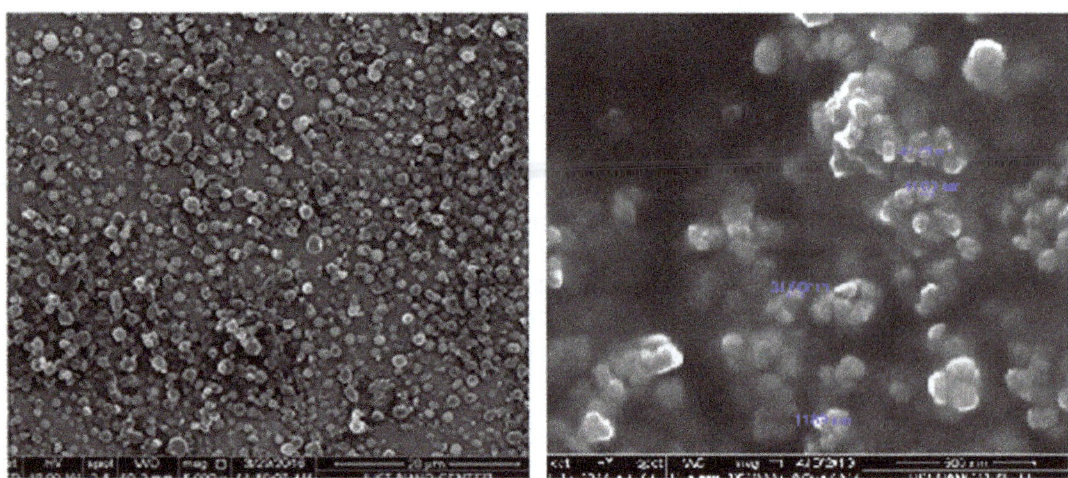

Figure 4. SEM images of synthesized SNPs using *rosemary* leaves aqueous extract

Percentage of seed germination was significantly affected by the interaction of SNPs. The results the control has shown the 75% germination. Cucumber seeds treated with SNPs have shown an increase in germination at different concentrations, viz., 100 ppm shows 90%, 200 ppm-600 ppm show a 100% in germination, Table 1. Control showed statistically significant difference and could not improve root and stem lengths.

Table 1. Growth charactestics of cucumber at different SNPs concentrations within 12 days

SNPs (ppm)	SG (%)	Root length (cm)	Stem length (cm)
Control	75	4	10
100	90	5	10.6
200	100	5.6	10.9
300	100	6.1	11.6
400	100	8.2	13.4
600	100	8.4	13.5

In the present study, the sulfur nanoparticles showed an increase in root and stem lengths with increasing the concentration of SNPs. At low concentrations 100 ppm of SNPs showed less effect on root and stem. The control showed 75% germination. The increase in root and stem growth at higher doses may attributed to the importance of sulfur in building chlorophyll, proteins, amino acids, vitamins. Also SNPS helps the plant's resistance to diseases and helping the plant's growth.

The stem fresh and dry weight was found to be influenced by different concentrations of SNPs, Table 2. Figures 5 and 6 showed the effect of different concentrations of SNPs on cucumber root growth and increasing number of seminal roots. Sulfur nanoparticles can stimulate cucumber's growth.

Table 2. Effect of sulfur nanoparticles (SNPs) on seed germination of cucumber

Treatment with SNPs (ppm)	Root fresh Wt. (g)	Root dry Wt. (g)	Stem fresh Wt. (g)	Stem dry Wt. (g)
Control	0.006	0.0034	0.33	0.0123
100	0.012	0.0054	0.43	0.0169
200	0.016	0.0057	0.49	0.0172
300	0.019	0.0060	0.53	0.0180
400	0.024	0.0068	0.59	0.021
600	0.026	0.0069	0.62	0.23

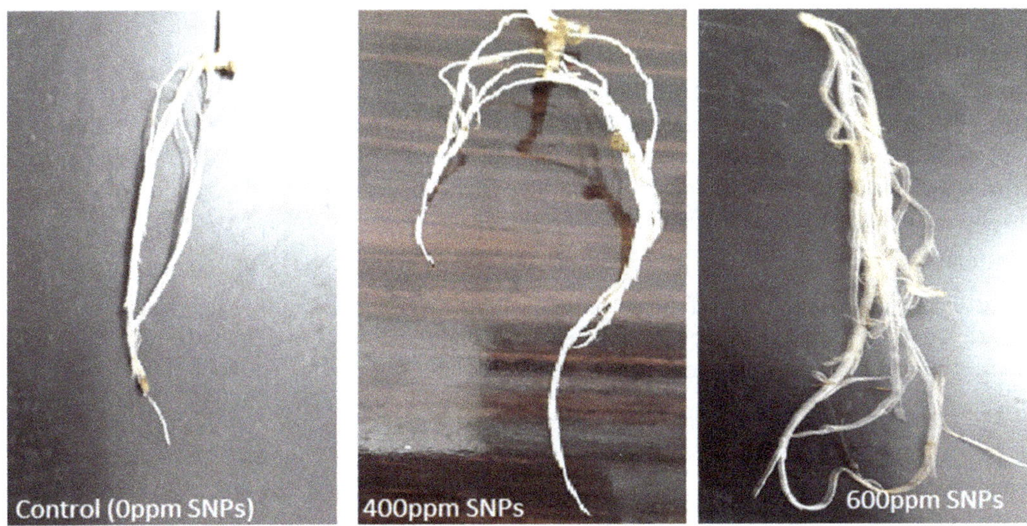

Figure 5. Photos show the positive effect of SNPs on root growth compared with control

Figure 6. A photographs for cucumber plant showed the difference in root growth, the one seeds treated with 200 ppm SNPs and the control 0 ppm

4. Conclusion

In this research paper the synthesized SNPs by rosemary leaves extract were characterized by different techniques for determining the crystalline size, particles size, and morphology. Different concentrations (100-600 ppm) of synthesized sulfur nanoparticles effect on cucumber seed germination and the root and shoot lengths. These results indicated that sulfur nanoparticles are necessary in the process of formation of proteins, amino acids, enzymes, chlorophyll, and resistance to disease, therefore aids in cucumber growth.

Acknowledgments

This research work was supported by funding program from Scientific Research Support Fund (SRF), Jordan. No. (Agr/2/13/2013). All Thanks to the Royal Scientific Society and the University of Jordan, Jordan for given all facilities to carry out this research work.

References

Canivet, L., Dubot, P., Garcon, G., & Denayer, F. O. (2015). Effects of engineered iron nanoparticles on the bryophyte physcomitrella patents (Hedw.) Bruch & Schimp, after foliar exposure. *Ecotoxicology and Environmental Safety, 113*, 499-505. http://dx.doi.org/10.1016/j.econv.2014.12.035

Gao, J., Xu, G., Qian, H., Liu, Y., Zhao, P., & Hu, Y. (2013). Effects of nano-TiO_2 on photosynthetic characterestics of *Ulmus elongate* seedlings. *Environ. Pollution, 176*, 63-70. http://dx.doi.org/10.1016/j.envpol.2013.01.027

Hojjat, S. S. (2015). Impact of silver nanoparticles on germinated fenugreek seed. *Inter. J. Crop Sci., 8*, 627-630. Retrieved from http://ijagcs.com/wp-content/uploads/2015/05/627-630.pdf

Jayarambabu, N., Kumari, B. S., Rao, K. V., & Prabhu, Y. T. (2014). Germination and growth characteristics of mungbean Seeds (*Vigna radiata* L.) affected by synthesized zinc oxide nanoparticles. *Inter. J. Current Eng. and Technol., 4*, 3411-3416. Retrieved from http://inpressco.com/category/ijcet

Juhel, G., Batisse, E., Hugues, Q., Daly, D., van Pelt, F. N. A. M., O'Halloran, J., & Jansen, M. A. K. (2011). Alumina nanoparticles enhance growth of *Lemna minor*. *Aquatic Toxicology, 105*, 328-336. http://dx.doi.org/10.1016/j.aquatox.2011.06.019

Klug, H. P., & Alexander, L. E. (1954). *X-ray diffraction procedure for polycrystalline and amorphous materials*. Wiley, New York. Retrieved from http://www.wiley.com/wileycda/wileytitle/productcd

Kumar, S., Patra, A. K., Datta, S. C., Rosin, K. G., & Purakayastha, T. J. (2015). Phytotoxicity of nanoparticles to seed germination of plants. *Inter. J. Adv. Res., 3*, 854-865. Retrieved from http://www.journalijar.com

Lin, D., & Xing, B. (2007). Phytotoxicity of nanoparticles: Inhibition of seed germination and root growth. *Environ. Pollution, 150*, 243-250. http://dx.doi.org/10.1016/j.envpol.2007.01.016

Moon, Y.-S., Parka, E.-S., Kimb, T.-O., Lee, H.-S., & Lee, S. E. (2014). SELDI-TOF MS-based discovery of a biomarker in *Cucumis sativus* seeds exposed to CuO nanoparticles. *Environ. Toxic. and Pharmac., 38*, 922-931. http://dx.doi.org/10.1016/j.etap.2014.10.002

Ngo, Q. B., Dao, T. H., Nguyen, H. C., Tran, X. T., Nguten, T. V., Khuu, T. D., & Huynh, T. H. (2014). Effects of nanocrystalline powders (Fe, Co and Cu) on the germination, growth, crop yield and product quality of soybean (Vietnamese species DT-51). *Adv. Nat. Sci.: Nanosci. Nanotechnol., 5*, 1-7. http://dx.doi.org/10.1088/2043-6262/5/1/015016

Nithiya, P., Chakra, C. S., & Ashok, C. (2015). Synthesis of TiO_2 and ZnO nanoparticles by facile polyol method for the assessment of possible agents for seed germination. *Mater. Today Proc., 2*, 4483-4488. http://dx.doi.org/10.1016/j.matpr.2015.10.056

Parveen, A., & Rao, S. (2015). Effect of nanosilver on seed germination and seedling growth in *Pennisetum glaucum*. *J. Cluster Sci., 26*, 693-701. http://dx.doi.org/10.1007/s10876-014-0728-y

Raskar, S. V., & Laware, S. L. (2014). Effect of zinc oxide nanoparticles on cytology and seed germination in onion. *Inter. J. Curr. Microbiol. App. Sci., 3*, 467-473. Retrieved from http://www.ijcmas.com

Razzaq, A., Ammara, R., Jhanzab, H. M., Mahmood, T., Hafeez, A., & Hussain, S. (2016). A novel nanomaterial to enhance growth and yield of wheat. *J. Nanosci. and Nanotechnol., 2*, 55-58. Retrieved from http://www.jacsdirectory.com/jnst

Salem, N. M., Albanna, L. S., Abdeen, A. O., Ibrahim, Q. I., & Awwad, A. M. (2016). Sulfur Nanoparticles Improves Root and Shoot Growth of Tomato. *J. Agri. Sci., 8*, 179-185. http://dx.doi.org/10.5539/jas.v8n4p179

Salem, N. M., Albanna, L. S., Abdeen, A. O., Ibrahim, Q. I., & Awwad, A. M. (2016). Green Synthesis of Nano-Sized Sulfur and Its Effect on Plant Growth, *J. Agri. Sci., 8*, 188-194. http://dx.doi.org/10.5539/jas.v8n1p188

Samadi, N., Yahyaabadi, S., & Rezayatmand, Z. (2014). Effect of TiO_2 and TiO_2 nanoparticle on germination, root and shoot length and photosynthetic pigments of *Mentha piperita*. *Inter. J. Plant & Soil Sci., 3*, 408-418. http://dx.doi.org/10.9734/IJPSS/2014/7641

Siddiqui, M. H., & Al-Whaibi, M. H. (2014). Role of nano-SiO_2 in germination of tomato (*Lycopersicum esculentum*) seeds Mill. *Saudi J. Biolog. Sci., 21*, 13-17. http://dx.doi.org/10.1016/j.sjbs.2013.04.005

Silva, S., Oliveira, H., Craveiro, S. C., Calado, A. J., & Santos, C. (2016). Pure anatase and rutile + anatase nanoparticles differently affect wheat seedlings. *Chemosphere, 151*, 68-75. http://dx.doi.org/10.1016/j.chemosphere.2016.02.047

15

Helminth Parasites of Lane Snapper, *Lutjanus synagris* from Santiaguillo Reef, Veracruz, Mexico

Jesús Montoya-Mendoza[1], María del Refugio Castañeda-Chávez[1], Fabiola Lango-Reynoso[1] & Salvador Rojas-Castañeda[1]

[1] Instituto Tecnológico de Boca del Río, División de Estudios de Posgrado e Investigación, Veracruz, México

Correspondence: Jesús Montoya-Mendoza, Laboratorio de Investigación Acuícola Aplicada, División de Estudios de Posgrado e Investigación, ITBOCA, Km 12 Carretera Veracruz-Córdoba, Boca del Río, Veracruz C.P. 94290, México. E-mail: jesusmontoya@itboca.edu.mx

The research is financed by Tecnológico Nacional de México, Dirección General de Educación Superior Tecnológica, México, Clave 5242.14-P.

Abstract

In 51 specimens of lane snapper, *Lutjanus synagris*, captured in Santiaguillo Reef, Veracruz Reef National Park System, State of Veracruz, in the Southern Gulf of Mexico, a total of 25 helminth species were recovered, as follows: 9 digeneans (8 adults, and 1 metacercaria), 7 monogeneans, 6 nematodes (4 adults, and 2 larvae), 2 cestodes (both larvae), and 1 acanthocephalan (juvenile). Out of the 25 species, 11 are new host records; 2 have prevalence > 50%, and mean intensity > 4.7; *Haliotrematoides cornigerum* (monogenean) had the highest prevalence, 94.11%, followed by *Euryhaliotrema tubocirrus* with prevalence of 66.67%. Richness (S = 25) and diversity (Shannon index H' = 2.13) at component community, and endoparasites infracommunity level (S = 6.27 ± 2.5, Brillouin index H = 1.07 ± 0.42), and ectoparasites infracommunity level (S = 3.6875 ± 1.87, Brillouin index H = 0.74 ± 0.4), were similar to those found in other marine fish. Results suggests that the host feeding habits determine the endoparasites composition, while the ectoparasites composition is associated to the environmental conditions.

Keywords: *Lutjanus synagris*, richness, diversity, parasites

1. Introduction

Demersal species known as lane snapper is distributed from North Carolina in the US, to Brazil, including the Gulf of Mexico and the Caribbean Sea (Allen, 1985). It has high biological and commercial relevance (Freitas et al., 2011), with growing demand for its value for regional fisheries in Mexico (Jiménez-Badillo et al., 2006; Arreguín-Sánchez & Arcos-Huitrón, 2011), and the Caribbean (Landínez et al., 2009). Considering only the Southeast US and the Colombian Caribbean, 48 parasite species have been recorded for *Lutjanus synagris*, a figure similar to that found in *L. griseus* for the same region (Argaéz-García et al., 2010). Helminth species registered for *L. synagris*: trematodes, *Hamacreadium mutabile*, *Helicometrina nimia*, *Lecithochririum floridense*, *L. microstomum*, *L. parvum*, *Metadena globose*, and *Stephanostomum casum* in the southeast USA (Overstreet et al., 2009); *Siphodera vinaledwardsii*, *M. globosa*, *Paracryptogonimus neoamericanus*, *S. casum*, *Lepocreadium trulla*, *H. mutabile*, *H. gullela*, *Prosogonotrema bilabiatum*, *Aponurus laguncula*, and other as Didymozoidae, *Pseudopecoelus* sp., *Xystretum* sp., *Lasiotocus* sp., *Megalomyzon* sp., for the Colombian Caribbean (Velez, 1987; Cortés et al., 2009a); the monogeneans: *Haliotrematoides longihamus*, *H. magnigastrohamus*, *H. heteracantha*, *H. cornigerum*, *Euryhaliotrema longibaculum*, *E. tubocirrus*, *E. torquecirrus* in hosts from Cuba (Kritsky & Boeger, 2002; Kritsky, Yang, & Sun, 2009), and the nematodes: *Capillaria* sp., *Contracaecum* sp., *Cucullanus* sp., *Raphidascaris* sp., in hosts from Colombia Caribbean (Cortés et al., 2009b). Parasite species found in lane snapper are unknown in México, and only *Mesostephanus appendiculatoides* has been recorded (Pérez-Ponce de León et al., 2007). This is why, we are addressing lane snapper parasites in this paper, as the composition, richness, and diversity characteristics of parasite communities are unknown, contrasting with other tropical marine fish where some baseline data has been published (Sánchez-Ramírez & Vidal-Martínez, 2002; Aguirre-Macedo et al., 2007; Montoya-Mendoza et al., 2014b).

Considering the previous parasite records for Lutjanids, it was expected that lane snappers would have a helminth community as rich and diverse as that of other reef-associated fish (Rohde & Heap, 1998; Justine et al., 2012). Herein we are describing the helminth community of *L. synagris* in terms of species richness and diversity.

2. Material and Methods

2.1 Sampling Procedures

From October, 2012 to March, 2013, 51 specimens of *L. synagris* were collected for helminthological examination. Fish were captured with fishing-baited hooks and longlines at 10-20 m depth in Santiaguillo Reef (19°08′30.00″ N y 95°48′00.41″ W), located in the Veracruz Reef National Park System, state of Veracruz, Mexico, in the Southern Gulf of Mexico. Fish specimens were kept in plastic containers with ice and transported to the lab for examination within 24 hours post-capture. Tissues and organs were reviewed using a stereomicroscope. The external examination included skin, scales, fins, gills, eyes, nostrils, mouth, and anus. Gills were removed and analyzed separately in Petri dishes with seawater. Internal examination included mesenteries, liver, kidney, and gonads, and the whole digestive system was placed in Petri dishes with 0.75% saline for examination. Helminths were fixed with hot 4% formalin and preserved in 70% ethyl alcohol.

For taxa identification, monogeneans, digeneans, cestodes, and acanthocephalans were stained using either Mayer's paracarmine, Gomori's triple stain, or Erlich's hematoxylin and then dehydrated in a graded alcohol series, cleared with clove oil, and mounted whole in Canada balsam. Nematodes were studied on temporary slides and cleared in glycerin, after which they were preserved in 70% alcohol. In order to study sclerotized structures, some specimens of monogeneans were fixed with ammonium picrate (Vidal-Martínez et al., 2001). Voucher specimens were deposited at the National Helminths Collection (*Colección Nacional de Helmintos*) (CNHE), Institute of Biology of the National Autonomous University, Mexico City. Prevalence (percentage of infected hosts) and mean intensity (mean number of parasites per infected fish) were calculated following Bush et al. (1997).

2.2 Sample Size

Helminth communities were analyzed at the component community (all helminths in all individuals of *L. synagris* examined), and infracommunity (helminths in each single fish examined) levels (Holmes & Price, 1986). Helminth species richness observed was one measure of component community structure adopted. Sampling adequacy for the component community was evaluated with a similar procedure at the helminth parasites community as *L. campechanus* (Montaya-Mendoza et al., 2014b), and it consisted in using a randomized (100x) sample-based species accumulation curve computed in EstimateS (version 8.0 RK Colwell, http://viceroy.eeb.unconn.edu/estimates) (Moreno & Halffter, 2001). For the component community, we examined the asymptotic richness based on the Clench's model equation (Soberon & Llorente, 1993), as well as the final slope of the randomized species accumulation curve (Jiménez-Valverde & Hortal, 2003). Clench's model is described by the following function:

$$V2 = (a \times V1)/[1 + (b \times V1)] \quad (1)$$

Where, $V2$ is the observed richness, $V1$ is the number of hosts examined, and a and b are curve parameters, a equals the new species adding rate, and b is a parameter related to the curve shape. These values were calculated iteratively using the EstimateS and Statistica (StatSoft, Inc., Tulsa, Oklahoma) software as in Jiménez-Valverde and Hortal (2003). The slope of the cumulative species curve was calculated as $a/(1 + b \times n)^2$, where a and b are parameters above and n is the number of hosts examined from a given component. Clench's model equation allows estimating the total number of species in a component as a/b. To calculate the number of rare species missing at the component community level, the nonparametric species-richness estimator bootstrap was calculated from data observed, as recommended by Poulin (1998). The Shannon index of diversity (H'), was calculated for the component community as in Magurran (2004). Descriptors of infracommunities included the mean number of helminth species per fish, the mean number of helminth individuals per fish, and the mean value of the Brillouin's diversity index per fish (H).

3. Results

All 51 *L. synagris* were sexually mature adults (23 males and 28 females), total fish length 22.2-49.4 cm (33.8 ± 4.3); fish weight, 157-1074 g (552.6 ± 177.1), and all were parasitized. Twenty-five helminth species were recorded. The analysis of cumulative species curves for component community suggested that the inventory of the helminth species was near completed and the slope of cumulative species curve was 0.03. Thus, an asymptote was reached, and richness estimated by Clench's model was 26.18 species (a = 6.440828, b =

0.245964; a/b = 26.18); the value of the nonparametric species-richness estimator bootstrap (S_b = 26.79) confirms, indeed, that most, if not all, helminth species from the component community were recovered. Out of the 25 species of helminths, 9 were digeneans (8 adults, and 1 metacercaria), 7 were monogeneans, 6 were nematodes (4 adults, and 2 larvae), 2 were cestodes (both larvae), and one was acanthocephala (1 juvenile) (Table 1).

Two of the 25 helminth species appeared as frequent with prevalence values > 50% and mean intensity > 4.7 helminths per infected fish (Table I). Nine out of the 25 were rare with a prevalence < 6% and mean intensity < 1.9, and fourteen species hold an intermediate position. Monogeneans *H. cornigerum* (prevalence 94.11%), *E. tubocirrus* (66.67%), *H. heteracantha* (43.14%), and *E. longibaculum* (43.14%) reached the highest prevalences recorded in this study. Subtle differences were observed in helminth infections when comparing male and female fish; only in the number of parasites in males (n = 1121) it was higher than in females (n = 819). No significant correlation was found between the total number of species (S) or the total number of helminths (N), when compared to the host size (total host length *vs.* S, r = 0.039; *vs.* N, r = 0.099) and weight (weight *vs.* S, r = 0.037; *vs.* N, r = 0.055). However, a highly significant correlation (r = 0.77) was found when comparing prevalence and mean intensity of the helminth species.

Table 1. Prevalence, mean intensity, and site of infection of helminth parasites in lane snapper, *Lutjanus synagris*, from Santiaguillo Reef, Veracruz, Mexico

Species	CNHE	Site	n (% prevalence)	mi (±SD)	Range
Trematoda					
Lecithochirium floridense (Manter, 1934)	10209	sto	2 (3.92)	28±38.18	1-55
Preptetos trulla (Linton, 1907)	10210	int	12 (23.53)	3.41±2.87	1-10
Siphodera vinaledwardsii (Linton, 1901)	10211	int	20 (39.22)	5.75±8.13	1-35
Metadena adglobosa Manter, 1947*	10212	int	7 (13.73)	6±8.96	1-26
Metadena crassulata Linton, 1910*	10213	int	15 (29.41)	2.93±2.84	1-11
Lepocreadium sp.*	10214	int	2 (3.92)	1±0	1-1
Opechona sp.*	10215	int	2 (3.92)	1±0	1-1
Stephanostomum sp.	10216	int	1 (1.96)	5	-
Trematoda$^{(mt)}$	-	fins	9 (17.65)	2.56±2.12	1-7
Monogenea					
Haliotrematoides cornigerum (Zhukov, 1976)**	10217	gills	48 (94.11)	18.48±22.6	1-135
Haliotrematoides heteracantha (Zhukov, 1976)	10218	gills	22 (43.14)	2.68±1.89	1-8
Haliotrematoides longihamus (Zhukov, 1976)**	10219	gills	14 (27.45)	2.21±1.71	1-7
Haliotrematoides magnigastrohamus (Zhukov, 1976)**	10220	gills	16 (31.37)	3.43±3.71	1-15
Euryhaliotrema longibaculum (Zhukov, 1976)**	10221	gills	22 (43.14)	3.27±2.21	1-8
Euryhaliotrema tubocirrus (Zhukov, 1976)	10222	gills	34 (66.67)	4.74±4.11	1-19
Euryhaliotrema torquecirrus (Zhukov, 1976)	10223	gills	21 (41.18)	2.67±1.91	1-8
Cestoda					
Callitetrarhynchus sp.$^{(p)}$*	10224	int	11 (21.57)	3.27±3.07	1-10
Tetraphyllidea gen. sp.$^{(p)}$	10225	Int	24 (47.06)	7.21±13.15	1-54
Nematoda					
Cucullanus pargi González-Solís, Tuz-Paredes y Quintal-Loria, 2007*	10227	int	10 (19.61)	2.4±1.65	1-6
Hysterothylacium reliquens Norris y Overstreet, 1975*	10228	Int	19 (37.25)	2.42±1.54	1-6
Anisakis sp. $^{(l)}$*	10229	mes	1 (1.96)	1	-
Contracaecum sp.$^{(l)}$	10230	int	3 (5.88)	1±0	1-1
Procamallanus (*Spirocamallanus*) sp.*	10231	int	2 (3.92)	1.5±0.71	1-2
Phyllometridae gen. sp.*	10232	cau fn	1 (1.96)	1	-
Acanthocephala					
Gorgorhynchoides sp.$^{(j)}$*	10226	mes	2 (3.92)	1±0	1-1

Note. Life stages: *mt*, metacercaria; *p*, plerocercoid; *j*, juvenile; *l*, larva. *, New host record. **, New record for Mexico. Site: sto, stomach; int, intestine; mes, mesenteries, cau fn; caudal fin. n, number of hosts infected. mi, mean intensity.

A total of 1940 individual helminths were collected; infections ranged from 5 to 155 helminth individuals per infected host. Richness of the component community was $S = 25$ and the Shannon index diversity value was $H' = 2.13$. Richness in infracommunities ranged from 2 to 12 species of helminths per fish, three hosts were infected by 2, 4 had 3, 8 had 4, 6 had 5, 7 had 6, 4 had 7, 9 had 8, 6 had 9, 1 had 10, 2 had 11 and 1 host had the maximum of 12 parasites species. As it can be observed, about one half of examined hosts (n = 30) were infected with 6 to 12 helminth species. The average number of parasites species per individual host was 6.27 ± 2.5, while the average number of helminth individuals per host was 38.03 ± 29.47. The value of Brillouin's index for each infracommunity ranged from 0.14 to 1.79 with an average value of 1.07 ± 0.42. Fourteen helminth species inhabit the intestine (Table 1); intestines of only 3 hosts were free of parasites, i. e., 48/51 (94%) hosts had intestinal helminths. Twelve hosts (25%) had a single helminth species and 36 (75%) had concurrent infections of intestinal helminths. Up to a maximum of 7 intestinal helminth species co-occurred in the infracommunities. A total of 592 intestinal individual helminths were collected; the number of helminth individuals per host ranged from 1 to 68 with an average of 12.3 ± 15.54. The number of intestinal helminth species ranged from 1 to 7 per examined host and the average was 2.7 ± 1.5. Brillouin's index for intestinal species was from 0 to 1.13 with an average of 0.49 ± 0.36. Seven helminth species inhabit the gills (Table 1); the branchial of only 3 hosts were free of parasites, i.e., 48/51 (94%) hosts had branchial helminths. Five hosts (10.4%) had a single helminth species and 43 (89.4%) had concurrent infections of branchial helminths. Up to a maximum of 7 branchial helminth species co-occurred in the ectoparasite infracommunities. A total of 1321 branchial individual helminths were collected; the number of helminths per host ranged from 1 to 146 with average of 27.5 ± 27.4. The number of ectoparasite helminth species ranged from 1 to 7 per examined host and the average was 3.6 ± 1.8. Brillouin's index for branchial species was from 0 to 1.5 with average of 0.66 ± 0.44.

4. Discussion

Adding the previous known parasites records for *L. synagris* and those reported in the present investigation, the updated helminths inventory for this fish reached 59 helminth species. This research is adding 11 new host records and 4 new location records. The number of parasite species on *L. synagris* is higher if compared with the number of species found in other Lutjanids; for example, 44 species have been reported in *Lutjanus griseus* from the Gulf of Mexico and Caribbean region (Argaéz-García et al., 2010). This larger inventory of parasites covers a systematic sampling involving several locations and a much larger sample size. Nevertheless, an accumulation curve suggested that our parasites inventory is near completed concerning the site studied with 42.3% (25/59) in this helminths survey. Trematodes and nematodes species are the main components of intestinal helminth parasites of marine fish most studied in northern temperate zones (Zander et al., 1999; Fernández et al., 2005). The same taxonomic groups of helminths have also been recorded as the most frequent in marine fish in the southern Gulf of Mexico (Moravec et al., 1997; Sánchez-Ramírez & Vidal-Martínez, 2002; Rodríguez-González & Vidal-Martínez, 2008; Argáez-García et al., 2010; Espínola-Novelo et al., 2013; Montoya-Mendoza et al., 2014a, 2014b). The number of trematodes and nematodes species in *L. synagris* was larger than that of cestodes and acanthocephalans. Proportions among parasite groups indicate the relevance of intermediate hosts (Palm & Overstreet, 2000; 2002; Sánchez-Ramírez & Vidal-Martínez, 2002). Lane snapper are carnivorous, and feed on crustaceans such as stomatopods, penneids, and portunids, and other as gastropods (Doncel & Paramo, 2010; Rosa et al., 2015); consumption of these organisms facilitates parasitic infections using them as intermediate hosts (Deardorff & Overstreet, 1981; Aguirre-Macedo et al., 2007; Lagrue et al., 2011). Presence of living cestode, nematode, and acanthocephalan larvae suggest that lane snappers are at intermediate level in the marine food web, just as are other fish in the Western Atlantic (Luque & Poulin, 2004).

Host specificity seems to be an important ectoparasites trait in structuring the communities of helminths of *L. synagris*. Four out of the seven helminth parasite species reached maximum prevalence in this survey. The highest prevalence and mean intensity was recorded by *H. cornigerum*, *E. tubocirrus*, *H. heteracantha* and *E. longibaculum*, with previous records in different snapper such as *L. analis*, *L. apodus*, *L. cyanopterus*, *L. mohogoni*, *L. synagris*, *Ocyurus chrysurus*, and *Rhomboplites aurorubens* (Zhukov, 1976; Kritsky & Boeger, 2002; Kritsky, 2012). This number of monogeneans in *L. synagris*, is higher if compared to others Lutjanids (Zhukov, 1976), and similar to *Caranx hippos* from Veracruz (Montoya-Mendoza et al., 2008), Venezuela (Boada et al., 2012), and Brazil (Luque & Alves, 2001). Richness and diversity of ectoparasites infracommunity in *L. synagris*, can be explained particularly from host-parasite relationship and the environmental conditions prevailing in coral reef systems (Justine et al., 2012), and considering that reefs in southeastern Gulf of Mexico are part of the Barrera Reef, running from the Caribbean Sea to the coast of Veracruz (Jordán-Dahlgren & Rodríguez-Martínez, 2003), providing larger distribution areas for Lutjanids and their ectoparasites (Zhukov, 1976; Montoya-Mendoza et al., 2014a).

These data shows that the helminth community of *L. synagris* is as rich and diverse as those of other marine fish in temperate (Châari et al., 2015) and tropical zones (Luque & Poulin, 2007; Madhavi & Triveni Lakshmi, 2012). This community is also similar to others described from marine fish in the southern Gulf of Mexico and the Caribbean (Sánchez-Ramírez & Vidal-Martínez, 2002; Aguirre-Macedo et al., 2007; Espínola-Novelo et al., 2013; Montoya-Mendoza et al., 2014a, 2014b). Infracommunity richness and diversity were not different from those recorded for marine fish at temperate latitudes (Madhavi & Sai Ram, 2000; Châari et al., 2015), the tropical Atlantic (Sánchez-Ramírez & Vidal-Martínez, 2002; Espínola-Novelo et al., 2013; Montoya-Mendoza et al., 2014b), and the Australian tropic (Rohde & Heap, 1998; Justine et al., 2012). The composition, richness, and diversity of the helminth communities of *L. synagris* are mainly associated to their feeding habits as reef inhabitants. Now, the intermediate host distribution is a factor to be considered, as fish feed on the most abundant prey, together with environmental stability in tropical latitudes. Therefore, the fish feeding habits, and the intermediate host abundance, and the habitat continuum are important factors determining the parasite community structure. A further potential factor to explain the composition of these communities is host switching among sympatric, related species in this area, as 60% or more of intestinal parasite species found in this study have been registered in other Lutjanids (Kritsky & Boeger, 2002; Overstreet et al., 2009; Argáez-García et al., 2010; Kritsky, 2012; Montoya-Mendoza et al., 2014a, 2014b).

References

Aguirre-Macedo, M. L., Vidal-Martínez, V. M., González-Solís, D., & Caballero, P. L. (2007). Helminth communities of four commercially important fish species from Chetumal Bay, Mexico. *Journal of Helminthology, 81*, 19-31. http://dx.doi.org/10.1017/S0022149X0721209X

Allen, G. R. (1985). Snappers of the world: An annotated and illustrated catalogue of lutjanid species known to date. *FAO Fish Synopsis, 6*, 1-208. Retrieved June, 2016, from http://www.fao.org/docrep/009/ac481e/ac481e00.htm

Argáez-García, N., Guillén-Hernández, S., & Aguirre-Macedo, M. L. (2010). Intestinal helminths of *Lutjanus griseus* (Perciformes: Lutjanidae) from three environments in Yucatán (Mexico), with a checklist of its parasites in the Gulf of Mexico and Caribbean region. *Revista Mexicana de Biodiversidad, 81*, 903-912. http://dx.doi.org/10.7550/rmb.21538

Arreguín-Sánchez, F., & E. Arcos-Huitrón, E. (2011). La pesca en México: Estado de la explotación y uso de los ecosistemas. *Hidrobiológica, 21*, 431-462.

Boada, M., Bashirullah, A., Marcano, J., Alió, J., & Vizcaíno, G. (2012). Estructura comunitaria de ectoparásitos en branquias del jurel *Caranx hippos* (Linnaeus, 1776) en Santa Cruz y Carúpano, Estado Sucre, Venezuela. *Revista Científica, 22*, 259-272.

Bush, A. O., Lafferty, K. D., Lotz, J. M., & Shostak, A. W. (1997). Parasitology meets ecology on its own terms: Margolis et al. revisited. *Journal of Parasitology, 83*, 575-583. http://dx.doi.org/10.2307/3284227

Châari, M., Feki, M., & Neifar, L. (2015). Metazoan parasites of the mediterranean garfish *Belone belone gracilis* (Teleostei: Belonidae) as a tool for stock discrimination. *Open Journal of Marine Science, 5*, 324-334. http://dx.doi.org/10.4236/ojms.2015.53027

Cortés, J., Valbuena, J., & Manrique, G. (2009a). Determinación taxonómica de trematodos digéneos en las especies de pargos *Lutjanus synagris* (Linneaus, 1758) y *Lutjanus analis* (Cuvier, 1828), en las Bahías de Santa Marta y Neguanje, Parque Nacional Natural Tayrona, Caribe Colombiano. *Revista Médica Veterinaria y Zootecnia, 56*, 7-22.

Cortés, J., Valbuena, J., & Manrique, G. (2009b). Nemátodos parásitos de *Lutjanus synagris* (Linneaus, 1758) y *Lutjanus analis* (Cuvier, 1828) (Perciformes, Ludjanidae) en las zonas de Santa Marta y Neguanje, Caribe Colombiano. *Revista Médica Veterinaria y Zootecnia, 56*, 23-31.

Deardoff, T. L., & Overstreet, R. M. (1981). Larval *Hysterothylacium* (= *Thynnascaris*) (Nematoda: Anisakidae) from fishes and invertebrates in the Gulf of Mexico. *Proceedings of the Helminthological Society of Washington, 43*, 113-120.

Doncel, O., & Páramo, J. (2010). Hábitos alimenticios del pargo rayado, *Lutjanus synagris* (Perciformes: Lutjanidae), en la zona norte del Caribe colombiano. *Latin American Journal of Aquatic Research, 38*, 413-426. http://dx.doi.org/10.4067/S0718-560X2010000300006

Espínola-Novelo, J. F., González-Salas, C., Guillén-Hernández, S., & Mackenzie, K. (2013). Metazoan parasites of *Mycteroperca bonaci* (Epinephelidae) off the coast of Yucatan, Mexico, with a checklist of its parasites in the Gulf of Mexico and Caribbean region. *Revista Mexicana de Biodiversidad, 84*, 1111-1120. http://dx.doi.org/10.7550/rmb.27989

Fernández, M., Aznar, F. J., Montero, F. E., & Raga, J. A. (2005). Endoparasites of the blue whiting, *Micromesistius poutassou* from north-west Spain. *Journal of Helminthology, 79*, 15-21. http://dx.doi.org/10.1079/JOH2004269

Freitas, M. O., Moura, R. L., Francini-Filho, R. B., & Minte-Vera, C. V. (2011). Spawning patterns of commercially important reef fish (Lutjanidae and Serranidae) in the tropical western South Atlantic. *Scientia Marina, 75*, 135-146. http://dx.doi.org/10.3989/scimar.2011.75n1135

Jiménez-Badillo, M. L., Pérez-España, H., Vargas-Hernández, J. M., Cortés-Salinas, J. C., & Flores-Pineda, P. A. (2006). *Catálogo de especies y artes de pesca del Parque Nacional Sistema Arrecifal Veracruzano.* CONABIO, Universidad Veracruzana, Mexico, D. F.

Jiménez-Valverde, A., & Hortal, J. (2003). Las curvas de acumulación de especies y la necesidad de evaluar la calidad de los inventarios biológicos. *Revista Ibérica de Aracnología, 8*, 151-161.

Jordán-Dahlgren, E., & Rodríguez-Martínez, R. E. (2003). The Atlantic coral reefs of México. In J. Cortés (Ed.), *Latin American coral reefs* (pp. 131-158). Elsevier Press, Amsterdam. http://dx.doi.org/10.1016/b978-044451388-5/50007-2

Justine, J.-L., Beveridge, I., Boxshall, G. A., Bray, R. A., Miller, T. L., Moravec, F., ... Whittington, I. D. (2012). An annotated list of fish parasites (Isopoda, Copepoda, Monogenea, Digenea, Cestoda, Nematoda) collected from snappers and bream (Lutjanidae, Nemipteridae, Caesionidae) in New Caledonia confirms high parasite biodiversity on coral reef fish. *Aquatic Biosystems, 8*, 1-29. http://dx.doi.org/10.1186/2046-9063-8-22

Kritsky, C. D. (2012). Dactylogyrids (Monogenoidea: Polyonchoinea) parasitizing the gills of snappers (Perciformes: Lutjanidae): Revision of *Euryhaliotrema* with new and previously described species from the Red Sea, Persian Gulf, the eastern and Indo-west Pacific Ocean, and the Gulf of Mexico. *Zoologia, 29*, 227-276. http://dx.doi.org/10.1590/S1984-46702012000300006

Kritsky, D. C., & Boeger, W. A. (2002). Neotropical Monogenoidea. 41: New and previously described species of Dactylogyridae (Platyhelminthes) from the gills of marine and freshwater perciform fishes (Teleostei) with proposal of a new genus and a hypothesis on phylogeny. *Zoosystema, 24*, 7-40.

Kritsky, D. C., Tingbao, Y., & Yuan, S. (2009). Dactylogyrids (Monogenoidea, Polyonchoinea) parasitizing the gills of snappers (Perciformes, Lutjanidae): Proposal of *Haliotrematoides* n. gen. and descriptions of new and previously described species from marine fishes of the Red Sea, the eastern and Indo-west Pacific Ocean, Gulf of Mexico and Caribbean sea. *Zootaxa, 1970*, 1-51. http://dx.doi.org/10.11646/%25x

Lagrue, C., Kelly, D. W., Hicks, A., & Poulin, R. (2011). Factors influencing infection patterns of trophically transmitted parasites among a fish community: Host diet, host-parasite compatibility or both? *Journal of Fish Biology, 79*, 466-485. http://dx.doi.org/10.1111/j.1095-8649.2011.03041.x

Landínez, R., Ospina, S., Rodríguez, D., Arango, R., & Márquez, E. (2009). Análisis genético de *Lutjanus synagris* en poblaciones del Caribe Colombiano. *Revista Ciencias Marinas, 35*, 321-331.

Luque, J. L., & Alves, D. (2001). Ecologia das comunidades de metazoários parasitos, do xaréu, *Caranx hippos* (Linnaeus) e do xerelete, *Caranx tatus* Agassiz (Osteichthyes, Carangidae) do litoral do estado do Rio de Janeiro, Brasil. *Revista Brasileira Zoologia, 18*, 399-410. http://dx.doi.org/10.1590/S0101-81752001000200011

Luque, J. L., & Poulin, R. (2004). Use of fish as intermediate hosts by helminth parasites: A comparative analysis. *Acta Parasitologica, 49*, 353-361.

Luque, J. L., & Poulin, R. (2007). Metazoan parasite species richness in Neotropical fishes: hotspots and the geography of biodiversity. *Parasitology, 134*, 865-878. http://dx.doi.org/10.1017/S0031182007002272

Madhavi, R., & Sai Ram, B. K. (2000). Community structure of helminth parasites of the tuna, *Euthynnus affinis*, from the Visakhapatnam coast, Bay of Bengal. *Journal of Helminthology, 74*, 337-342. http://dx.doi.org/10.1017/S0022149X00000494

Madhavi, R., & Triveni Lakshmi, T. (2012). Metazoan parasites of the Indian mackerel, *Rastrelliger kanagurta* (Scombridae) of Visakhapatnam coast, Bay of Bengal. *Journal of Parasitic Diseases, 35*, 66-74. http://dx.doi.org/10.1007/s12639-011-0028-5

Magurran, A. E. (2004). *Measuring biological diversity.* Blackwell Publishing, Oxford, U. K.

Montoya-Mendoza, J., Jiménez-Badillo, M. L., & Salgado-Maldonado, G. (2014a). Helminths of *Ocyurus chrysurus* from coastal reefs in Veracruz, Mexico. *Revista Mexicana de Biodiversidad, 85*, 957-960. http://dx.doi.org/10.7550/rmb.43343

Montoya-Mendoza, J., Jiménez-Badillo, M. L., Salgado-Maldonado, G., & Mendoza-Franco, E. F. (2014b). Helminth parasites of the red snapper, *Lutjanus campechanus* (Perciformes: Lutjanidae) from the reef Santiaguillo, Veracruz, Mexico. *Journal of Parasitology, 100*, 868-872. http://dx.doi.org/10.1645/13-429.1

Montoya-Mendoza, J., Salgado-Maldonado, G., & Mendoza-Palmero, C. A. (2008). Monogenean parasites of Carangidae and Sciaenidae marine fish on the Alvarado. *Zootaxa, 1843*, 47-56. http://dx.doi.org/10.11646/%25x

Moravec, F., Vidal-Martínez, V. M., Vargas-Vázquez, J., Vivas-Rodríguez, C., González-Solís, D., Mendoza-Franco, E., ... Güemez-Ricalde, J. (1997). Helminth parasites of *Epinephelus morio* (Pisces: Serranidae) of the Yucatan Peninsula, southeastern Mexico. *Folia Parasitologica, 44*, 255-266.

Moreno, C. E., & Halffter, G. (2001). On the measure of sampling effort used in species accumulation curves. *Journal of Applied Ecology, 38*, 487-490. http://dx.doi.org/10.1046/j.1365-2664.2001.00590.x

Overstreet, R. M., Cook J. O., & Heard, R. (2009). Trematoda (Platyhelminthes) of the Gulf of Mexico. In D. W. Felder, & D. K. Camp (Eds.), *Gulf of Mexico-Origins, Waters, and Biota. Volume 1. Biodiversity* (pp. 419-486). Texas A&M University Press, College Station, Texas, USA.

Palm, H. W., & Overstreet, R. M. (2000). New records of trypanorhynch cestodes from the Gulf of Mexico, including *Kotorella pronosoma* (Stossich, 1901) and *Heteronybelinia palliata* (Linton, 1924) comb. n. *Folia Parasitologica, 47*, 293-302. http://dx.doi.org/10.14411/fp.2000.051

Pérez-Ponce De León, G., García-Prieto, L., & Mendoza-Garfias, B. (2007). Trematode parasites (Platyhelminthes) of wildlife vertebrates in Mexico. *Zootaxa, 1534*, 1-247. http://dx.doi.org/10.11646/%25x

Poulin, R. (1998). Comparison of three estimators of species richness in parasite component communities. *Journal of Parasitology, 84*, 485-490. http://dx.doi.org/10.2307/3284710

Rodríguez-González, A., & Vidal-Martínez, V. M. (2008). Las comunidades de helmintos del lenguado (*Symphurus plagiusa*) en la costa de Campeche, México. *Revista Mexicana de Biodiversidad, 76*, 159-173. http://dx.doi.org/10.7550/rmb.5369

Rohde, K., & Heap, M. (1998). Latitudinal differences in species and community richness and in community structure of metazoan endo- and ectoparasites of marine teleost fish. *International Journal for Parasitology, 28*, 461-474. http://dx.doi.org/10.1016/S0020-7519(97)00209-9

Rosa, D. M., Vilar, C. C., & Musiello-Fernandes, J. (2015). Relative effect of seasonality and body size on the diet of juvenile *Lutjanus synagris* (Perciformes: Lutjanidae) at a sandy beach in Southeastern Brazil. *Boletim do Instituto de Pesca, 41*, 19-29.

Sánchez-Ramírez, C., & Vidal-Martínez, V. M. (2002). Metazoan parasite infracommunities of Florida pampano (*Trachinotus carolinus*) from the coast of the Yucatán Peninsula, Mexico. *Journal Parasitology, 88*, 1087-1094. http://dx.doi.org/10.1645/0022-3395(2002)088[1087:MPIOFP]2.0.CO;2

Soberón, M. J., & Llorente, J. (1993). The use of species accumulation functions for the prediction of species richness. *Conservation Biology, 7*, 480-488. http://dx.doi.org/10.1046/j.1523-1739.1993.07030480.x

Vélez, I. (1987). Sobre la fauna de tremátodos en peces marinos de la familia Lutjanidae en el mar Caribe. *Actualidades Biológicas, 16*, 70-84.

Vidal-Martínez, V. M., Aguirre-Macedo, M. L., Scholz, T., González-Solís, D., & Mendoza-Franco, E. (2001). *Atlas of the helminth parasites of cichlid fishes of Mexico.* Academia, Academy of Sciences of the Czech Republic, Prague, Czech Republic.

Zander, C. D., Reimer, L. W., & Barz, K. (1999). Parasite communities of the Salzhaff (Northwest Mecklenburg, Baltic Sea). I. Structure and dynamics of communities of littoral fish, especially small-sized fish. *Parasitology Research, 85*, 356-372. http://dx.doi.org/10.1007/s004360050562

Zhukov, E. V. (1976). New monogenean species of the genus *Haliotrema* Johnston and Tiegs, 1922, from the Gulf of Mexico fishes of the Fam. Lutianidae, in Fauna, systematics and phylogeny of Monogenoidea. *Proceedings, Institute of Biology and Pedology, Far-East Science Centre, Academy of Sciences of the U. S. S. R., New Series, 35*, 33-47.

16

Energy Consumption in Onion and Potato Production within the Province of El Hajeb (Morocco): Towards Energy Use Efficiency in Commercialized Vegetable Production

Khalil Allali[1], Boubaker Dhehibi[2], Shinan N. Kassam[3] & Aden Aw-Hassan[2]

[1] Département d'Economie Rurale, Ecole Nationale d'Agriculture de Meknes, Meknes-El Menzeh, Morocco

[2] Sustainable Intensification and Resilient Production Systems Program (SIRPSP), International Center for Agricultural Research in the Dry Areas (ICARDA), Amman, Jordan

[3] Sustainable Intensification and Resilient Production Systems Program (SIRPSP), International Center for Agricultural Research in the Dry Areas (ICARDA), Cairo, Egypt

Correspondence: Boubaker Dhehibi, Sustainable Intensification and Resilient Production Systems Program (SIRPSP), International Center for Agricultural Research in the Dry Areas (ICARDA), P.O. Box 950764, Amman 11195, Jordan. E-mail: b.dhehibi@cgiar.org

Abstract

Energy use efficiency is a key requirement for sustainability in agricultural production, but often overlooked. The aim of this study was to quantify the amount and efficiency of energy consumed in the production of onions and potatoes in El Hajeb province of Morocco. These estimates are of significant importance in informing contemporary policy discourse related to energy subsidy reform in Morocco, and more specifically within an ongoing national strategy for 'modernizing' the agricultural sector under the 'Green Morocco Plan'. Data were collected through the administration of a direct questionnaire with 60 farmers and analyzed using PLANETE. Our results indicate that total energy consumption in onion production is 107483 MJ ha^{-1} with butane (79.5%) as the main source of direct energy. Chemical fertilizers (61.53%) and water for irrigation (30%) were main sources of indirect energy. Energy indices related to energy efficiency ratios, energy profitability and energy productivity were estimated at 0.78, -0.22 and 0.54 kg MJ^{-1}, respectively. Total energy consumption in potato production was estimated at 74,270 MJ ha^{-1}, with direct energy consumption of 28,521 MJ ha^{-1} stemming from butane (70%) and diesel (19.14%) as primary sources. Indirect energy consumption was estimated at 45749 MJ ha^{-1} and generated principally through the use of fertilizers (60%). Energy indices (efficiency, profitability and productivity) were estimated at 1.54, 0.54, and 0.45 kg MJ^{-1}, respectively. GHG emissions were found to be 3.47 t CO_{2eq} ha^{-1} in the production of onions and 3.63 t CO_{2eq} ha^{-1} for potatoes. We find that within the study area, increases in the size of production plots are not necessarily consistent with increases in energy use efficiency.

Keywords: energy analysis, energy consumption, onions, potatoes, energy indices, GHG emissions

1. Introduction

A continued desire to ensure food security in the face of sustained population growth has been one driving force in the process of agricultural innovation. At the same time, however, agricultural technology development and adoption have led to an increase in energy-dependence, and particularly so in terms of fossil fuels. In environmental terms, issues of sustainability in energy resource use lie in their non-renewable nature, and in the negative externalities generated by their use, particularly in terms of contribution to global warming through the emissions of greenhouse gases (GHGs). It has been estimated that 24% of 2010 global greenhouse gas emissions emanated from the agriculture sector (cultivation of crops and rearing of livestock) and exacerbated by deforestation. This estimate does not include the CO_2 that ecosystems beneficially remove from the atmosphere through the sequestration of carbon in biomass, dead organic matter and soils, and through which, approximately 20% of emissions from the agricultural sector are offset (Hillier et al., 2011; Vermeulen et al., 2012; Thornton, 2012).

Energy is a critical issue in food security, but one which is often overlooked. Energy efficiency is defined as "the ability of producing the same level of output with minimum used resources" (Sherman, 1988 quoted

Mousavi-Avval et al., 2012). While this is of critical importance for profitability at the farm gate, national level concerns related to energy efficiency use naturally include a desire for reduction in imports, with implications for foreign exchange, mitigation of greenhouse gas emissions and more specifically for this paper, competitiveness of the agricultural sector through a reduction in production costs.

Energy related challenges are of both historical and contemporary concern to Morocco given that the Kingdom imports 97% of its energy needs. Recognizing this vulnerability, Morocco has recently launched a number of national plans and programs aimed at promoting energy use efficiency in all economic activities, including agriculture, and through encouraging and incentivizing the use of renewable energy sources. Studies analyzing energy consumption, and evaluating energy efficiency of agricultural production in Morocco, are essential in order to inform the discovery of avenues for generating and disseminating knowledge on (energy) efficient and sustainable agricultural production practices which embody desired economic incentives. This study is placed within this context, with an overall objective to assess energy consumption (*i.e.*, testing if farmers producers are using energy unnecessarily excessively and if so, why are they using energy excessively/or inefficiently, and finally what measures can be taken to increase energy efficiency use and reduction of GHG emissions by these systems?) in the production of onions and potatoes two nationally important agricultural commodities produced within the province of El Hajeb.

2. Methodological Approach

2.1 Crops Selection and Study Area

This study was undertaken within the province of El Hajeb, which is well known within Morocco for its historical importance in the production of onions and potatoes. Indeed, these crops represent 41% and 32% respectively of the area under vegetable production within the province, with a corresponding share of 52% and 33% in terms of market production for all vegetables. More broadly, areas under cultivation of onions and potatoes are found to be 60% and 38% of the total area these crops at the regional level (Meknès-Tafilet) and 16% and 8% nationally (DSS, DRA Meknès-Tafilalet, DPA El Hajeb, 2014). For the kingdom as a whole, onions and potatoes represent 13.33% and 21% of the areas devoted to vegetable production. Similarly, at the regional level, these two crops combined command 35% and 44% of the market shares for all cash crops.

2.2 Data Collection and Calculation Method

While data related to the area under onions and potatoes is available for public consumption, data on the number of producers is conspicuously absent. We therefore opt for sampling at a single level of stratification, and more specifically at the level of district within the province of El Hajeb. More specifically, our stratification employs aggregated data from 3 districts within the study area: Ain Taoujtate, Agourai and El Hajeb (Note 1), with the number of growers chosen within each district determined by:

$$N_i = (S_i/S) \times N \quad (1)$$

Where,

N_i: Number of growers to investigate in district i; S_i: area under vegetables within district i; S: area under vegetables within El Hajeb province; N: total number of growers to investigate.

A survey of 60 vegetable growers, with individuals randomly drawn from each district (El Hajeb, Ain Taoujtate and Agourai) was administered within the 2012/2013 cropping season. The questionnaire was designed in a manner such that contextual practices and limited time in administration were considered in the collection of data on energy inputs and outputs for each crop. Analysis of data was undertaken through utilization of PLANET balance (Method of Energy Analysis of Operations). As a tool designed for measuring fossil energy consumed directly or indirectly through inputs into a system, PLANET has generally been utilized for estimating GHG emissions at the scale of a farm (Figure 1) and taking into account both crop and livestock systems.

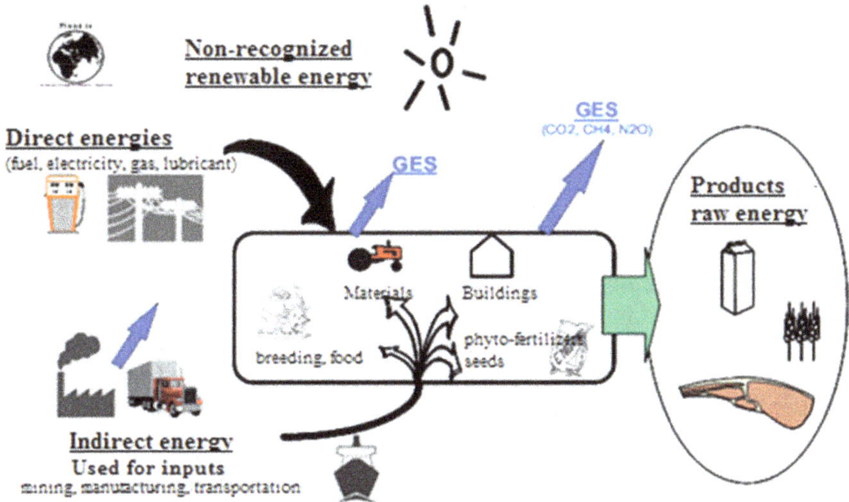

Figure 1. General diagram of PLANET balance

Source: Bochu (2007).

The underlying principle is one of converting physical quantities of inputs and outputs into units of MJ (Mega Joule) and evaluated on the basis of energy consumed and emitted per hectare. These coefficients and factors are consistent with international standards of life cycle analysis and environmental audits (Bochu, 2007). Conversion calculations were undertaken according to the following formula:

$$E_i/o = Q_i \times EE_i \tag{2}$$

Where,

E_i/o: Energy of specific input or output (MJ); Q_i: Physical quantity of the specific input or output (unit); EE_i: Equivalent unit of energy of the specific input or output (MJ/unit).

The specific sequence of steps of energy analysis followed were:

- Analysis of energy consumption (direct and indirect energy);
- Establishment of the energy balance (energy consumption and production);
- Assessment of energy performance; and
- GHG emission estimation and evaluation of global warming potential.

This analysis was conducted for each crop taking into account the class of producer. For the production of onions and potatoes, energy inputs are divided into direct and indirect sources of energy. Direct energy (ED) sources are those consumed on the production site, while indirect energy (EI) is that consumed during manufacturing process and transportation of inputs; in other words, EI corresponds to energy incorporated into factors which are generally outside of the farm gate.

Direct energy includes petroleum products and electricity used to produce each studied crop. Petroleum products include diesel (EG) utilized by agricultural machinery for the implementation of various agricultural operations (land preparation, phytosanitary treatments, etc.); butane (EB) used in irrigation water pumping; and gasoline (EE) used as a fuel for engines in the application of phytosanitary treatments (largely in spraying). Electricity (EEL) is also used to pump irrigation water within the area of study and included within the category of direct energy.

Indirect energy includes those embodied in seeds (ES), irrigation water (ACS), manure (EF), mineral fertilizers (EEM), phytosanitary treatments (FTEs) and agricultural machinery (EM). Energy utilized in irrigation is considered indirect when it relates to the conveyance of water to the plot as well as in the maintenance of equipment and irrigation canals (cleaning, maintenance, etc.). For mineral fertilizers, known to be major consumers of energy in the production process, we consider nitrogen, phosphorus, potassium and sulfur. Energy coefficients take into account mining, formulation, packaging, transport and distribution of these nutrients (Bochu, 2007).

Energy sequestered in agricultural machinery represents energy incorporated in commodities, manufacturing energy, repair and maintenance (Mousavi-Avval et al., 2012). Its equivalent energy matches the energy payback timeline (AE) which is measured through the following equation:

$$EE = AE = (M \times EA)/CTA \tag{3}$$

Where,

AE: Energy amortization schedule; M: The mass of the machine in kg; EA: Equipment annual energy; CTA: The annual work capacity h/year.

Energy balance consists of energy inputs which are comprised of consumed energy (direct and indirect) and energy outputs or emitted energy. For onion and potato production, outputs include yields of bulb (onion) and tubers (potato) which are converted to mega-joules. Based on obtained energy equivalents, we proceed with the evaluation of performance of energy utilized in the production of onions and potatoes through the creation of relevant indices: energy ratio, energy productivity, energy intensity, energy profitability and net energy gain.

The energy ratio (ER), also called Energy Efficiency Index, is the ratio of energy emitted to energy sequestered in production factors. This index indicates the influence of inputs expressed in the energy unit of energy output. The energy ratio in a system can be improved by reducing the energy sequestered in inputs and/or through reducing losses by increasing production yields (Kitani, 1999 quoted by Mousavi-Avval et al., (2012).

$$ER = \frac{\text{Total produced energy (MJ}\cdot\text{ha}^{-1})}{\text{Total consumed energy (MJ}\cdot\text{ha}^{-1})} \tag{4}$$

Energy productivity (ECD) is a measure of the amount of product obtained per unit of input energy. It is therefore calculated as the ratio between the energy expended in directly producing crop yield and total consumed energy:

$$ECD\ (\text{kg}\cdot\text{MJ}^{-1}) = \frac{\text{Crop yield (kg}\cdot\text{ha}^{-1})}{\text{Total consumed energy (MJ}\cdot\text{ha}^{-1})} \tag{5}$$

Specific energy (SE) is the inverse of energy productivity and corresponds to the energy value required for an emitted unit of output:

$$SE\ (\text{kg}\cdot\text{MJ}^{-1}) = \frac{1}{ECD\ (\text{kg}\cdot\text{MJ}^{-1})} = \frac{\text{Consumed energy (MJ}\cdot\text{ha}^{-1})}{\text{Crop yield (kg}\cdot\text{ha}^{-1})} \tag{6}$$

Net energy gain (NEG) is the difference between the energy expended in delivering output the total energy required in producing output:

$$NEG\ (\text{MJ}\cdot\text{ha}^{-1}) = \text{Total produced energy (MJ}\cdot\text{ha}^{-1}) - \text{Total consumed energy (MJ}\cdot\text{ha}^{-1}) \tag{7}$$

Energy profitability (EPB) is defined as:

$$EPB = \frac{NEG\ (\text{MJ}\cdot\text{ha}^{-1})}{\text{Total consumed energy (MJ}\cdot\text{ha}^{-1})} \tag{8}$$

GHG emissions in the production of onion and potato are produced through both direct and indirect consumption of energy with CO2 and N2O as principal components. Quantifying emissions of each of these gases is undertaken on the basis of global warming potential and representing the total weight of estimated gas quantities:

$$1 \text{ ton } CO_2 = 1 \text{ equivalent tons of } CO_2 \tag{9}$$

$$1 \text{ ton of } N_2O = 310 \text{ equivalent tons of } CO_2 \tag{10}$$

3. Results and Discussion

3.1 Energy Analysis in the Production of Onions

3.1.1 Direct Energy Consumption

Based on energy equivalence, the average direct energy consumption for all respondent onion producers is approximated at 72135 MJ/ha. Taking each energy item separately, respondent producers consumed an average of 57334 MJ/ha of butane, 7424 MJ/ha of diesel, 6920 MJ/ha of electricity and 457.2 MJ/ha of gasoline.

Based on these empirical results, the main component of direct energy consumption in the production of onions is Butane. It represents 79.48% of total direct energy consumption. This is followed by diesel (10.29%), electricity (9.59%) and lastly gasoline (0.63%). The high proportion occupied by butane in total direct energy consumption is due to its ubiquitous use in the pumping of water for irrigation, but incentivized by significant subsidies which have been targeted for household cooking purposes. In Aval and Moghaddam (2013) and Barber (2004), research undertaken in Iran and New Zealand respectively, direct energy utilized in the production of onion are diesel and electricity. Butane, which is the main source in direct (subsidized) energy consumption for onion production in Morocco, is not generally utilized in other countries and signals a need for policy dialogue in Morocco in terms of both subsidy reform, as well as in terms of moving towards cleaner uses of energy.

Irrigation, as a source of direct energy, consumes an average of 68212 MJ/ha or 94.56% of total consumption. This is followed by the process of land preparation, consuming 3209 MJ/ha; and thereafter phytosanitary treatment processes which utilize 714.96 MJ/ha. In terms of production costs, irrigation maintains a significant proportion (20%); hence the need for rationalizing irrigation practices in technical terms (speed, duration, frequency) so as to ensure a better allocation of both water and energy resources for both productivity and efficiency purposes.

Our results indicate that total direct energy consumption follows an upward trend with increasing plot size. Indeed, 'small' onion producers with plots of less than 2 hectares consume the least amount of energy (41895 MJ/ha), with those between 2 and 5 hectares consuming 66180 MJ/ha; and 67897 MJ/ha in consumption for producers in the 5 to 10-hectare plot size range. 'Large' producers with parcels exceeding 10 hectares consume in excess of 2.5 times smaller producers, with estimates as high as 104065 MJ/ha within the sample. We suggest that these very large variances are most likely a reflection of the different weights of energy utilized in pumping irrigation water and in terms of conveyance systems utilized in the delivery of water to the crop.

3.1.2 Indirect Energy Consumption

Indirect energy utilized in the production of onions consists of energy sequestered in the factors of production; namely seeds, mineral fertilizers, phytosanitary treatments, water irrigation and farm equipment. The average consumption of indirect energy for all surveyed onion producers was estimated at 35348 MJ/ha and by component: 21749 MJ/ha in mineral fertilizers, 10624 MJ/ha in irrigation water, 2020 MJ/ha in farm equipment, 839 MJ/ha in phytosanitary treatments and 116 MJ/ha in seeds. However, the main sources of indirect energy consumed in the production of onions are mineral fertilizers (61.53%) and irrigation water (30.05%). These are followed by agricultural equipment (5.71%), phytosanitary treatment (2.37%) and seeds (0.33%).

With respect to mineral fertilizers, nitrogen is most prominent and with consumption estimated at 17062 MJ/ha or 78.45% of total fertilizer consumption. This is followed by phosphorus at 3778 MJ/ha or 17.37% of total fertilizer consumed, with energy consumption of potassium and sulfur not of (relatively) significant value, but estimated at 715.1 MJ/ha and 19.66 MJ/ha respectively. In comparing the results of the use of indirect energy obtained in this study with those reported by Aval and Moghaddam (2013) and Barber (2004) in Iran and New Zealand, there is general consistency in a claim that mineral fertilizers are key consumers of indirect energy in vegetable production; together with a more specific claim that Moroccan producers are relatively greater consumers of energy in the application of mineral fertilizer in onion production.

In terms of energy consumption in the conveyance and application of irrigation water, Moroccan producers would appear to consume greater quantities relative to Iranian producers of onions and potatoes. Indeed, the level of consumption of the indirect energy is 12174 MJ/ha in Morocco and 10624 MJ/ha in Iran (Aval and Moghaddam, 2013). The government of Morocco has placed much policy interest on providing incentives to farmers to adopt drip irrigation as a water saving practice, together with ancillary benefits related to improved quality and yield. While a significant number of farmers have adopted drip irrigation, anecdotal evidence from the field would suggest that farmers continue to apply significantly more water than is technically in the production of both onions and potatoes. One explanation for this relates to heavy subsidies on butane, to support households in lowering the cost of cooking fuels, but in practice equally important for farm households in the pumping of irrigation water. Another relates to efficacy in extension service provision and knowledge dissemination. While a significant national strategy for reforming extension services in Morocco has been designed and is in process, there have been notable delays in implementing the plan; and concerns related to overlapping of mandates across a number of organizations which may lead to institutional conflict (Note 2).

Regarding the energy embedded within the category of phytosanitary treatments, Iranian producers consume 1192 MJ/ha while Moroccan producers use approximately 839 MJ/ha (Aval & Moghaddam, 2013). In large part, this is due to a greater number of treatments, with Iranian producers applying an average of 6.2 relative to 3.7 for Moroccan producers. The breakdown of energy consumed according to the type of treatment indicates that Iranian farmers apply more herbicides and insecticides while Moroccan farmers tend to use more fungicides. In New Zealand, onion growers consumed 12050 MJ/ha in phytosanitary treatments (Barber, 2004) for the

2000/2001 cropping season. This figure, however, should be viewed with caution as the author reports that the season under study was exceptionally unique given heavier than average rainfall and significant humidity.

On average, Barber (2004) suggests that a normal use of fungicides is 20 to 25 percent of the calculated amount. The vagaries of weather, therefore, may play a key role in the amount of energy utilized in phytosanitary treatments. Equally important are on farm storage and drying practices which may result in the greater use of fungicides. To be sure, a standard practice on Moroccan farms, within the area of study, is to dry onions under stone mounds which are covered with straw. The incidence of mould and fungus is therefore potentially high in the absence of controlled environments for storage, and thereby resulting in greater need for applications of phytosanitary treatments.

For producers with land areas of less than 2 ha indirect energy consumption was estimated at 29460 MJ/ha. For producers with plot sizes between 5 and 10 ha, consumption was estimated at 46342.58 MJ/ha. Producers with land holdings of greater than 10 ha consumed approximately 12% in energy consumption relative to the middle class of producers. In all cases, mineral fertilizer consumption was the key input into indirect energy consumption and accounting for close to 62% of total indirect energy consumed.

3.1.3 Energy Balance

Average energy consumed across the sample was estimated at 107483 MJ/ha for onions with the amount of energy emitted at 84269 MJ/ha. Direct energy accounted for 67.11% of admissions with indirect energy accounting for the remainder of 32.89% of total energy consumed.

Of the total energy consumed, on average across all sampled onion producers, 53.34% was generated from the use of butane and 20.24% from the use of fertilizer. Initiatives aimed at reducing energy use in the production of onions within the area of study would achieve much impact, therefore, if concentration was placed upon efficient and rational use of these two inputs. A need for policy dialogue on existing subsidies is clearly evident from these results, particularly in terms of butane subsidy, but equally important in terms of delivering more effective knowledge and information to farmers on the effective use of fertilizers through both reduction as well as in more effective water saving irrigation technologies.

Table 1. Energy balance in the production of onions within the Meknes region

Items		< 2 ha	[2,5] ha	[5,10] ha	≥ 10 ha	Total sample	% energy input
Energy inputs							
Direct Energy	Diesel	7 212.94	4 891.27	11 581.00	4 748.33	7 424.01	6.91
	Gasoline	138.33	329.04	520.64	937.21	457.19	0.43
	Butane	34 544	60 960.00	50 319.71	69 951.60	57 333.93	53.34
	Electricity	0.00	0.00	5 475.27	28 428.33	6 920.14	6.44
	Total	*41 895.28*	*66 180.30*	*67 896.62*	*104 065.48*	*72 135.27*	*67.11*
Indirect Energy	Seeds	88.20	98.70	144.33	148.23	115.40	0.11
	Fertilizers	16 037.98	16 535.42	32 495.78	28 528.30	21 749.48	20.24
	Phytosanitary	609.52	710.39	990.17	1 174.67	839.41	0.78
	Irrigation water	11 179.20	11 120.62	10 556.81	9 193.60	10 623.76	9.88
	Materials	1 545.36	1 894.95	2 155.49	2 166	2 020.01	1.88
	Total	*29 460.27*	*30 360.09*	*46 342.58*	*41 210.79*	*35 348.05*	*32.89*
Total energy consumed		71 355.55	96 540.39	114 239.20	145 276.27	107483.31	100
Total energy emitted		74 755.56	77 678.57	82 254.55	108 629.17	84 269.17	-

Source: Own elaboration from Survey (2015).

3.1.4 Energy Performance Indices

The calculation of energy performance indices for the survey sample indicates that the energy ratio is less than 1, with negative profitability. One MJ of energy converts to 0.54 kg of onion or put differently, one kilogram of onion requires 1.85 MJ of energy. Analysis by plot size reveals that smaller producers have the best values of these indices and is likely a reflection of lower input levels as well as relatively low levels of mechanization.

Scale of farm operation, therefore, is likely to increase energy consumption per kilogram of output rather than inducing efficiency in energy utilization.

3.1.5 GHG Estimation

Key greenhouse gases emitted in the production of onion are carbon dioxide (CO_2) and Nitrous oxide (N_2O) with emission values of 2.68 t/ha and 2.54 kg/ha respectively. From empirical findings, 79% of CO_2 emissions accrue from the use of mineral fertilizers, 9% from the combustion of butane in the pumping of irrigation, 6% in the manufactory of agricultural equipment, 3% in the use of electricity for irrigation, 2% in the combustion of diesel (land preparation, irrigation and spraying of phytosanitary treatment) and 1% when in the application of pesticides.

In the case of mineral fertilizers, almost all of the emission (more than 99%) is a result of energy consumed in the manufacturing stage as opposed to application on farm. Transforming these emission values into global warming potential (GWP), expressed as CO_2 equivalent (CO_{2eq}), our results indicate that onion production has a potential for atmospheric warming of approximately 3.467 t CO_{2eq}/ha within the study area. Carbon dioxide (CO_2) contributes 77.29% to the potential for global warming with an average of 2.68 CO_{2eq} /ha while Nitrous oxide (N_2O) contributes approximately 22.71% in GWP through its emissions of 0.78 t CO_{2eq}/ha. As expected, global warming potential increases with larger plot sizes.

3.2 Energy Analysis in the Production of Potatoes

3.2.1 Direct Energy Consumption

An analysis of the use of direct energy in potato production indicates a total consumption of 28521 MJ/ha. The average consumption of each component was found to be: 19964 MJ/ha for butane, 5458 MJ/ha for diesel, 2608 MJ/ha for electricity, and 490 MJ/ha for gasoline. In terms of consumption as percentage of total energy consumption: Butane (70%), followed by gas oil (19.14%), electricity (9.14%), and finally gasoline (1.72%).

From the perspective of agricultural production, we find that the most energy-intensive activity is irrigation (84.35%), followed by land preparation (11.56%), phytosanitary treatment (3.05%) and finally the maintenance of the cultivation and harvesting equipment with 0.52% each. Comparing the results of this study with those of Mohammadi et al. (2008), Ghahderijani et al., (2013), Barber (2004) and Pimentel et al. (2002), cited by Barber, 2004), it would appear that butane is not used by Iranian, New Zealand and US potato producers in the pumping of irrigation water, while it represents the main direct energy consumption item for Moroccan producers. For Iran, the only direct conventional power/energy source considered in relatively recent studies is diesel (Mohammadi et al., 2008; Ghahderijani et al., 2013). This source of energy is largely consumed during the land preparation, planting, irrigation, application of phytosanitary treatments, fertilization and harvesting.

In the Iranian province of Ardabil, diesel consumption was estimated at 12897 MJ/ha and in Esfahan province at 5638 MJ/ha (Ghahderijani et al., 2013). In Morocco, average diesel consumption was estimated at 5458 MJ/ha but is not the only choice of fuel for pumping of water in potato production. In New Zealand and the United States, the sources of direct energy used in potato production are diesel and electricity, with diesel used in all cultivation operations. Its consumption for 294 L/ha in New Zealand is 11966 MJ/ha, for 424 L/ha in the United States is 17257 MJ/ha and for 134 L/ha in Morocco is 5458 MJ/ha. These results indirectly reflect the degree of mechanization of the production of potatoes in each country. With respect to electricity, consumption in New Zealand was estimated by the authors (ibid) at 360 KWh/ha or 3456 MJ/ha; in the US, it is 47 kWh/ha or 451 MJ/ha; and in Morocco we estimate this to be 272 KWh/ha or 3 973 MJ/ha. Several variables generate these differences in energy consumption. These include: the source of irrigation water (water pumped and surface water), annual rainfall, farmers' practices, and the depth of groundwater, among others.

As expected, direct consumption of energy increases with increasing plot size. Indeed, the producers of small producers, with plot sizes under 2 hectares consume 16449 MJ /ha, while, those with plots between 2 and 5 hectares consuming 29049 MJ/ha, and 34083 MJ/ha for those with plots sizes of between 5 and 10 hectares. Farm sizes exceeding 10 hectares consumes 46561 MJ/ha of direct energy. The energetic study undertaken by Ghahderijani et al. (2013) in Iran was also conducted according to the field sizes, with small producers classified on the basis of less than 1 ha, medium producers with an areas between 1 and 5 ha and large producers with an areas greater than 5 hectares. Their analysis in Iran indicate a contrary observation: energy consumption decreases with an increase of field size. Indeed, small producers reach average consumption levels of 50549 MJ/ha, the medium 44796 MJ/ha and the large ones consume about 42963 MJ/ha. Iran, therefore, has an economy of scale in energy consumption with increasing field sizes while Morocco is experiencing dis-economies of scale at this level.

3.2.2 Indirect Energy Consumption

Sources of indirect energy in the production of potato are; seeds, irrigation water, fertilizer (chemical fertilizers and organic manure), phytosanitary treatments, agricultural machinery, straw and plastic (for pre-germination treatment). Average indirect energy consumption was estimated at 45749 MJ/ha. The main energy consumption items are: fertilizers (27241 MJ/ha; 59.54%), seeds (9682 MJ/ha; 21.16%) and irrigation water (4672 MJ/ha; 10.21%).

Energy sequestered in fertilizers is largely due to the use of nitrogen (81.59%) containing 22227 MJ/ha, followed by phosphate 12.65% used up to 3445 MJ/ha, then potassium (4.72%) and finally manure (0.8%) and sulfur (0.2%) with quantities that are relatively small (228 MJ/ha and 58 MJ/ha, respectively).

Comparing the results of this study with those of Iran, New Zealand and the United States, it appears that in all these countries, the largest single consumer of indirect energy is fertilizer. Seeds were not considered in the study of Barber (2004) for New Zealand while it occupied the second largest item of indirect energy consumption for studies in the US and Iran. The energy sequestered in irrigation water was also not included in studies in New Zealand and for the USA. For Iran, energy consumption in the conveyance and application of irrigation water was placed in third after fertilizers and seeds. Nitrogen represented the key element within all studies in relation to the category of fertilizer.

Indirect energy consumption increases with an increase in plot size as expected. Indeed, the smaller holdings (less than 2 ha) consume 41801 MJ/ha, while those between 2 and 5 ha consume 44815 MJ/ha; larger areas between 5 and 10 ha use 49353 MJ/ha and those exceeding 10 hectares consume 53328 MJ/ha.

3.2.3 Energy Balance

We find that average energy consumption for the entire sample per hectare of potatoes is in the range of 74270 MJ/ha and the amount of emitted energy per hectare estimated at 114634 MJ/ha (Table 2). The structure of energy consumption is dominated by indirect energy representing 61.60% of total consumption with the balance of 38.40% as consumption of direct energy.

Table 2. Energy balance in the production of potatoes within the Meknes region

Items		< 2 ha	[2,5] ha	[5,10] ha	≥ 10 ha	Total sample	% energy input
Energy inputs							
Direct Energy	Diesel	3 034.00	7 063.49	4 431.78	6 410.25	5 458.43	7.35
	Gasoline	226.36	468.95	617.89	1 037.50	490.45	0.66
	Butane	13 189.53	20 147.28	22 487.47	32 004.00	19 964.40	26.88
	Electricity	0.00	1 369.98	6 546.00	7 110.23	2 608.06	3.51
	Total	*16 449.89*	*29 049.70*	*34 083.13*	*46 561.98*	*28 521.34*	*38.40*
Indirect Energy	Seeds	9 545.45	9 600.00	9 888.89	10 000.00	9 681.82	13.04
	Fertilizers	23 488.52	26 690.86	30 220.19	33 602.66	27 240.53	36.68
	Phytosanitary	329.78	436.22	574.81	621.23	454.77	0.61
	Irrigation water	5 054.01	4 506.36	4 724.19	4 329.90	4 671.79	6.29
	Materials	1 423.56	1 610.35	2 006.34	2 720.90	1 730.71	2.33
	Total	*1 520.59*	*1 529.28*	*1 504.50*	*1 593.00*	*1 527.83*	*2.06*
	Seeds	439.09	441.60	434.44	460.00	441.18	0.59
	Fertilizer	41 801.01	44 814.68	49 353.37	53 327.69	45 748.63	61.60
Total energy consumed		58 250.90	73 864.37	83 436.50	99 889.67	74 269.97	100
Total energy emitted		107 263.64	115 575.00	119 983.33	118 162.50	114 634.09	-

Source: Own elaboration from Survey (2014).

We note that 76.59% of the average energy consumption of all interviewed potato producers are divided between "fertilizers" (36.68%), "butane "(26, 88%), and seeds (13.04%). These are therefore areas which are of key target for inducing energy-saving. Analysis by class of producers globally shows that the inputs and outputs increase with the size of the plots and consistent with our results.

3.2.4 Energy Performance Indices

The net energy gain of the entire sample is positive and equals 40364 MJ/ha. One MJ of energy per hectare is required to produce 0.45Kg of potatoes, or put differently, 1 kilogram of potato emitted requires 2.24MJ of energy. Analysis by plot class reveals that small producers have the best values of these indices relative to larger producers. This is largely due to their relatively lower use of agricultural inputs and relatively low level of mechanization.

3.2.5 GHG Estimation

The main greenhouse gases emitted in the production of potatoes are carbon dioxide CO_2 and nitrous oxide N_2O, with emission estimates of 2.61 t/ha and 3.30 kg/ha, respectively. Based on the empirical results, approximately 84.67% of CO_2 emissions come from mineral fertilizers, 6.63% from agricultural equipment, 3.37% from butane, 2.89% from organic manure, 1.46% from diesel combustion, 0.05% from phytosanitary products, and 0.46% from plastic used to stimulate the germination of potato tubers.

By transforming its emissions in global warming potential (GWP), expressed as CO_2 equivalent (CO_{2eq}), the results indicate that potato production has potential for atmospheric warming of approximately 3.628 CO_{2eq} /ha. Carbon dioxide (CO_2) contributes 71.82% in the potential of global warming with an average of 2.61 CO_{2eq} /ha while N_2O Nitrous oxide contributes 28.17% in GWP with emissions of 1.02 CO_{2eq} /ha. Comparing between different classes of the producers on GWP, we note that GWP reported per hectare increases with the increase in the size of the plots as expected.

4. Concluding Remarks and Recommendations

Energy is essential for food security and development. However, food production and current energy use patterns may no longer be viable in the long term given current rates of natural resource extraction. Given that energy use in agriculture is related to crop choices and management practices within the production process, a diagnosis such as that undertaken within this study may allow decision makers and farmers to consider mutually consistent incentives in the adoption of energy saving practices; thereby contributing to direct improvements in energy efficiency and (indirectly induced) reductions in greenhouse gas emissions stemming from agricultural production.

Studies on energy analysis of agricultural productions in Morocco are limited; reflecting a significant shortfall in research. Vegetable crops are among the most intensive agricultural production inputs including energy resources. This study has analyzed energy consumption for two vegetable crops: onion and potato in El Hajeb province of Meknes.

The results of energy analysis in onion production suggest that total energy consumption is 107483 MJ/ha and divided into direct energy (67.11%) and indirect energy (32.89%), with butane (79.48%) as the main source of direct energy consumption. Significant indirect energy consumption items were found to be mineral fertilizers (61.53%) and energy used in the pumping of irrigation water (30.05%). Total energy emitted was estimated at 84269 MJ/ha. Empirical findings from energy analysis in the production of potatoes indicate total energy consumption of 74270 MJ/ha. This is divided into direct energy (38.40%) and indirect energy (61.60%) with Butane (70%) as the main source of direct energy. Significant indirect energy was found to be consumed in the form of fertilizers (59.54%), seeds (21.16%) and pumping of irrigation water (10.21%). Energy production (emitted) in potato production was estimated at 114634 MJ/ha.

An analysis of energy performance indices indicates that energetic efficiency, profitability and productivity are 0.78, 0.22 and 0.54 kg/MJ, respectively for onion and 1.54, 0.54 and 0.45 kg/MJ for potato. The analysis of GHG emissions suggests that global climate warming potential is about 3.47 CO_{2eq}/ha for the production of onion and 3.63 CO_{2eq}/ha for potato production. Fertilizers, and particularly nitrogen, were found to be significant emitters of greenhouse gas emissions. One key learning is that for both onion and potato, energy consumption increases with an increase in the size of farm plots and standardized on a per hectare basis. This is different from the experience of Iranian producers, as detailed within the literature, and signals that there may be little savings in energy through expansion in the scale of production. It appears, therefore, that larger farm operations within the study area are not necessarily more energy efficient relative to smaller production units. This is an important policy related finding, and one which requires more in depth understanding of the nature of incentives provided through subsidies such as butane, as well as within a general movement towards consolidating fragmented land parcels.

Acknowledgements

This work was undertaken as part of, and funded by, the CGIAR Research Program on Dryland Systems (http://drylandsystems.cgiar.org/) led by the International Center for Agricultural Research in the Dry Areas

(ICARDA: http://www.icarda.org). The opinions expressed here belong to the authors, and do not necessarily reflect those of Dryland Systems, ICARDA, or CGIAR.

References

Aval, F. H., & Moghaddam, P. R. (2013). Energy Efficiency Evaluation and Economical Analysis of Onion (*Allium cepa* L.) Production in Khorasan Razavi Province of Iran. *Iranian Journal of Applied Ecology, 2*(3), 1-11.

Barber, A. (2004). *Seven Case Study Farms: Total Energy & Carbon Indicators for New Zealand Arable & Outdoor Vegetable Production*. AgriLINK New Zealand Ltd. http://www.agrilink.co.nz/Files/Arable%20Vegetable%20Energy%20Use%20Main%20Report.pdf

Bochu, J. L. (2007). *Planete result analyze, energy consumption and greenhouse gas emission from French frams*. ADEME, SOLAGRO.

Ghahderijani, M., Pishgar-Komleh, S. H., Keyhani, A., & Sefeedpari, P. (2013). Energy analysis and life cycle assessment of wheat production in Iran. *African Journal of Agricultural Research, 8*(18), 1929-1939. http://dx.doi.org/10.5897/AJAR11.1197

Hillier, J., Hawes, C., Squire, G., Hilton, A., Wale, S., & Smith, P. (2011). The carbon footprints of food crop production. *International Journal of Agricultural Sustainability, 7*(2), 107-118. http://dx.doi.org/10.3763/IJAS.2009.0419

Mohammadi, A., Tabatabaeefar, A., Shahan, S., Rafiee, S., & Keyhani, A. (2008). Energy use and economic analysis of potato production in Iran a case study: Ardabil Province. *Energy Conversion and Management, 49*, 3566-3570. http://dx.doi.org/10.1016/j.enconman.2008.07.003

Mousavi-Avval, S., Rafiee, H., Jafari, S., & Mohammadi, A. (2012). Optimization of energy consumption for soybean production using Data Envelopment Analysis (DEA) approach. *Applied Energy, 88*(11), 3765-3772. http://dx.doi.org/10.1016/j.apenergy.2011.04.021

Thornton, P. (2012). Recalibrating Food Production in the Developing World: Global Warming Will Change More Than Just the Climate. *CCAFS Policy Brief* (No. 6). CGIAR Research Program on Climate Change, Agriculture and Food Security.

Vermeulen, S. J., Campbell, B. M., & Ingram, J. S. I. (2012). Climate change and food systems. *Annual Review of Environment and Resources, 37*, 195-222. http://dx.doi.org/10.1146/annurev-environ-020411-130608

Notes

Note 1. El Hajeb province is administratively divided into 16 districts, of which one district is also named El Hajeb.

Note 2. Based on information gathered in the field and through interviews with key officials within existing public extension services.

17

Lipometabolic Alteration in Mice Feeding Eatable Tissues of Chinese Mitten Crab

Jun Jing[1,†], Wenhui Wu[1,2,†], Xinfeng Xiao[1], Yu Zhou[1], Xiaoyu Wang[1], Shangqiao Chen[1], Shujun Wang[2], Yongxu Cheng[3], Xugan Wu[3] & Bin Bao[1]

[1] Department of Marine Pharmacology, College of Food Science and Technology, Shanghai Ocean University, Shanghai, China

[2] Huaihai Institute of Technology, Lianyungang, China

[3] Key Laboratory of Exploration and Utilization of Aquatic Genetic Resources, Shanghai Ocean University, Shanghai, China

Correspondence: Bin Bao, Department of Marine Pharmacology, College of Food Science and Technology, Shanghai Ocean University, Shanghai 201306, China. E-mail: bbao@shou.edu.cn

Xugan Wu, Key Laboratory of Exploration and Utilization of Aquatic Genetic Resources, Shanghai Ocean University, Shanghai 201306, China. E-mail: xgwu@shou.edu.cn

† *These authors contributed equally to this research.*

Abstract

Objective: Chinese mitten crab is a famous aquatic species in eastern Asian region, but their edible parts, particularly hepatopancreas and gonads, generally contain very high levels of lipids that may have negative effects on human health. This study investigated the effects of different edible parts of Chinese mitten crab on the body weight and lip metabolism for Kunming mice.

Method: The mice were fed with diets containing one part of an Chinese mitten crab or the mixture of parts of an Chinese mitten crab for 4 weeks. There were 9 treatments. The triacylglycerol (TG), total cholesterol (TC), high-density lipoprotein cholesterol (HDL-C) and low density lipoprotein cholesterol (LDL-C) were enzymatically determined using commercial kits (purchased from Nanjing Jiancheng Bioengineering Institute, China). The arteriosclerosis index (AI) was calculated by the equation: AI = (TC – HDL-C)/HDL-C. The levels of fatty acid syntheses (FAS), the 3-hydroxy-3-methyl-glutaryl-coenzyme A reductase (HMG-CoA) and lipoprotein lipase (LPL) were measured using commercially available kits according to the manufacturer's instructions. The significant differences between the groups were further analyzed by Bonferronis's t-test.

Results: Our results showed that the crab hepatopancreas, gonads and the mixed male crab-edible parts increased blood lipids in some experiment group of mice corresponding to a change in the nutrition-related liver enzymes. It shows that addition of the Chinese mitten crab has an adverse effect on the blood lipid levels in mice. The FFH, FFMI and FMMI groups had significantly higher weight than the FN group ($P < 0.05$). The crab hepatopancreas, crab gonads and the mixed male crab-edible parts cause an increase in the blood lipid levels. The crab mixture significantly affected the AI value of male and female mice ($P < 0.01$). The level of FMMI group was significantly higher than the FN group ($P < 0.05$). Other groups showed no significant difference. The level of the FFMI group was significantly lower than the FN group ($P < 0.05$), and levels in the MMM and MFMI groups were significantly lower than the MN group ($P < 0.05$).

Conclusion: It clearly showed that long-term feeding with the Chinese mitten crab has an adverse effect on the blood lipid levels in mice. One the one hand, the weight, liver index and fat index of experimental mice were changed than normal mice. On the other hand, the crab diet affects the level of TC, TG, AI and FASN on increasing. It is suggested that the special diet has affected lip metabolic alteration associated with contents of serum lipids and metabolic enzymes. But according to a certain regular feeding, there would be no adverse effect on mice. On the contrary, it may adjust the blood lipid in mice

Keywords: blood lipid, Chinese mitten crab, liver enzyme, mice

1. Introduction

Chinese mitten crab has delicious taste with a unique and pleasant aroma. It also has good nutritive value including high levels of fatty acids and amino acids (Guo, Gu, Wang, Zhao, & Zheng, 2014). There are three quality ranks to evaluate the value of a Chinese mitten crab according to Chinese National Standard GB/T 19957-2005, and they are separated mainly based on weight [General Administration of Quality Supervision, Inspection and Quarantine of the People's Republic of China (2005) GB/T 19957-2005 Product of geographical indication-Yangcheng Lake Chinese mitten crab. Standards Press of China, Beijing]. These include special, first, and second classes (Gs, G1, and G2), for which male/female crabs should weigh over 200/150 g, 150/125 g, and 125/100 g, respectively. They differ in nutrition contents and flavors.

Table 1. Proximate composition (% of wet weight) in hepatopancreas, mature gonads and muscle of adult Chinese mitten crab (Wu, 2007)

	Yield (%)	Moisture (%)	Protein (%)	Total lipid (%)	Ash (%)
Male crab muscle	23-28	75-80	17-20	0.8-1.2	1.2-1.5
Male crab hepatopancreas	3.5-6	52-65	9-14	9-16	1.6-2.3
Male crab gonads	2.5-4.5	70-73	16-19	1.5-2	2.4-2.8
Female crab muscle	21-25	72-78	18-20	1.5-2.5	2.1-2.4
Female crab hepatopancreas	4.0-7.5	37-48	10-17	17-36	1.4-2
Female crab gonads	8.0-13.0	40-50	30-35	16-19	2-2.5

In general, Chinese mitten crab contains 18.9% crude protein, and about 80% of the protein is in the crab meat portion. About 90% of the fat is in the viscera (Chen, Zhang, & Shrestha, 2007) (Table 1). The Chinese mitten crab has an elevate highly unsaturated fatty acid (HUFA) levels in their gonad and hepatopancreas (Guo, Gu, Wang, Zhao, & Zheng, 2014; Wu et al., 2007). It has been reported that fatty acid profiles, especially of essential fatty acids (EFAs), are closely related to nutrient quality. The EFAs consist of a-linolenic acid (ALA, 18:3n-3) and linoleic acid (LA, 18:2n-6). ALA is a precursor of n-3 fatty acids, including eicosapentaenoic acid (EPA, 20:5n-3) and docosahexaenoic acid (DHA, 22:6n-3), while LA is a precursor of arachidonic acid (AA, 20:4n-6) (Gil, 2002).

AA and DHA are major components of cell membrane phospholipids and abundant in the central nervous system (He et al., 2014; Singh, 2005). EPA and DHA are important in infant brain growth (Wu et al., 2007) and have cardio-protective (Lee, O'Keefe, Lavie, Marchioli, & Harris, 2008; Ross, Lombardo, & Chicco, 2010) and anti-cancer (Alberts & Greenspan, 1984; Donaldson, 2004) properties. Balancing the nutritive material with risk is important.Crab meat is also an excellent source of minerals, particularly calcium, iron, zinc, potassium and phosphorus (Naczk, Williams, Brennan, Liyanapathirana, & Shahidi, 2004).

One study showed that the feeding gonad and hepatopancreas of Chinese mitten crab enriches fat in mice quickly with normal blood lipids (Su, Li, Ouyang, & Liu, 1995). However, high levels of fat intake may promote pathogenesis of many diseases including Crohn's disease (Calder, 2006) and inflammatory diseases (Gil, 2002). This reports suggests that eating a proper amount of gonad and hepatopancreas of Chinese mitten crab can play an important role in nutrition and biological activities.

The aim of this study was to measure the effect of eating crab gonad and hepatopancreas on blood lipid levels. We will study the weight-gaining effect of Chinese mitten crab consumption and thus determine whether the harm of crab-eating outweighs its merits. The study was 4 weeks and we use serum biochemical indices in mice.

2. Materials and Methods

These studies were approved by the university committee for animal experiments.All the experiments followed Chinese legislations on the use and care of laboratory animals.

2.1 Animals

Healthy Kunming specie mice, weighing 18-22 g, were purchased from shanghai Slac Laboratory Animal (SCXK2012-0002). The animals were maintained under standard conditions (12 h day/night cycle, 22±2 °C, 50%-60% humidity) with free access to food and tap water. Mice body weights were recorded weekly. Animals were allowed to acclimate to the environment for at least 1 week before use in the described experiments. Then the female mice were randomly divided into nine groups (n = 10), each group includes 10 mice; Group 1: The normal control group (FN) was fed a normal diet (composed of wheat (30%), ginglly oil cake (25%), black gram husk (29%), soybean meal (15%) and mineral mixture (10%)). Group 2: The FFG group was fed female crab gonads (showing in Table 2, following same) with a normal diet. Group 3: The FMG group was fed male crab gonads with a normal diet. Group 4: The FFM group was fed female crab muscle with a normal diet. Group 5: The FMM group was fed male crab muscle with a normal diet. Group 6: The FFH group was fed female crab hepato-pancreas with a normal diet. Group 7: The FMH group was fed male crab hepatopancreas with a normal diet. Group 8: The FFMI group was fed with a mixture of edible parts from female crab with the normal diet. Group 9: The FMMI group was fed a mixture of edible parts from male crab with a normal diet.The male mice were used and they were under the same treatment. The experimental diet was lasted for 4 weeks.

Table 2. Each group mice feed formulation

Groups	Normal diet (%)	Male crab hepatopancreas (%)	Male crab muscle (%)	Male crab gonads (%)	Female crab hepatopancreas (%)	Female crab muscle (%)	Female crab gonads (%)
MMMI FMMI	95.18	0.73	3.65	0.44	0	0	0
FMH MMH	99.27	0.73	0	0	0	0	0
FMM MMM	96.35	0	3.65	0	0	0	0
FMG MMG	99.56	0	0	0.44	0	0	0
MN	100	0	0	0	0	0	0
MFMI FFMI	95.82	0	0	0	0.63	2.4	1.15
FFH FMH	99.37	0	0	0	0.63	0	0
FFM MFM	97.6	0	0	0	0	2.4	0
FMG FMG	98.85	0	0	0	0	0	1.15
FN	100	0	0	0	0	0	0

Then the mice were sacrificed, and the liver were excised, weighed, and then homogenized for enzymatic analysis.

2.2 Estimation of Plasma Lipid Profile

At the end of the experiment, animals were fasted overnight (14 h) and euthanized under diethyl ether anesthesia in the morning by withdrawing blood from the abdominal vena using a vacuum tube. The blood was clotted and plasma was harvested by centrifugation at 4 °C (1800 ×g, 10 min). The triacylglycerol (TG), total cholesterol (TC), high-density lipoprotein cholesterol (HDL-C) and low density lipoprotein cholesterol (LDL-C) were enzymatically determined using commercial kits (purchased from Nanjing Jiancheng Bioengineering Institute, China) with Beckman coulter chemistry analyzer AU5800 Series. The arteriosclerosis index (AI) was calculated by the equation: AI = (TC – HDL-C)/HDL-C.

2.3 Enzymatic

The livers were quickly removed and homogenized in 50mM Tris-HCl, pH 7.4 (1/10, w/v). The homogenate was centrifuged at 4 °C (2400 ×g, 15 min). The levels of FAS, HMG-CoA and LPL were measured using commercially available kits according to the manufacturer's instructions.

2.4 Statistical Analysis

All data were represented as mean±SD from 10 samples per group. The mean values were statistically analyzed using one-way analysis of variance (ANOVA). The significant differences between the groups were further analyzed by Bonferronis's t-test. Analyses were performed using the SPSS 19.0 software. *P* values less than 0.05 were considered as significant. Plots were made with package ggplot2 (Wickham, 2009) of R (Team, 2014).

3. Results and Discussion

3.1 Effect of Chinese Mitten Crab on Mice Weight

Body weight and liver tissue weight for each experimental diet group are shown in Tables 3 and 4. In female mice, Chinese mitten crab consumption led to a higher body weight veraus normal control diets. This was true both for FFH, FFMI and FMMI groups—they had significantly higher body weight than the FN group ($P < 0.05$).

All female mice groups had significantly higher body weight gain than the FN group except the FFG group ($P < 0.05$). However, there was no significant difference in the liver tissue weights ($P > 0.05$).

Table 3. Effect of Chinese mitten crab on growth indicators of female mice

Group	Initial body weight(g)	Final body weight (g)	Body weight gain (g)	Liver tissue weights (g/100 g)
FN	24.38±1.18	39.76±5.34a	15.38±2.34a	4.78±0.50
FFG	24.28±1.82	42.32±3.46ab	18.04±2.13ab	4.70±0.51
FMG	23.70±1.28	43.07±3.11ab	19.23±2.23b	4.79±0.48
FFM	24.72±2.74	43.97±2.44ab	19.25±1.21b	4.55±0.37
FMM	24.34±1.89	43.76±2.19ab	19.42±1.54b	5.03±0.54
FFH	24.66±1.12	45.57±2.59b	20.91±1.86b	4.73±0.62
FMH	23.90±1.55	44.57±2.78ab	20.67±2.13b	4.6±0..47
FFMI	24.64±0.72	46.00±2.73b	21.36±1.34b	4.72±0.38
FMMI	24.48±1.97	45.76±3.14b	21.28±1.24b	4.91±0.32
MMMI	23.54±1.44	37.27±4.48	13.73±2.38	5.61±0.56b

In male mice, Chinese mitten crab consumption had only a minimal effect on weight. Both the final body weight and the body weight gain showed no significant difference between the groups ($P > 0.05$). Liver tissue weights, however, were significantly higher in the MFH, MFMI and MMMI groups than in the MN group ($P < 0.05$).

Table 4. Effect of Chinese mitten crab on growth indicators of male mice

Group	Initial body weight (g)	Final body weight (g)	Body weight gain (g)	Liver tissue weights (g/100 g)
MN	23.76±1.09	33.31±4.53	9.56±2.39	4.98±0.28ab
MFG	24.20±0.84	32.06±4.34	7.86±3.46	6.22±0.32a
MMG	24.24±1.18	33.58±2.67	9.34±2.13	5.01±0.57ab
MFM	24.00±0.77	32.79±4.11	8.79±2.61	5.49±0.55a
MMM	23.60±1.05	32.92±3.87	9.32±2.42	5.32±0.43ab
MFH	23.60±1.50	35.69±4.13	12.09±2.18	5.75±0.66ab
MMH	24.11±1.09	34.71±3.71	10.6±2.27	5.20±0.58ab
MFMI	23.80±1.43	37.43±4.05	13.63±2.32	6.20±0.35b
MMMI	23.54±1.44	37.27±4.48	13.73±2.38	5.61±0.56b

3.2 Effect of Chinese Mitten Crab on TC, TG, HDL-C, LDL-C, and AI

The levels of TC, TG, HDL-C and LDL-C from treatment plasma were displaying in the part of Tables 5 and 6.

Total cholesterol level in the FFG, FFMI, and FMMI groups were significantly higher than that of the FN group ($P < 0.05$). Compared to the normal control group, TC levels in the FMG and FFH groups were elevated, and FFM, FMM, and FMH groups were reduced but not significantly ($P > 0.05$). The triglyceride level in the FFM, FMM, FFH and FMMI groups were significantly higher than that of the normal control group ($P < 0.05$); other groups were elevated, but these differences were not significant ($P > 0.05$). Chinese mitten crab does not affect the levels of HDL-C and LDL-C. All groups were normal.

The total cholesterol level in the MFG, MMG, MFH and MMMI groups were significantly higher than that in the normal control group ($P < 0.05$). Furthermore, MFMI group was markedly increased versus the normal control group ($P < 0.01$). Except for the MMM group, other groups were higher than the normal control group. However, these differences were not significant ($P > 0.05$). In terms of the triglyceride level, the MFG and MFMI groups were significantly higher than the normal control group ($P < 0.05$). The HDL-C and LDL-C level in the MFG, MMG, MFH, and MFMI were significantly higher than the normal control group ($P < 0.05$). The MFH group was very significant ($P < 0.01$). However, the MMM and MFM groups were significantly lower than the normal control group ($P < 0.05$). The LDL-C level in the MFMI group was significantly higher than the normal control group ($P < 0.05$).

The total cholesterol was inclined to elevate alteration in male or female mice treatment because the total cholesterol of main resource is diet including reach cholesterol (Leontowicz, 2011). The triglyceride level of plasma was raising for hyperlipidemia in few treatment groups from the male or female groups. The alteration of HDL-C and LDL-C level was observed in male treatment group. This is consistent with the general principle (Rony, 2014). The eatable tissue of hepatopancreas or gonads was added to the diet which were can raising the lipometabolite level than muscle including lowly lipids amount in mice treatment groups. The lipometabolete level of male mice than the female mice was easily affected by hepatopancreas or gonads or mixed eatable tissue of Chinese mitten crab.

The AI level in the FFMI and FMMI groups were significantly higher than the normal control group ($P < 0.01$). Other groups were higher than the normal control group, but these differences were not significant ($P > 0.05$).

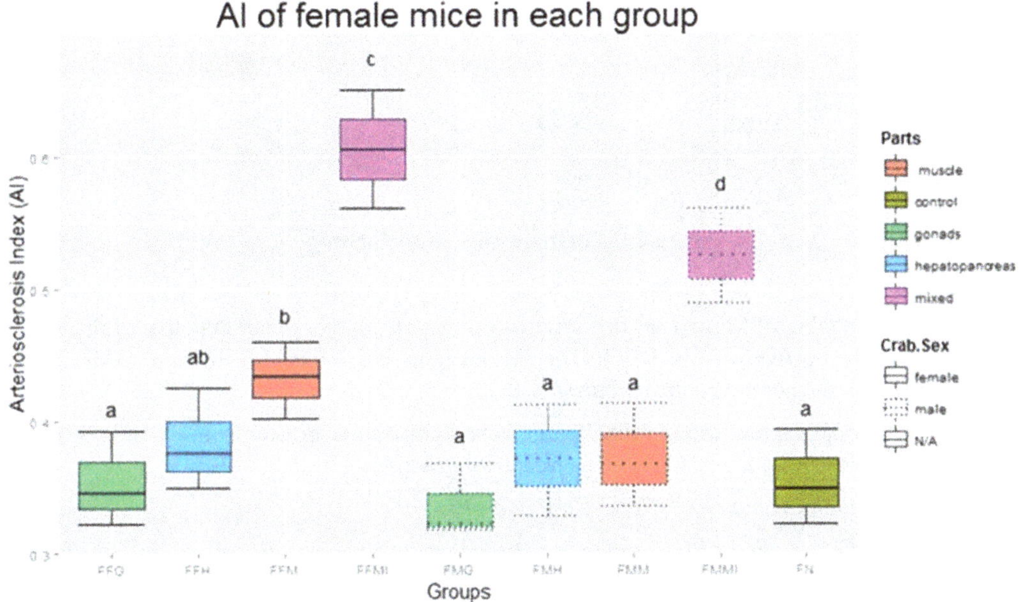

Figure 1. AI of female mice in each group

The AI level (Figure 1) in the MFG, MMM, MMH, MFMI and MMMI groups were significantly greater than that of the normal control group ($P < 0.01$). Other groups were higher than that of the normal control group, but these differences were not significant ($P > 0.05$).

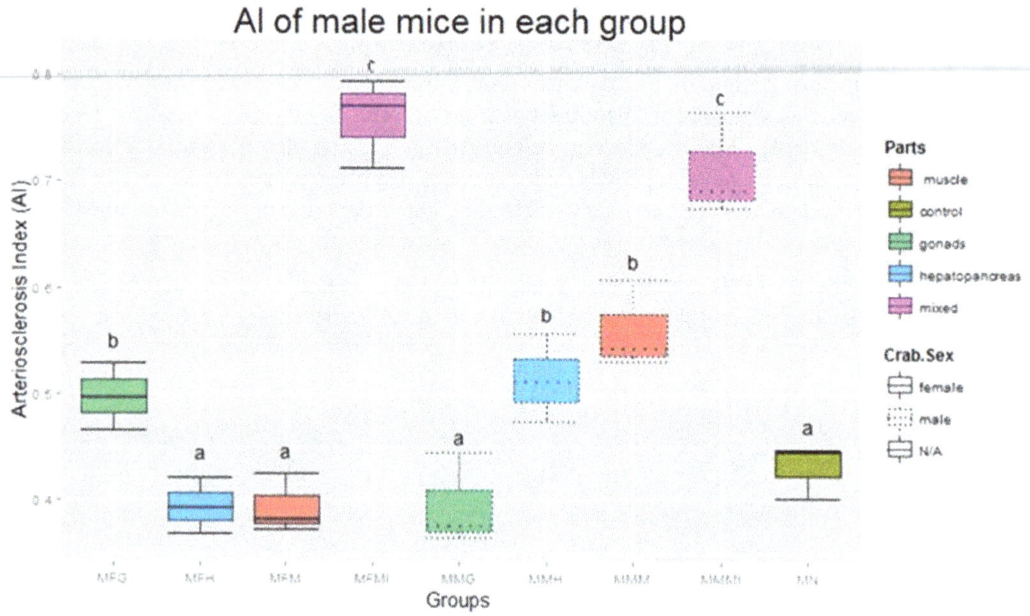

Figure 2. AI of male mice in each group

The AI level (Figure 1) in the MFMI and MMMI groups were significantly greater than that of the MN group by 75.93% and 64.54%, respectively ($P < 0.01$). The AI level in the MFG, MMM and MMH groups were significantly greater than the normal control group ($P < 0.05$).

The AI of treatment groups as few eating mixed parts were significantly greater in male mice or female mice. The edible gonads have caused a lowly AI value on treatment groups.

3.3 Effect of Chinese Mitten Crab on FAS, HMG-CoA, and LPL

The levels of FAS, HMG-CoA, and LPL from the liver tissue were displaying as well in the part of Tables 5 and 6.

Table 5. Effect of Chinese mitten crab on FAS, HMG-CoA, LPL, TC, TG, HDL-C and LDL-C of female mice

Group	FAS (nmol/g protein)	HMG-CoA reductase (ng/g protein)	LPL (U/g protein)	TC (mmol/L)	TG (mmol/L)	HDL-C (mmol/L)	LDL-C (mmol/L)
FN	2.93±0.07cd	116.88±6.82abcd	70.15±3.36ce	1.64±0.12abc	0.85±0.10a	1.17±0.12ab	0.34±0.08abc
FFG	2.97±0.22cd	109.08±3.59ab	63.61±7.31abcd	1.99±0.12de	0.85±0.07a	1.33±0.16b	0.42±0.07bc
FMG	2.87±0.15c	113.47±5.40abc	66.71±5.53bcd	1.76±0.20de	0.98±0.11ab	1.28±0.27ab	0.43±0.06bc
FFM	2.64±0.14abc	120.43±2.01cd	67.89±3.59bcd	1.41±0.23a	1.20±0.13b	1.02±0.22ab	0.31±0.08ab
FMM	2.52±0.14ab	119.95±7.36cd	64.75±4.35abcd	1.48±0.19ab	1.21±0.19b	0.96±0.20a	0.31±0.06ab
FFH	2.81±0.22bc	126.58±5.28d	60.61±5.03ab	1.88±0.16cde	1.07±0.20ab	1.35±0.14b	0.45±0.05c
FMH	2.46±0.21q	118.36±2.56bcd	61.46±2.04abc	1.54±0.14cde	1.01±0.10ab	1.02±0.12ab	0.31±0.06ab
FFMI	3.21±0.25de	107.21±5.35a	56.58±5.18a	2.06±0.11de	1.05±0.09ab	1.10±0.11ab	0.28±0.05a
FMMI	3.44±0.21e	115.34±7.91abc	70.54±2.17d	2.10±0.21e	1.16±0.09b	1.19±0.19ab	0.29±0.07a

The FAS of FMMI group (Table 5) was significantly higher than the FN group ($P < 0.05$). In the male mice (Table 6), the level of FAS in MFG, MMG, MFH, MFMI and MMMI groups were significantly higher than that of the MN group ($P < 0.05$). Other groups showed no significant difference. The triglyceride stored in adipose tissue originates either from the esterification of FFAs provided mainly from the diet or from denovo synthesis. The activity of the lipogenic pathway in adipose tissue is highly dependent on nutritional conditions (Rossi, 2010). Fatty acid synthetase (FAS) is the designation given to the enzyme system that catalyzes the synthesis of long-chain fatty acids from a short-chain acyl-CoA primer (Alberts & Greenspan, 1984).

Table 6. Effect of Chinese mitten crab on FAS, HMG-CoA, LPL, TC, TG, HDL-C and LDL-C of male mice

Group	FAS (nmol/g protein)	HMG-CoA reductase (ng/g protein)	LPL (U/g protein)	TC (mmol/L)	TG (mmol/L)	HDL-C (mmol/L)	LDL-C (mmol/L)
MN	2.91±0.09cde	113.51±5.89ab	65.25±8.27ab	1.65±0.22abc	0.83±0.16abc	1.25±0.12b	0.27±0.06a
MFG	3.16±0.38def	118.64±5.64b	60.24±7.21ab	1.95±0.18cd	1.27±0.11e	1.45±0.08c	0.33±0.04ab
MMG	2.76±0.06bcd	108.30±9.05ab	64.87±3.25ab	1.89±0.23bcd	1.02±0.17cde	1.43±0.12c	0.29±0.06ab
MFM	2.38±0.11b	104.99±9.98ab	60.25±2.11ab	1.59±0.13ab	0.73±0.09ab	0.94±0.07a	0.34±0.10ab
MMM	2.36±0.34b	102.73±8.27a	55.98±4.25a	1.34±0.13a	0.62±0.17a	0.88±0.06a	0.34±0.04ab
MFH	2.70±0.28bc	102.10±6.59a	65.26±3.14ab	1.98±0.20cd	0.94±0.17bcd	1.64±0.09d	0.30±0.09ab
MMH	1.92±0.23a	103.17±4.83a	58.77±6.79ab	1.81±0.16bcd	0.70±0.16ab	1.22±0.11b	0.28±0.05ab
MFMI	3.45±0.14f	118.72±5.20b	55.98±4.25a	2.47±0.13e	1.13±0.22de	1.62±0.09d	0.41±0.04b
MMMI	3.29 ±0.12ef	108.17±7.81ab	66.90±2.53b	1.99±0.17d	1.02±0.06cde	1.24±0.11b	0.35±0.09ab

As for HMG-CoA, the level of this reductase in MFG and MFMI were significantly higher than that in the MMM, MFH, MMH groups ($P < 0.05$). However, both female and male mice belong to treatment groups showed no significant difference compared with the normal group ($P > 0.05$). People with higher levels of hydroxymethylglutaryl coenzyme A (HMG-CoA) reductase expression, as indicated by their serum mevalonate or cholesterol levels, appear to be more resistant to dietary cholesterol (Ness, 2015). Cholesterol biosynthesis localized in the endoplasmic reticulum (ER) starts with the rate limiting enzyme (HMG-CoA) reductase whose activity is strongly controlled by several feed-back mechanisms involving endogenous pathways and exogenous cholesterol intake by nutrition (Reynolds et al., 1984).

The level of LPL in the FFMI group was significantly lower than that in the FN group ($P < 0.05$); levels in the MMM and MFMI groups was significantly lower than that in the MN group ($P < 0.05$).The lipoprotein lipase (LPL) plays an important role in lipid metabolism by hydrolyzing triglycerides in chylomicrons and very low density lipoproteins (Goldberg, 1996). An increasing number of studies have suggested an association of LPL gene variants with the risk of cardiovascular and cerebrovascular diseases (Munshi et al., 2012). Several structural mutations in the LPL gene have been documented (Mead, Irvine, & Ramji, 2002). These have been associated with various lipid traits like hypertriglyceridemia and reduced HDL levels (two polymorphisms in the human lipoprotein lipase (LPL) gene).

4. Conclusion

The lipids contents are different in the eatable tissues like gonads, muscle, hepatopancreas which were associated with male and female of Chinese mitten crab (Wu, 2007). The Different lipids contents as an eatable tissues maker of diets were fed with 8 experiment groups including of FFG, FMG, FFM, FMM, FFH, FMH, FFMI and FMMI for female mice. The similar experiment design is in male mice. The list of highest to lowest on the lipids content about diets were postfix MI group, postfix G group, postfix H group and postfix M group (data no showing). The special diet containing of eatable tissues of Chinese mitten crab was affected on the lipometabolite showing in the blood biochemistry index and metabolism enzyme of mice.

Chinese mitten crab was including to the diet has an adverse effect on the blood lipid levels in mice. (1) Mice feed the diet containing eatable tissue of Chinese mitten crab which the body weight was more increased. The FFH, FFMI and FMMI groups had significantly higher body weight than the FN group ($P < 0.05$); (2) the crab-edible parts was included in the special diet had a greater adverse impacts on the male mice. The crab hepatopancreas, crab gonads and the mixed male crab-edible parts caused an increase in the blood lipid levels. (3) The eatable Crab part increased the lipid index in male and female mice. The crab mixture significantly affected the AI value of male and female mice ($P < 0.01$). (4) The level of FAS in MFG, MMG, MFH, MFMI and MMMI groups were significantly higher than that of the MN group ($P < 0.05$). The level of FMMI group was significantly higher than the FN group ($P < 0.05$). Other groups showed no significant difference. (5) The level of of LPL in the FFMI group was significantly lower than the FN group ($P < 0.05$), and levels of LPL in the MMM and MFMI groups were significantly lower than the MN group ($P < 0.05$).

Hyperlipidemia is a major cause of build-up in coronary atherosclerosis (Steinberg, 2005). Whether TC or TG levels increased, or both increased, they are referred to as hyperlipidemia. We found that feeding mice with diets rich in lipids resulted in increased TC, TG and LDL cholesterol levels. The results was in accordance with the the report belong Engelking. Fatty acid synthetase (FAS) is the designation given to the enzyme system which catalyzes the synthesis of long-chain fatty acids from a short-chain acyl-CoA primer (Cole & Kramer, 2016;

Watkins, 2013). The concentration of FAS enzymes in liver affects the concentration of TC and TG in plasma (Myant, 1990). Feeding with the Chinese mitten crab may increase the FAS enzyme level to increase TC and TG. Moreover, HDL is directly anti-androgenic and it is believed to remove cholesterol from the developing lesions. LDL is a risk factor and plays a role in several steps of atherosclerosis.

However, to understand the implications of these findings on humans, we need further research to understand other potential health hazards.

References

Alberts, A. W., & Greenspan, M. D. (1984). Chapter 2: Animal and bacterial fatty acid synthetase: Structure, function and regulation. In N. Shosaku (Ed.), *New Comprehensive Biochemistry* (pp. 29-58). Elsevier. https://doi.org/10.1016/s0167-7306(08)60120-9

Calder, P. C. (2013). Chapter 4: Omega-6 and Omega-3 Polyunsaturated Fatty Acids and Inflammatory Bowel Diseases. In R. R. W. R. Preedy (Ed.), *Bioactive Food as Dietary Interventions for Liver and Gastrointestinal Disease* (pp. 55-79). Academic Press, San Diego. https://doi.org/10.1016/b978-0-12-397154-8.00014-2

Chen, D.-W., Zhang, M., & Shrestha, S. (2007). Compositional characteristics and nutritional quality of Chinese mitten crab (*Eriocheir sinensis*). *Food Chemistry, 103*, 1343-1349. https://doi.org/10.1016/j.foodchem.2006.10.047

Cole, L., & Kramer, P. R. (2016). Chapter 1.3: Sugars, Fatty Acids, and Energy Biochemistry. *Human Physiology, Biochemistry and Basic Medicine* (pp. 17-30). Academic Press, Boston. https://doi.org/10.1016/b978-0-12-803699-0.00019-0

De la Cruz-García, C., López-Hernández, J., González-Castro, M. J., Rodríguez-Bernaldo De Quirós, A. I., & Simal-Lozano, J. (2000). Protein, amino acid and fatty acid contents in raw and canned sea urchin (*Paracentrotus lividus*) harvested in Galicia (NW Spain). *Journal of the Science of Food and Agriculture, 80*, 1189-1192. https://doi.org/10.1002/1097-0010(200006)80:8%3C1189::AID-JSFA618%3E3.0.CO;2-7

Donaldson, M. S. (2004). Nutrition and cancer: A review of the evidence for an anti-cancer diet. *Nutrition Journal, 3*, 19-21. https://doi.org/10.1186/1475-2891-3-19

Engelking, L. R. (2015). Chapter 65: VLDL, IDL, and LDL. *Textbook of Veterinary Physiological Chemistry* (3rd ed., pp. 416-420). Academic Press, Boston. https://doi.org/10.1016/b978-0-12-391909-0.50065-7

Gil, Á. (2002). Polyunsaturated fatty acids and inflammatory diseases. *Biomedicine & Pharmacotherapy, 56*, 388-396. https://doi.org/10.1016/s0753-3322(02)00256-1

Goldberg, I. J. (1996). Lipoprotein lipase and lipolysis: Central roles in lipoprotein metabolism and atherogenesis. *Journal of Lipid Research, 37*, 693-707.

Guo, Y.-R., Gu, S.-Q., Wang, X.-C., Zhao, L.-M., & Zheng, J.-Y. (2014). Comparison of fatty acid and amino acid profiles of steamed Chinese mitten crab. *Fisheries Science, 80*, 621-633. https://doi.org/10.1007/s12562-014-0738-6

He, J., Wu, X., Li, J., Huang, Q., Huang, Z., & Cheng, Y. (2014). Comparison of the culture performance and profitability of wild-caught and captive pond-reared Chinese mitten crab (*Eriocheir sinensis*) juveniles reared in grow-out ponds: Implications for seed selection and genetic selection programs. *Aquaculture, 434*, 48-56. https://doi.org/10.1016/j.aquaculture.2014.07.022

Kong, L., Cai, C., Ye, Y., Chen, D., Wu, P., Li, E., & Song, L. (2012). Comparison of non-volatile compounds and sensory characteristics of Chinese mitten crabs (*Eriocheir sinensis*) reared in lakes and ponds: Potential environmental factors. *Aquaculture, 364*, 96-102. https://doi.org/10.1016/j.aquaculture.2012.08.008

Lee, J. H., O'Keefe, J. H., Lavie, C. J., Marchioli, R., & Harris, W. S. (2008). Omega-3 Fatty Acids for Cardioprotection. *Mayo Clinic Proceedings, 83*, 324-332. https://doi.org/10.4065/83.3.324

Leontowicz, H., Leontowicz, M., Jesion, I., Bielecki, W., Poovarodom, S., Vearasilp, S., ... Gorinstein, S. (2011). Positive effects of durian fruit at different stages of ripening on the hearts and livers of rats fed diets high in cholesterol. *European Journal of Integrative Medicine, 3*, 169-181. http://dx.doi.org/10.1016/j.eujim.2011.08.005

López-Miranda, J., Pérez-Martinez, P., & Pérez-Jiménez, F. (2006). Health benefits of monounsaturated fatty acids. *Improving the Fat Content of Foods* (pp. 71-106). https://doi.org/10.1533/9781845691073.1.71

McDonagh, J. E. R. (1966). The Essential Amino Acids. *Protein: The Basis of All Life* (pp. 15-18). https://doi.org/10.1016/b978-1-4831-8038-0.50013-7

Mead, J., Irvine, S., & Ramji, D. (2002). Lipoprotein lipase: Structure, function, regulation, and role in disease. *J. Mol. Med., 80*, 753-769. https://doi.org/10.1007/s00109-002-0384-9

Munshi, A., Babu, M. S., Kaul, S., Rajeshwar, K., Balakrishna, N., & Jyothy, A. (2012). Association of LPL gene variant and LDL, HDL, VLDL cholesterol and triglyceride levels with ischemic stroke and its subtypes. *Journal of the Neurological Sciences, 318*, 51-54. https://doi.org/10.1016/j.jns.2012.04.006

Myant, N. B. (1990). *LDL: Origin and Metabolism, Cholesterol Metabolism, Ldl, and the Ldl Receptor* (Chapter 7, pp. 184-232). Academic Press. https://doi.org/10.1016/b978-0-12-512300-6.50012-0

Naczk, M., Williams, J., Brennan, K., Liyanapathirana, C., & Shahidi, F. (2004). Compositional characteristics of green crab (*Carcinus maenas*). *Food Chemistry, 88*, 429-434. https://doi.org/10.1016/j.foodchem.2004.01.056

Ness, G. C. (2015). Physiological feedback regulation of cholesterol biosynthesis: Role of translational control of hepatic HMG-CoA reductase and possible involvement of oxylanosterols. *Biochimica et Biophysica Acta (BBA): Molecular and Cell Biology of Lipids, 1851*, 667-673. https://doi.org/10.1016/j.bbalip.2015.02.008

Reynolds, G. A., Basu, S. K., Osborne, T. F., Chin, D. J., Gil, G., Brown, M. S., ... Luskey, K. L. (1984). HMG CoA reductase: A negatively regulated gene with unusual promoter and 5′ untranslated regions. *Cell, 38*, 275-285. https://doi.org/10.1016/0092-8674(84)90549-x

Rony, K. A., Ajith, T. A., Nima, N., & Janardhanan, K. K. (2014). Hypolipidemic activity of Phellinus rimosus against triton WR-1339 and high cholesterol diet induced hyperlipidemic rats. *Environmental Toxicology and Pharmacology, 37*, 482-492. http://dx.doi.org/10.1016/j.etap.2014.01.004

Rossi, A. S., Lombardo, Y. B., & Chicco, A. G. (2010). Lipogenic enzyme activities and glucose uptake in fat tissue of dyslipemic, insulin-resistant rats: Effects of fish oil. *Nutrition, 26*, 209-217. https://doi.org/10.1016/j.nut.2009.04.006

Singh, M. (2005). Essential fatty acids, DHA and human brain. *The Indian Journal of Pediatrics, 72*, 239-242. https://doi.org/10.1007/bf02859265

Steinberg, D. (2005). Thematic review series: The Pathogenesis of Atherosclerosis. An interpretive history of the cholesterol controversy, part III: Mechanistically defining the role of hyperlipidemia. *Journal of Lipid Research, 46*, 2037-2051. https://doi.org/10.1194/jlr.R500010-JLR200

Su, X., Li, T., Ouyang, F., & Liu, P. (1995). Study on the nutritive compositions of Portunus trituberculatus. *[Ying Yang Xue Bao] Acta nutrimenta Sinica, 18*, 342-346.

Team, R. C. (2014). *R: A language and environment for statistical computing.* R Foundation for Statistical Computing, Vienna, Austria.

Vilasoa-Martínez, M., López-Hernández, J., & Lage-Yusty, M. A. (2007). Protein and amino acid contents in the crab, *Chionoecetes opilio*. *Food Chemistry, 103*, 1330-1336. https://doi.org/10.1016/j.foodchem.2006.10.045

Watkins, P. A. (2013). Fatty Acyl-CoA Synthetases A2. In W. J. Lennarz & M. D. Lane (Eds.), *Encyclopedia of Biological Chemistry* (pp. 290-295). https://doi.org/10.1016/B978-0-12-378630-2.00100-6

Wickham, H. (2009). *ggplot2: Elegant graphics for data analysis.* Springer Science & Business Media. https://doi.org/10.1007/978-0-387-98141-3

Wu, X., Cheng, Y., Sui, L., Yang, X., Nan, T., & Wang, J. (2007). Biochemical composition of pond-reared and lake-stocked Chinese mitten crab *Eriocheir sinensis* (H. Milne-Edwards) broodstock. *Aquaculture Research, 38*, 1459-1467. https://doi.org/10.1111/j.1365-2109.2007.01728.x

18

Factors Affecting Farmer's Chemical Fertilizers Consumption and Water Pollution in Northeastern Iran

Hosein Mohammadi[1], Abdolhamid Moarefi Mohammadi[2] & Solmaz Nojavan[1]

[1] Agricultural Economics Department, Agriculture Faculty, Ferdowsi University of Mashhad, Iran

[2] Department of Economics, University of Isfahan, Isfahan, Iran

Correspondence: Hosein Mohammadi, Agricultural Economics Department, Agriculture Faculty, Ferdowsi University of Mashhad, Iran. E-mail: hoseinmohammadi@um.ac.ir

Abstract

Pollution by fertilizers containing nitrogen is one of the most significant sources of water pollution, and agriculture sector has a considerable share in this type of pollution. In this study, factors affecting the level of contamination of surface and underground water resources by agricultural activities were examined. Data of 254 wheat farmers in the plain of Mashhad in Khorasan Razavi province in Iran were used for investigating the effect of some explanatory variables on the level of water pollution utilizing Ordered Logit Regression Model. The results show that main activity of farmers, years of experience, the level of education, awareness of organic farming, level of income, price of fertilizers and irrigation method have significant effect on the amount of fertilizers utilized by farmers and hence the level of water resources pollution. Efforts for decreasing pollution of water resources require strategies such as changing the main activity of farmers, increase the cost of using chemical fertilizers, create economic incentives for organic farming, and increase general information and knowledge of farmers.

Keywords: agriculture, chemical fertilizers, environment, nitrogen, ordered logit regression, water pollution

1. Introduction

Population growth and increasing water demand in various sectors of agriculture, industry and drinking water, has caused great pressure on groundwater resources. More than one billion people in the world have limited access to safe water and the major source of 80% of water in developing countries is from poor quality and unhealthy water resources. Usually in developing countries, the priority is finding suitable and clean underground aquifers to supply water for drinking and agricultural activities, while in developed countries; more attention is drawn to the issue of water quality (Seldon & Song, 1994).

Agricultural activities might have negative impacts on the environment. Water pollution from agriculture is defined as non-point source of pollution. Actors involved are many and it is quite a complex matter to define how and when polluting agent enters the water bodies and who the polluters are (Semaan et al., 2007). In the last decades, concerns about groundwater pollution caused by excessive utilization of nitrogen fertilizers in agricultural areas have increased (de Paz & Ramos, 2004). Nutrient pollution is one of the most widespread, costly and challenging environmental problems, and is caused by excess nitrogen and phosphorus in the air and water. When too much nitrogen and phosphorus enters the environment, usually from a wide range of human activities, the air and water can be polluted (EPA, 2016). Collins et al. (2016) emphasize that controlling excessive emissions of pollutants into water and air is a major policy challenge in several countries.

Majority of fertilizers contain nitrogen (N) because nitrogen is an essential nutrient for the growth of plants and its deficiency in the soil can affect the growth of plants. Nevertheless, if these fertilizers are not properly use, N can enter through rainfall or irrigation into surface and groundwater resources and it can contaminate water resources. To maximize production and minimize the environmental damages resulting from the use of N, studying the factors affecting consumption of these fertilizers is necessary. Excess amount of nitrogen in ground and surface waters is a serious problem in intensive agriculture areas, largely due to poor efficiency in using N and this subject has several reasons, including not timing N application to crop requirements and excess N application. Chemical fertilizers producing industries that provide the nitrogen needed to grow plants in the form of ammonium (NH_4^+) or nitrate (NO_3^-) are the main sources that produce nitrogen pollutants. Another form of N is found in urea fertilizers which when used in the soil, is rapidly converted to ammonium by natural enzymes. Compound fertilizers containing N such as ammonium, nitrate and urea are known as quick-release fertilizers because in these fertilizers, N dissolves rapidly in water to be utilized for plants. Excess nitrogen in the environment can contributes to many health and environmental problems such as coastal dead zones and fish kills, groundwater pollution, reduced crop productivity and global climate change (UNEP, 2014). Because the rapid dissolution of N can lead to environmental problems, the use of N fertilizers should be at standard levels.

Some companies produce and market slow release or controlled released fertilizers that released N with delays. Nevertheless, N fertilizers with quick or controlled rate of dissolution can be equally harmful to the environment if not properly utilized (Shober, 2014).

Emission of nitrogen into water resources has been a continuous process in different parts of the world including Europe, despite different regulations that have been enacted. Part of this problem is as a result of lack of enforcement and implementation of laws and the main part of the problem is due to excessive pollutants entering the environment. In these regards, the agriculture sector has a significant contribution in reducing the quality of water resources and about 55% of the nitrogen entering into water resources in the Europe is due to activities of the agricultural sector (Bouraoui & Grizzetti, 2013).

Agriculture is the single largest user of freshwater resources, using a global average of 70% of all surface water supplies. Agriculture is both cause and victim of water pollution. It is a cause via its discharge of pollutants and it is a victim via the use of wastewater and polluted surface and groundwater, which contaminate crops and transmit disease to consumers and farm workers (FAO, 1996).

Groundwater resources are mainly polluted by nitrates. In several areas, the groundwater is polluted to an extent that it is no longer fit to be used as drinking water according to present standards. Agriculture also makes a substantial contribution to the total atmospheric nitrogen loading into the North and the Baltic Seas. This amounts to 65 and 55% respectively (FAO, 1996).

Due to increase in world population and change in eating habits, it is anticipated that until 2020, consumption of nitrogen and phosphorus fertilizers will reach up to $114 \times 10^6 tN$ and $21 \times 10^6 tP$ respectively (Bumb & Baanate, 1996).

Peña-Haro et al. (2010) by utilizing the hydro-economic modeling framework, consider the relationship between fertilizer application and concentration of nitration on the ground water resources at different control sites while maximizing net economic benefits. Their results demonstrate that a high fertilizer price would be needed to decrease nitrate concentrations in groundwater below the standard of 50 mg/l.

Between 1983 and 2005, the average usage of chemical fertilizers per hectare has increased from 169 to 390 kg and Two-thirds of these fertilizers are nitrogen-containing fertilizers. In this period, average cereal production has increased from 3.7 to 5.3 tons per hectare. The results of other studies had demonstrated a high positive correlation between annual production of food products and consumption of fertilizers in recent years (Zhu & Chen, 2002). Increasing use of fertilizers containing nitrogen imposed various costs on the society. In most fertile land, the rate at which nitrogenous fertilizers are consumed is very high and it not only increases production costs and decreases efficiency, but also has various negative impacts on air and water quality (Zhou et al., 2010).

In many fertile areas of Iran, annual consumption of nitrogen base fertilizers are more than 350 kg per hectare which is much more than standard levels (Mahvi et al., 2005). Khamadi et al. (2015) showed that increasing application of N fertilizers in wheat farms from 160 to 360 kg per hectare, would decrease the efficiency of nitrogen fertilizers in Iran. According to Wang et al. (2004) excessive application of N fertilizer for wheat production has resulted in environment pollution. These researchers found that the optimal N application rate is 180-225 kg N hm^{-2} for wheat which resulted in the highest yield with standard pollution. Sharasbi et al. (2016) showed that in Iran conditions, the maximum grain yield and biological yield were obtained in 150 and 225 kg N hm^{-2} under normal irrigation condition. Nevertheless, nitrogen application up to optimum level enhanced grain yield, yield components and water productivity, but as water stress intensified, the positive effect of nitrogen fertilizer was diminishes. Seyed Sharifi et al. (2016) also showed that in Iran, the maximum yield, yield components, rate and grain filling period for wheat belonged to application of 180 kg N hm^{-2}.

It seems that several factors affect the amount of nitrogen fertilizers used by farmers such that by identifying the most important and influential factors, we can offer suggestions for reducing nitrogen fertilizers application and hence reduce water resources pollution.

This study attempts to identify the factors influencing the consumption of nitrogen-containing fertilizers by farmers in the plains of Mashhad. These factors are those other than soil type, terrain, weather and factors that are beyond human controls and can affect the amount of nitrogen absorption on water resources. These factors include; main activity of farmers, farming experience, farmers education, farmers income, price of fertilizer, irrigation method, fertilizing method, awareness of organic farming and awareness of the dangers of water pollution. Mashhad, which is the top city in Khorasan Razavi province, has about 90,000 hectares of cultivation areas and around 55% of them are irrigated by water and 45% are rain-fed cultivation. Around 66% of cereal cultivation in Mashhad Plain is devoted to irrigated wheat (Ansari et al., 2014).

The N requirement depends on many factors such as soil type and the cultivars, the climate conditions, crop type and other conditions that they are beyond the objective of this research. Therefore, by selecting Mashhad plains that have similar conditions for soil and temperature and by selecting only single product, our objective was to investigate the effect of human factors on the use of nitrogen base fertilizers and therefore water pollution.

2. Methods

To analyze the factors affecting the level of water resources pollution caused by the amount of chemical fertilizers consumed by farmers, Ordered Logit Model was employed. OLM is based on latent variable and in this study; it is the real amount of water pollution caused by the consumption of chemical fertilizers. latent variable is defined as follows:

$$y_i^* = X_i\beta + \varepsilon_i \qquad -\infty < y_i^* +\infty \tag{1}$$

Where, y_i^* is continuous variable that shows the real amount of water pollution by nitrogenous fertilizers. β is a vector of parameters that should be estimated. X_i is a vector of explanatory variables and ε_i is error term that has logestic distribution. Since y_i^* is not observable, the standard regression technique such as Ordinary Least Square (OLS) is not applicable (Long, 1997).

If y_i be discrete and observable variable that exhibits the level of pollution, the relationship between unobservable variable y_i^* and y_i can be obtain by ordered logit model as follows:

$$\begin{aligned} y_i &= 1 \quad if -\infty \le y_i^* < \tau_1 \quad i = 1,\dots,n \\ y_i &= 2 \quad if\ \tau_1 \le y_i^* < \tau_2 \quad i = 1,\dots,n \\ y_i &= 3 \quad if\ \tau_2 \le y_i^* < \tau_3 \quad i = 1,\dots,n \\ y_i &= 4 \quad if\ \tau_3 \le y_i^* < \tau_4 \quad i = 1,\dots,n \\ &\dots \end{aligned} \tag{2}$$

In these relationships, n is sample size and τ's are thresholds that define observable discrete responses and should be estimated. The probability that $y_i = J$ can be calculated as Equation (3):

$$\Pr(y_i = J) = \Pr(\varepsilon_i < \tau_J - X_i\beta|X_i) - \Pr(\varepsilon_i < \tau_{J-1} - X_i\beta|X_i) = F(\tau_J - X_i\beta|X_i) - F(\tau_{J-1} - X_i\beta|X_i) \tag{3}$$

In this equation, F is cumulative distribution function (CDF) for ε and β is a column vector of parameters and X_i is a vector of independent variables. One important test in OLM is the assumption of parallel regression. In this test, we assume equal β for each level of dependent variable. Parallel regression test (Brant and Likelihood ratio), assesses the rightness of parameters equality for all groups of dependent variable. Null hypothesis in this test is that β is the same for all groups.

Model Equation (3) can be estimated by using Maximum Likelihood (ML) method and marginal effects of each variable usually calculated in the means of other variables and summation of marginal effects for each variable is zero. The marginal change in the probability can be computed as follows:

$$\frac{\partial \Pr(y = m|X)}{\partial x_k} = \frac{\partial F(\tau_m - X\beta)}{\partial x_k} - \frac{\partial F(\tau_{m-1} - X\beta)}{\partial x_k} \tag{4}$$

Which is the slope of the curve relating x_k to Pr(y = m|X), holding all other variables constant (Long, 1997).

As previously mentioned, the aim of this study is to evaluate factors affecting water pollution by agricultural activities. To achieve this goal, we apply the OLM and our research model is as follows:

$$Y_i = \alpha_0 + \beta_1 act + \beta_2 exp + \beta_3 sat + \beta_4 edu + \beta_5 awa + \beta_6 org + \beta_7 way + \beta_8 inc + \beta_9 wat + \beta_{10} price + \varepsilon_i \tag{5}$$

In Equation (5), Y_i is dependent variable that shows the level of pollution entering into water resources by farmer *i*. This variable is divided into three categories and its description is shown in Table 1.

Classification of dependent variable is based on the level of chemical fertilizers containing nitrogen used by farmers during planting periods. We have three categories for dependent variables. In the first category, farmers use less than 180 kg N hm^{-2}, in the second category, farmers use standard level (between 180-225 kg) of N hm^{-2} while in the third category, farmers use more than 225 kg N hm^{-2} in planting season.

Data and information for this study were collected by completing questionnaires from 254 wheat farmers in the plains of Mashhad. Simple random sampling method was used for selecting a random sample of farmers in this area. About one million hectare of lands in Khorasan Razavi province are cultivated by irrigated and rain-fed crops and this area is the third regarding the area under cultivation in Iran.

3. Results

Descriptive statistics of data are reported in Tables 1 and 2. In Table 1, the frequency of each category of dependent variable reported. Due to the use of nitrogen fertilizers, pollution of water resources by farmers is classified into three groups of; low pollution (1), medium (standard) pollution (2) and high pollution (3).

Table 1. Level and frequency of dependent variable

Level of Dependent Variable	Frequency	Percentage Frequency	Cumulative Frequency
1	30	11.8	11.8
2	89	35	46.8
3	135	53.2	100
Total	254	100	

Table 2 shows the descriptive statistics of all variables.

Table 2. Descriptive statistics of variables

Variable	Mean	Standard Error	Min	Max	Description
Pollution	2.2	0.45	1	3	Level of pollution by nitrogen fertilizers
Main Activity	0.52	0.502	0	1	Agriculture = 1 and Other = 0
Experience	29.14	7.5	8	60	Farmers experience in year
Cultivation levels	1.52	1.04	0.3	4	Levels of cultivation in terms of hectare
Education	7	4	0	15	Farmers education in years
Awareness of Pollution damages	0.44	0.28	0	1	Aware = 1 and Not aware = 0
Awareness of Organic farming	0.47	0.37	0	1	Aware = 1 and Not aware = 0
Fertilizing Method	0.31	0.26	0	1	By hand = 0 and by machine = 1
Income	13	5.6	3.5	45	Annual farmers income
Fertilizers Price	0.57	0.23	0	1	Low = 0 and High = 1
Irrigation method	0.48	0.33	0	1	Drop = 0 and submerge = 1

The results of the ordered logit regression model are reported in Table 3. The results show that main activity of farmers, years of experience, the level of education, awareness of organic farming, level of income, price of fertilizers and irrigation method have significant effect on the amount of nitrogen entering into the water resources and therefore water pollution. Main activity, experience, level of income and irrigation method have direct and positive effect on water pollution; on the other hand, education, awareness of organic farming and price of fertilizers have indirect and negative effect on water resource pollution.

Table 3. Ordered logit regression model results

Variable	Coefficient	Standard Deviation	Z statistic	Prob.
Main Activity	0.82*	0.43	1.91	0.06
Experience	0.068**	0.031	2.18	0.031
Cultivation levels	0.25	0.19	1.3	0.19
Education	-0.53*	0.31	-1.7	0.09
Awareness of Pollution	0.26	0.48	0.58	0.55
Awareness of Organic	-0.97*	0.55	1.76	0.08
Fertilizing Method	0.01	0.5	0.02	0.98
Income	0.28**	0.14	2.02	0.04
Price	-0.46*	0.25	1.84	0.06
Irrigation method	0.81*	0.47	1.72	0.08
τ_1	4.85**	2.62	1.85	0.06
τ_2	4.21*	1.88	2.24	0.04
Log likelihood = -97.57		LRchi2(10) = 21.28		
$R_{McF}^2 = 0.298$		Prob > chi2 = 0.011		

Note. $^{**}p < 0.05$, $^{*}p < 0.1$.

The Likelihood Ratio (LR) chi-square test compares a given model to the constraint model in which all slope coefficients are equal to zero. As shown by the results, we can reject the hypothesis that all coefficients except the intercept are zero.

Since the main activity has positive and significant effect on the level of dependent variable, farmers that take agriculture as their main occupation are more likely to cause pollution of water resources. Farmers who don't have any other occupation other than agriculture depend on this main activity for their lives and livelihoods. They try to earn more money from this core business and therefore with excessive use of chemical fertilizers, in an effort to increase their incomes, unleash more pollution into water resources. Thus, efforts to reduce water pollution should include policies that change farmer's main activity to other activities that do not impose high social costs such as environmental pollution to the community.

Farmers experience also has positive and significant effect on the level of dependent variable. In other words, farmers with more experience in agriculture, find that with extra usage of chemical fertilizers, they can achieve higher production and income levels with no expenses which are detrimental to the environment as well. On the other hand, the prices of fertilizer have negative and significant effect on the level of dependent variable. Thus increasing the price of fertilizers can diminish the amount of fertilizers used by farmers. Hence raising the cost of utilizing chemical fertilizers containing nitrogen with economic tools, can reduce the consumption of these fertilizers and reduce environmental damage.

Education has negative effect on the probability of increasing pollution of water resources. Farmers that have greater education utilize lower level of fertilizers, and hence unleash less pollution on water resources. Farmers with more education utilized more environmental friendly and scientific solutions for increasing production and reducing their costs in farming. Moreover, educated farmers mainly know the exact time for using chemical fertilizers to achieve maximum productivity (Noshad, 2014). Therefore, employing educated farmers in agricultural activities can have a significant impact on the reduction of pollution.

Awareness of farmers about organic farming is another variable that has negative impact on the level of water pollution. When farmers are aware of the benefits and techniques of organic farming, they will use less chemical fertilizers and pesticides and hence the chance of water resources pollution is reduced. This result is consistent with other studies in this field. For instance, Pitmentel et al. (2011) stated that organic practices lead to increased water retention and using organic methods and materials can allow for a slower release of chemicals into water resources. Agricultural organic products due to their superior quality could be a good candidate for nonorganic products and by improving the general health of the society, can reduce the contamination of water and soil pollution. Efforts to increase awareness of organic farming and its benefits as well as creating rational economic returns could reduce water resources pollution.

Farmer's income has a positive and significant effect on the level of water pollution because when farmer's income increases, they could buy more fertilizers and pesticides and even use more water in their farms, thereby increasing water resources pollution. Using economic instruments for reducing the use of chemical fertilizers can control the level of pollution to the standard levels.

Finally, irrigation method has significant effect on the level of water resources pollution and farmers that apply submerge method for irrigating their farms are more likely to cause more pollution to water sources. According to FAO (2011), poor irrigation has one of the worst impacts on water quality, whereas precision irrigation is one of the least polluting practices as well as decreasing the net cost of supplied water. Developing sprinkler and drip irrigation methods can decrease the possibility of contamination of water resources.

Table 4 presents the results of ordered logit regression model in terms of marginal effects. It should be noted that variables such as main activity, awareness of organic farming, price of fertilizers and irrigation method, are discrete and dichotomous variables and when these variables change (with all other variables held at their mean constant), the possibility that the dependent variable stand in each of the three levels, will change as indicated in Table 4.

Table 4. Marginal effects of ordered logit model

Marginal effect of Dependent Variable	Group 1 (Low Pollution)	Group 2 (Medium Pollution)	Group 3 (High Pollution)
Main Activity*	-0.1887	0.0516	0.1371
Experience*	-0.0157	0.0042	0.0115
Cultivation levels	-0.058	0.047	0.011
Education*	0.122	-0.033	-0.089
Awareness of Pollution	-0.062	0.018	0.044
Awareness of Organic*	0.21	-0.03	-0.18
Fertilizing Method	-0.002	0.0006	0.0014
Income*	-0.0667	0.0181	0.0486
Price*	0.121	-0.037	-0.084
Irrigation method*	-0.189	0.052	0.137

For instance, the results in Table 4 indicates that when the main activity of farmers is agriculture, the possibility that farmers stand on group 1 with low pollution reduces by 0.19% and the possibility that they stand on group 2 with medium pollution and in the group 3 with high pollution increase by 0.05 and 0.14% respectively. Again, the results of Table 3 are repeated in Table 4 in terms of marginal effects and for three level of dependent variable.

For other continuous variables such as experience, education and income, it says that when these variables change by one percent, the possibility that dependent variable lies in each group of dependent variable, changes by the percent reported in Table 4. For example, when the income of farmers increases by one percent, the possibility that they stand in group 1 with low level of pollution reduces by 0.066% and the possibility that farmers stand on group 2 with medium pollution or stand on group 3 with high pollution, increases by 0.018 and 0.048% respectively.

Table 5 presents the results of parallel regression test in ordered logit model. One of the assumptions in the ordered logit model is that for any independent variable, the coefficient of parameters (β_s) in all groups of dependent variable are the same. Hence, the null hypothesis in the parallel regression test is that for all groups of dependent variable, parameters are identical (Long, 1997).

Table 5. The results of parallel regression test

Test	Chi-square	df	Prob.
Brant	8.32	9	0.502
LR	9.08	9	0.42
WALD	7.89	9	0.54

The results in Table 5 indicate that the null hypothesis is not rejected and so the ordered logit model is accepted and it can show the relationships between explanatory variables and the groups of dependent variable as well. If H_0 is rejected, we can use other methods such as generalized ordered logit models.

4. Discussion

In this study, factors affecting the level of consumption of fertilizers by farmers, and thus absorption of nitrogen in water resources were investigated in Mashhad plain area. The results from ordered logit model show that explanatory variables such as main activity, farmers experience, farmers education, awareness of organic farming, income, price of fertilizers and irrigation methods have significant effect on the level of dependent variable.

Efforts to reduce water pollution should include policies that change farmer's main activity to other activities such as complementary jobs in agricultural sector that do not impose high social costs to the community. Furthermore, increasing the price of fertilizers to an optimal level can reduce the amount of fertilizers utilized by farmers, thereby reducing environmental pollution.

Developing formal and informal educational centers for farmers and employing educated farmers in agricultural activities can have a significant effect on the reduction of pollution and when farmers are aware of the benefits and techniques of organic farming, they will use less chemical fertilizers and pesticides, thereby reducing the contamination of water resources. Efforts to increase awareness about organic farming as well as government support from these groups of farmers and finally creating rational economic returns could reduce the pollution of water resources. Developing sprinkler and drip irrigation methods also can decrease the possibility of contaminating water resources.

Finally, it is proper to mention that the solutions proffered in this study for reducing water pollution will be more effective when they are carried out simultaneously and together. For example, by raising the cost of chemical fertilizers, rational economic incentives for organic farming are create and the general knowledge and education of farmers increase. Secondly, the solutions should fit into the overall culture and the level of development of a region or a country. Strategies mentioned in this article are mainly for developing countries like Iran, where their main priority is earning more income from agricultural activities rather than considering environmental issues. Therefore, the results of other similar studies especially in developed countries may somewhat vary with the results of the present study.

References

Ansari, H., Salarian, M., Takarli, A., & Bayram, M. (2014). Determining Optimum Irrigation Depth for Wheat and Tomato Crops Using Aqua crop Model (A Case study in Mashhad). *Iranian Journal of Irrigation and Drainage, 8*, 86-95.

Bouraoui, F., & Grizzetti, B. (2013). Modeling mitigation options to reduce diffuse nitrogen water pollution from agriculture. *Science Total Environment, 15*, 468-469. https://doi.org/10.1016/j.scitotenv.2013.07.0662013

Bumb, B. L., & Baanate, C. A. (1996). *World Trends in Fertilizer Use and Projections to 2020*. International Food Policy Research Institute. Washington DC. Retrieved from http://www.ifpri.org/2020/briefs/number38.htm

Collins, A. L., Zhang, Y. S., Winter, M., Inman, A., Jones, J. I., Johnes, P. J., … Noble, L. (2016). Tackling agricultural diffuse pollution: What might uptake of Farmer-preferred measures deliver for emissions to water and air? *Science of the Total Environment, 547*, 269-281. https://doi.org/10.1016/j.scitotenv.2015.12.130

De Paz, J. M., & Ramos, C. (2004). Simulation of nitrate leaching for different nitrogen fertilization rates in a region of Valencia (Spain) using a GIS-GLEAMS system. *Agriculture, Ecosystems and Environment, 103*, 59-73. https://doi.org/10.1016/j.agee.2003.10.006

Khamadi, F., Mesgarbashi, M., Hosaibi, P., Enaiat, N., & Farzaneh, M. (2015). The effect of crop residue and nitrogen fertilizer levels on soil biological properties and nitrogen indices and redistribution of dry matter in wheat. *Agronomy Journal, 28*(4), 149-157.

Long, S. J. (1997). *Regression Models for Categorical and Limited Dependent Variables*. SAGE Publication, London.

Mahvi, A. H., Nouri, J., Babaei, A. A., & Nabizadeh, R. (2005). Agricultural activities impact on groundwater nitrate pollution. *Int. J. Environ. Sci. Tech., 2*, 41-47. https://doi.org/10.1007/BF03325856

Noshad, H. (2014). *The amount and timing of nitrogen application in sugar beet cultivation*. Organization of Research, Education and Promoting of Agriculture, Iran.

Ongley, E. D. (1996). *Control of water pollution from agriculture, GEMS/Water Collaborating Centre, Canada Centre for Inland Waters*. Food and Agriculture Organization of the United Nations, Rome.

Peña-Haro, S., Llopis-Albert, C., Pulido-Velazquez, M., & Pulido-Velazquez, D. (2010). Fertilizer standards for controlling groundwater nitrate pollution from agriculture: ElSalobral-Los Llanos case study, Spain. *Journal of Hydrology, 392*, 174-187. https://doi.org/10.1016/j.jhydrol.2010.08.006

Pimentel, D., Hepperly, P., Hanson, J., Douds, D., & Seidel, R. (2011). Environmental, Energetic, and Economic Comparisons of Organic and Conventional Framing Systems. *Bio Science, 55*(7), 573-580. https://doi.org/10.1641/0006-3568(2005)055[0573:EEAECO]2.0.CO;2

Seldon, T. M., & Song, D. (1994). Environmental quality and development: Is there a Kuznets Curve for air pollution emissions. *Journal of Environmental Economics and Management, 27*(2), 147-52. https://doi.org/10.1006/jeem.1994.1031

Semaan, J., Flichman, G., Scardigno, A., & Steduto, P. (2007). Analysis of nitrate pollution control policies in the irrigated agriculture of Apulia Region (Southern Italy): A bio-economic modeling approach. *Agricultural Systems, 94*, 357-367. https://doi.org/10.1016/j.agsy.2006.10.003

Seyed Sharifi, R., Ganbari, P., Khavazi, K., & Kamari, H. (2016). Study of interaction between nitrogen and biofertilizers on yield, grain growth of wheat and fertilizer use efficiency. *Journal of Soil Biology, 4*(1), 1-14.

Sharasbi, S., Emam, Y., Ronaghi, A., & Pirasteh Anoshe, H. (2016). Effect of drought stress and nitrogen fertilizer on grain yield and nitrogen use efficiency of wheat in Fars province, Iran conditions. *Iranian Journal of Crop Science, 17*(4), 349-363.

Shober, A. L. (2014). *Nitrogen in the Home Landscape*. Retrieved from http://edis.ifas.ufl.edu

United Nations Environment Programme. (2014). *UNEP year book, emerging issues in our global environment*. Retrieved from http://www.epa.gov/nutrientpollution

United Nations of Food and Agriculture Organization. (2011). *Fertilizers as Water Pollutants*. Natural Resources Management and Environment Department. Retrieved from http://www.fao.org/docrep/w2598e/w2598e06.htm

Wang, D. J., Liu, Q., Lin, Q., Lin, J. H., & Sun, R. J. (2004). Optimum Nitrogen Use and Reduced Nitrogen Loss for Production of Rice and Wheat in the Yangtse Delta Region. *Environmental Geochemistry and Health, 26*(2), 221-227. https://doi.org/10.1023/B:EGAH.0000039584.35434.e0

Zhou, Y., Yang, H., Mosler, H. J., & Abbaspour, K. C. (2010). Factors affecting farmers' decisions on fertilizer use: A case study for the Chaobai watershed in Northern China. *The Journal of Sustainable Development, 4*, 80-102. https://doi.org/10.7916/D8C24W3R

Zhu, Z. L., & Chen, D. L. (2002). Nitrogen fertilizer use in China–Contributions to food production, impacts on the environment and best management strategies. *Nutrient Cycling in Agroecosystems, 63*, 117-127. https://doi.org/10.1023/A:1021107026067

19

Comparison of Two Harvest Methods for Lettuce Production in an Aquaponic System

Gaylynn E. Johnson[1], Karen M. Buzby[2], Kenneth J. Semmens[3,4] & Nicole L. Waterland[1]

[1] Division of Plant and Soil Sciences, West Virginia University, Morgantown, WV, USA

[2] Department of Civil and Environmental Engineering, West Virginia University, Morgantown, WV, USA

[3] Division of Animal and Nutritional Sciences, West Virginia University, Morgantown, WV, USA

[4] Aquaculture Research Center, Kentucky State University, Frankfort, KY, USA

Correspondence: Nicole L. Waterland, Division of Plant and Soil Sciences, West Virginia University, Morgantown, WV, 26506, USA. E-mail: nicole.waterland@mail.wvu.edu

The research is financed by the U.S. Department of Agriculture Special Grant Agreement 2010-34386-21745 and West Virginia University Agriculture and Forestry Experimental Station (Hatch Grant WVA00640). Scientific Article No. 3291 of the West Virginia Agricultural and Forestry Experiment Station, Morgantown.

Abstract

Aquaponics is an integrated food production technology of aquaculture and hydroponics. Lettuce (*Lactuca sativa* L.) is an economically important vegetable crop that can be grown aquaponically. In addition to selecting the right choice of lettuce cultivars, developing an optimal harvest strategy could increase lettuce production. Lettuce production using two harvest methods, Cut-and-Come-Again (CC) and Once-and-Done (OD), was evaluated using 'Red Sails' lettuce in a flow-through aquaponic system rearing trout. With the CC method continual harvesting was possible on a weekly basis after the initial harvest, while it took five weeks for each harvest using the OD method. The total yield of lettuce by the CC method was 6.7 kg from 9 trays, while 22.6 kg of lettuce was harvested by the OD method using 54 trays. In harvests by the OD method, 6 times as many seeds were sown compared to the CC method. The average yield per tray harvested by the CC method (744.4 g/tray) was 78% higher than that by the OD method (418.5 g/tray) because the CC method used 6 times less trays. Productivity, calculated by the average yield per growing week, of the two harvest methods at the first harvest was similar, but 4.8 times higher in the CC method than in the OD method at the second harvest due to the shorter harvest time. However, visual and decay ratings of lettuce harvested by the CC method began to decline afterwards. Together, the OD method after two consecutive harvests by the CC method would help growers to obtain increased yield of quality lettuce.

Keywords: aquaculture, aquaponics, flow-through system, harvest method, high tunnel, lettuce, trout

1. Introduction

Aquaponics is a synergistic combination of fish farming (aquaculture) and soilless plant production in a nutrient solution (hydroponic) and is a potentially sustainable method of food production (Love et al., 2014). In an aquaponic system, the effluent containing nutrients generated from fish production passes through the root zone of the plants (Rakocy, 2007; Buzby & Lin, 2014). In turn, plants are able to utilize the soluble nutrients in the aquaculture effluent for growth. As the fish digests proteins, ammonia (NH_3) is excreted into the water. During a process called nitrification, bacteria utilize this nutrient and convert it from ammonium (NH_4^+) into nitrite (NO_2^{-1}) and then to nitrate (NO_3^{-1}) (Rakocy, Masser, & Losordo, 2006; Boyd & Tucker, 2014).

Many economically important vegetables and flowering plants can utilize NH_4^+ and NO_3^{-1} for growth (Rana et al., 2011). Among the most commonly grown plants are leafy greens such as lettuce and herbs (J. Oziel & T. Oziel, 2013). Lettuce and herbs make excellent crops because of their high market prices and short production cycles (Rakocy et al., 2006). These crops also have modest nutritional requirements and are well adapted to aquaponic systems (Blidariu & Grozea, 2011).

In this study, a cold water (14-15 °C) flow-through aquaponic system (FTS) was utilized. An FTS involves the continuous flow of water through a tank or a raceway (Adler, Harper, Takeda, Wade, & Summerfelt, 2000; Snow, Anderson, & Wootton, 2012; Soderberg, 1995). These systems are the most commonly used for producing salmonids such as rainbow trout (*Oncorhynchus mykiss*) (Snow et al., 2012). Clean water is typically directed through a channel where the fish are retained and fed (Blidariu & Grozea, 2011). Waste products are carried away in the flow and the effluent is discharged directly into a receiving body of water (Viadero, Cunningham, Semmens, & Tierney, 2005). Unlike recirculating aquaculture systems, flowing water aquaculture systems do not concentrate nutrients through water reuse so the effluent is characterized by high volume and low nutrient content (Hinshaw & Fornshell, 2002).

To produce crops in nutrient limited water, a selection of crops requiring a low level of nutrient for growth is important. One of the low nutrient demanding vegetables is lettuce. Lettuce is also a cool season crop with an optimal temperature ranging from 13 to 16 °C, but can tolerate temperatures as low as -2 °C (Borrelli, Koenig, Jaeckel, & Miles, 2013). As a cool season crop, lettuce is a good candidate for production in a cold water FTS. The cultivar 'Red Sails' was selected in this study because it performed well in previous study utilizing an FTS (Buzby & Lin, 2014; Buzby, West, Waterland, & Lin, 2016b).

In addition to crop selection for FTS, harvest methods can affect the harvestable yield and the timing of harvest. Lettuce is usually destructively harvested with the entire above ground portion of the plant being removed. In order to obtain a second crop the plants have to be resown and the process starts over again. Alternatively, multiple harvests with one time sowing would allow speedy harvest and conservation of resources and time. In this study, two harvest strategies were evaluated: Cut-and-Come-Again (CC) and Once-and-Done (OD). With the CC strategy, all plants in the system are sown at one time, but multiple harvests are made until the crop is no longer productive. In the CC method, only the aerial parts of the lettuce are harvested, leaving the meristem intact, allowing for continued growth. Because there is no need to constantly resow seeds in CC, the CC method could be potentially beneficial for growers by reducing the space needed for production, labor and material costs (seed, media, etc). In contrast, the OD strategy requires more seeds that need to be planted for each harvest.

Development of an effective harvest strategy can help to increase crop yield and efficiency by reducing the period of time it takes to grow a crop to harvestable size. With the CC method, the intact meristem remained after the initial harvest enabled lettuce to regrow quicker than lettuce from the new batch of seeds. The objective of this research was to evaluate lettuce production under two harvest methods (CC and OD) in an FTS. To the best of our knowledge, this is the first study evaluating harvest methods of lettuce production in an FTS.

2. Materials and Methods

2.1 Facility

The experiment was conducted in a high tunnel (HT; 7.9 m wide × 12.2 m long) located at Wardensville, West Virginia (394'32″N, 7835'40″W) between May 16th and July 25th, 2013. The HT housed 13 beds (1.2 m wide × 2.4 m long × 26.7 cm deep per bed) constructed with 1.9 cm plywood. Each bed was subdivided into three channels (38.1 cm wide × 2.4 m long per channel). The channels were lined with 20 mil white polyethylene liner (Raven Industries' Dura-Skrim R-Series, Model R20WW; Sioux Falls, SD, USA). Each channel held three floating trays (34.3 cm wide × 66.7 cm long) (Speedling, Inc., Model 32; Ruskin, FL, USA). Aquaculture effluent entered through a valve at the inlet of each channel and drained at the outlet. Water depth was maintained at 23 cm via a standpipe. Influent velocity was adjusted to 0.012 m^3 min^{-1} or 10 L min^{-1} with 9.5 mm aperture as described in Dyer (2006). There were sixteen fish tanks (3.82 m^3 per tank) in the aquaculture facility. The total number and weight of trout (*Oncorhynchus mykiss*) were 5,520 and 2.8 tons, respectively, during the period of experiment. The average fish density was 90 fish/m^3.

2.2 Growing Environment and Water Sample Analyses

Environmental conditions in the HT were continually measured over the course of the experiment. The average day and night air temperature inside the HT was 25.2/17.7 ± 4.0/3.3 °C day/night (mean ± SD). A black shade cloth was installed to reduce the effect of elevated temperatures and excessive light exposure in the HT. Average light intensity was 310.3 μmol m^{-2} s^{-1}. Water temperature was also monitored with probes submerged in the water. Average water temperature was 19.2 ± 3.7 °C (mean ± SD). Electrical conductivity (EC), pH and dissolved oxygen (DO) were measured every two weeks. Average pH, EC and DO were 7.0 ± 0.08, 0.13 ± 0.01 mS cm^{-1} and 8.0 ± 0.7 mg L^{-1}, respectively.

Water samples for mineral analysis and total suspended solids (TSS) were also taken every two weeks and analyzed according to methods delineated by the American Public Health Association (APHA, 1995) for nutrient content (4500-NH_3, phenate method, 4110-NO_2 ion chromatography with direct conductivity detection and 4500-P ascorbic acid method) and TSS (2540D). Nutrients measured included total ammonia nitrogen (TAN), nitrite (NO_2^{-1}), nitrate (NO_3^{-1}), and phosphate (PO_4^{-3}). Total suspended solids consisted of nominal particle size of $\geq$ 1.2 μm.

2.3 Plant Material

Seeds of red leaf lettuce 'Red Sails' (*Lactuca sativa* L. var. *crispa*) were purchased from Johnny's Selected Seeds (Winslow, ME, USA) and were vacuum-seeded in 128-cell trays (Speedling, Inc., Model 32; Ruskin, FL, USA) filled with vermiculite (Therm-O-Rock East Inc., Grade 3A; New Eagle, PA, USA) as substrate.

2.4 Harvest Methods

Seeds for the first harvests of CC and OD were sown at the same time (Week 0). The CC method was sown only once (CC), while seeds for OD (OD 1-6) were sown six times every week for six weeks to coincide harvest time with CC (Figure 1). OD 2-6 were sown in a staggered method by one week intervals. Lettuce was harvested six times each in both methods until the lettuce in the CC treatment was no longer productive. A total of 24 channels in eight beds were used; three channels for a control, another three channels for the CC harvest method, and 18 channels for the OD harvest method. Each channel contained three trays and each treatment (CC and OD 1-6) had a total of nine trays for three replications (n = 3). The trays in control channels contained only vermiculite without plants to evaluate nutrient removal by the vermiculite substrate. Trays were rotated every week to avoid preferential nutrient uptake by lettuce placed closest to the inlet (Alder et al., 2000). All lettuce from a set of nine trays were harvested at the same time when the average height of lettuce reached 12.7-15.24 cm tall as measured from the media level to the top of the tallest leaf (United States Department of Agriculture, 2013). Lettuce for the CC method was cut 12.7 mm above the meristem (25.4 mm above the tray) to allow for lettuce to grow continuously, while lettuce for the OD method were cut at the base of the plant. For the CC method, the first harvest was carried out five weeks after seeds were sown and the sequential harvests were made every week for a total of six harvests.

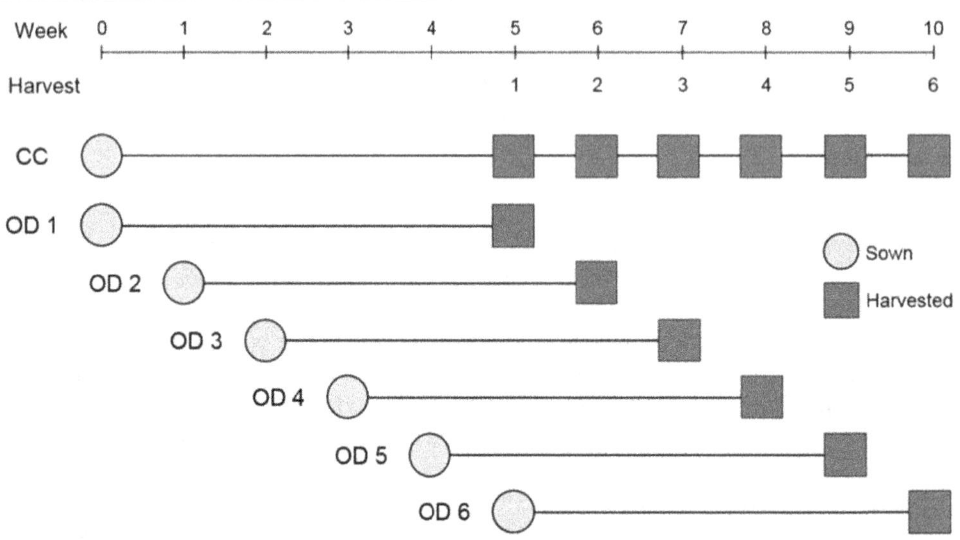

Figure 1. A timeline of the two harvest methods, Cut-and-Come-Again (CC) and Once-and-Done (OD)

Note. For the CC method, lettuce was first harvested five weeks after sowing and subsequently harvested every week for the next five weeks. For the OD method, lettuce was harvested once every five weeks. Seeds were sown every week for six weeks in the trays for the OD method so that the harvests for the OD method could coincide with the CC harvests. Six separate harvests for the OD method are labeled (OD 1-6). Circles and squares indicate sowing and harvesting time, respectively. The numbers above the line on top of the diagram shows the number of weeks after the first seeds were sown and the numbers below the line are harvest numbers.

2.5 Stand Establishment, Average Yield, Productivity

Stand establishment (SE) described the ability of the seedlings to survive after germination and was determined as the percentage of two-week old seedlings in a tray. SE of trays for the CC method was determined once from the initial trays. SE for the OD method was the average of six sets of individual SE. The average yield and productivity of 'Red Sails' lettuce harvested by Cut-and-Come-Again (CC) and Once-and-Done (OD) methods were determined. The average yield was determined by the average harvest fresh weight of the aerial part of the lettuce per tray. Lettuce productivity was determined by the average yield per week.

2.6 Quality Analysis

The deleterious effect of multiple harvests on the lettuce quality ratings of leaf color, texture, decay, and visual quality were determined according to Kader, Lipton, and Morris (1973) and the description of market quality rating scale was provided in Table 1.

Table 1. Market quality rating scale of lettuce from Kadar et al. (1973)

Score[z]	Visual quality description
5	Excellent, essentially free from defects
4	Good, minor defects, not objectionable
3	Fair, slight to moderate objectionable defects, lower limit of sales appeal
2	Poor, excessive defects, limit of salability
1	Extremely poor, not usable/salable
Score	Color description
5	Dark green leaves/heavy redness of leaves
4	Green leaves/slight to moderate redness of leaves
3	Dull green leaves/no redness in leaves
2	Light leaves, premature or dying
1	Yellow/dead leaves, not usable/salable
Score	Leaf texture description
3	Crispy, ruffed (crunchy – romaine, iceberg; smooth, soft – butterhead)
2	Wilted, poor texture, not usable/salable
1	Dead, not usable/salable
Score	Decay description
5	None
4	Slight, slightly objectionable, may impair salability
3	Moderate, objectionable, definitely impairs salability
2	Severe, salvageable, but normally not salable
1	Extreme, not usable

Note. [z] Ratings are based on presumed consumer acceptance and a rating of one is least desirable.

2.7 Statistical Analysis

Values are the means of three replications (n = 3). Analysis of variance (ANOVA) was performed with PROC GLM (generalized linear model) by SAS version 9.3 (SAS Institute, Inc., Cary, NC). Differences among treatment means were determined using Tukey's significance test at $P \leq 0.05$.

3. Results

3.1 Nutrient Removal

The concentrations of TAN, NO_2^{-1}, NO_3^{-1}, PO_4^{-3} and total suspended solids consist of particulates ≥ 1.2 µm (TSS) in the influent and effluent of all channels was measured. There was no difference in nutrient or TSS removal by lettuce grown for CC or OD method (Table 2). The control contained only vermiculite substrate and did not remove nutrient or reduce TSS (Table 2).

Table 2. Level of total ammonia nitrogen (TAN), nitrite (NO_2^{-1}), nitrate (NO_3^{-1}), and phosphate (PO_4^{-3}) in effluent of a cold water flow-through aquaponic system

Date	Week	TAN			NO_2^{-1}			NO_3^{-1}			PO_4^{-3}		
		Control[z]	CC	OD	Control	CC	OD	Control	CC	OD	Control	CC	OD
		mg L^{-1}											
30-May	2	0.39Aa[y]	0.39Aa	0.39Aa	3.90Aa	4.06Aa	4.00Aa	0.28Ca	0.28Aa	0.28BCa	0.15Ba	0.15Ba	0.15Ba
13-Jun	4	0.20Da	0.20Da	0.20Ea	1.33Ba	1.53Ba	1.85Ba	0.32Aa	0.30Aa	0.31Aa	0.18Aa	0.18Aa	0.18Aa
27-Jun	6	0.38ABa	0.37Ba	0.37Ba	4.00Aa	4.00Aa	4.00Aa	0.30Ba	0.29Aa	0.29Ba	0.17Aa	0.17Aa	0.18Aa
11-Jul	8	0.35Ba	0.36Ba	0.35Ca	4.00Aa	4.00Aa	4.00Aa	0.29BCa	0.29Aa	0.30Ba	0.15Ba	0.15Ba	0.15Ba
25-Jul	10	0.31Ca	0.31Ca	0.31Da	3.67Aa	3.33Aa	3.67Aa	0.26Da	0.27Aa	0.26Ca	0.15Ba	0.15Ba	0.15Ba

Note. [z] Control, vermiculite only; CC, Cut-and-Come-Again; OD, Once-and-Done.

[y] Values are the means of three replications (n = 3).

Different upper case letters down the column indicate significant difference among sampling dates within a treatment by Tukey's significance test at $P \leq 0.05$.

Different lower case letters across the row indicate significant difference among treatments by Tukey's significance test at $P \leq 0.05$.

3.2 SE, Average Yield and Productivity

The average SE of lettuce for the CC and OD harvest methods were 81.6 and 83.3%, respectively. A total of six harvests were made for each harvest method during the ten week growing period. The total yield or harvest biomass of the six harvests combined by the CC and OD harvest methods were 6.7 and 22.6 kg, respectively.

The average yield (harvest biomass per tray) between CC and OD methods at each harvest and among harvests for each method were compared (Table 3). The average yields of the first two harvests were lower than the rest of harvests for each harvest method and were not significantly different between the two harvest methods. However, the average yield from the CC method gradually declined from the third harvest and the average yield of the 5th and 6th harvest by the CC method were nearly 9 and more than 13 times lower than those by the OD method at the same harvest, respectively. On the contrary, the average yield of lettuce harvested by the OD method generally increased or showed no change (Table 3).

Because the growing period of the lettuce in the two harvest methods was different, productivity (average yield per week) at each harvest was also compared (Table 3). The productivity at the first harvest of the CC method was not different from the OD method, but after the first harvest the productivity of the CC method significantly increased (Table 3). The first four harvests of CC had productivity that was generally equivalent to or higher than that of the OD method. However, the productivity of the CC method declined significantly at 5th and 6th harvests, while the productivity of the newly sown lettuce in the OD method increased or stayed at the same level (Table 3).

Table 3. Lettuce yield and productivity from two harvest methods, Cut-and-Come-Again (CC) and Once-and-Done (OD) at each harvest

Harvest	Average yield (g/tray)			Productivity (yield/number of growing weeks)		
	Treatment[z]		Significance	Treatment		Significance
	CC	OD		CC	OD	
1	180.91 ABa[y]	221.02 Da	NS	36.18 ABa	44.20 Da	NS
2	225.78 Aa	232.97 Da	NS	225.78 Aa	46.59 Db	***
3	130.99 Bb	459.87 BCa	***	130.99 Ba	91.97 BCb	*
4	144.59 ABb	614.90 Aa	***	144.59 ABa	122.98 Aa	NS
5	46.36 Cb	406.59 Ca	***	46.36 Cb	81.32 Ca	**
6	42.70 Cb	572.42 ABa	***	42.70 Cb	114.48 ABa	***
Significance	***	***		***	***	

Note. [z] CC, Cut-and-Come-Again; OD, Once-and-Done.

[y] Values are the means of three replications (n = 3).

Different upper case letter down the column indicate significant difference among harvests within a treatment by Tukey's significance test at $P \leq 0.05$.

Different lower case letter across the row indicate significant difference between treatments by Tukey's significance test at $P \leq 0.05$.

NS, *, **, ***: Non-significant, significant at $P \leq 0.05$, 0.01, or 0.001, respectively.

3.3 Quality Analysis

The lettuce harvested by the OD method showed no decay in visual ratings (Figure 2). However, the lettuce harvested by the CC method showed decay and consequently visual ratings began to decline from the third harvest. In some cases, not only dead leaves, but redness along the ribs and leaf margins were visible. There was no difference in ratings of leaf color, and leaf texture among the lettuce harvested by CC or OD method (data not shown).

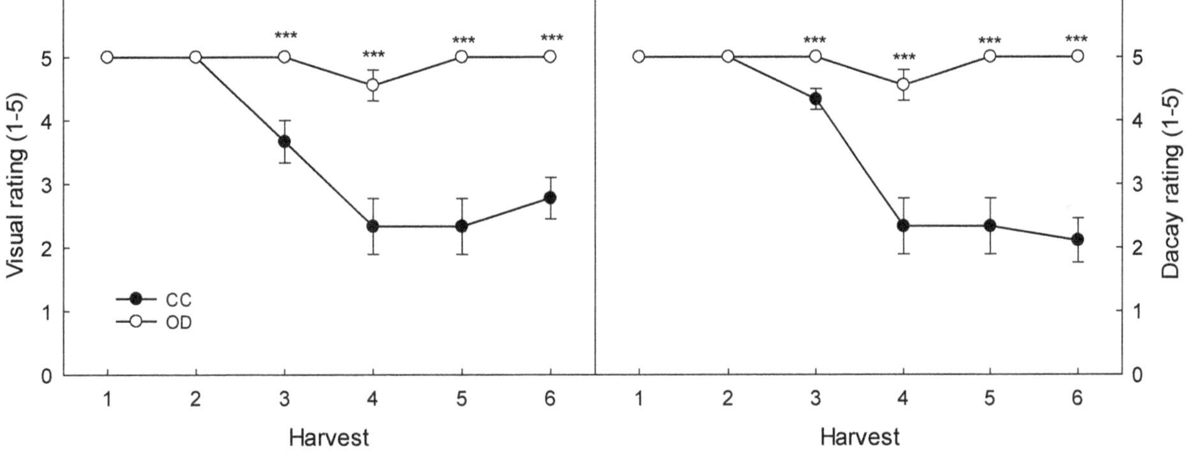

Figure 2. Visual (left) and decay (right) ratings of lettuce in Cut-and-Come-Again (CC) and Once-and-Done (OD) harvest methods at each harvest

Note. Vertical bars are standard errors of the means with three replications (n = 3). ***: Significant at $P \leq 0.001$.

4. Discussion

4.1 Nutrient Removal

There appeared to be no significant nutrient removal in any channel for control, CC and OD treatments during the ten weeks of the growing period in our study. Because the aquaculture effluent continuously flowed with a low level of nutrient in the FTS, it was difficult to determine the exact amount of nutrient removed by the lettuce. Even though low level of nutrient was provided through FTS, constant supply of those nutrients might have helped lettuce to grow without any nutrient deficiency symptoms. However, nutrient removal was observed in the previous report in an FTS (Buzby et al., 2016b). This discrepancy could be nutrient compensation effect of nutrients generated from the solids accumulated at the bottom of the channels. The solids might have decayed and released nutrient to compensate for the amount of nutrient removed by lettuce, resulting in no apparent removal. In the experiment performed by Buzby et al. (2016b), the accumulated solids were removed prior to water analysis.

4.2 SE, Average Yield, and Productivity

The feasibility of lettuce production in a cold water FTS has been demonstrated (Buzby et al., 2016b). However, an efficient harvest method to increase lettuce yield has not been evaluated. In our study, the harvest method designed to harvest multiple times from a single planting (CC) was compared to a method of a single harvest from a single planting (OD). The average SE was over 80% in both harvest methods. Total yield of lettuce obtained from all six harvests by the CC method was 6.7 kg compared to 22.6 kg by the OD method. The reason for this large discrepancy was because in the CC treatment lettuce was harvested repeatedly from the same plants sown once using nine trays, while total of 54 trays were used for the OD method (nine trays each harvest for six harvests). However, when the yield per tray was taken into consideration for the 10 week of growing period, 744.4 g of lettuce per tray were harvested by the CC method (6.7 kg/9 trays), while 418.5 g of lettuce per tray (22.6 kg/54 trays) was obtained by the OD method. Additionally, it would have taken 30 weeks to harvest, if lettuce were harvested every 5 weeks in the OD method. Thus, the CC method not only increased harvest biomass per tray and production efficiency, but also reduced the cost of resources such as labor for seeding, number of seeds and trays.

The average yields of lettuce measured from the 3rd through 6th harvest using the OD method were higher than the first two harvests. The first two harvests were made on June 20th and 27th. Nutrient analyses of the aquaponic water were initiated on May 30th, and repeated every two weeks until July 25th (Table 2). According to the nutrient analysis, significantly lower concentrations of TAN (0.2 mg L^{-1}) were detected in all treatments on June 13th, which was one week prior to the first harvest, compared to the concentration of TAN (0.31-0.39 mg L^{-1}) detected during all other time periods. On June 8th through 10th, fish were harvested. Prior to the fish harvest, feeding was discontinued temporarily and feeding resumed after the harvest was completed. Nutrient concentrations in the aquaculture water, especially TAN, decreased during this transition period before nutrient level increased to 0.37-0.38 mg L^{-1} in all treatment as observed on the second lettuce harvest (June 27th). The fish harvest might have influenced the nutrient level in the wastewater, consequently resulting in lower average yield of the first two harvests in both harvest treatments. This low level of TAN could have affected the 3rd harvest as well, while lettuce harvested after 3rd harvest might not have been affected as much, if not at all. Well-coordinated planning for fish and lettuce production is strongly recommended.

As observed in the last four harvests (harvests 3 through 6), the average yield using the CC method gradually declined to less than 50 g per tray at the 5th and 6th harvest, while the average harvest using the OD method was over 400 g per tray through the 6th harvest. In the CC method, not only average yield decreased, but also visual and decay ratings declined as well (Figure 2), resulting in only the first two harvests being marketable using the CC method which was comparable to the OD method.

Initially, it took five weeks until lettuce reached the marketable size of 12.7-15.24 cm. The productivity at the second harvest by the CC method was particularly high because it only took one more week to regenerate and reach harvestable size. In the CC method, when lettuce was harvested, the meristem remained intact to allow continued growth after the harvest. Because of this accelerated regeneration, the productivity of CC method was significantly higher than that of the OD method at the 2nd and 3rd harvest, and comparable at the 4th harvest even with the declining harvestable yield. However, the productivity of CC treatment significantly declined following the fourth harvest due to a damage from repeated harvests of the same lettuce as evidenced by the decay and visual ratings (Figures 2 and 3). On the other hand, the productivity of OD method was significantly higher at 5th and 6th harvest compared to that of the CC method and remained relatively high from 3rd through the last harvest because they were grown from a new set of seeds at every harvest (Table 3).

Figure 3. Representative photos of lettuce during the first, fourth, and sixth harvests by Cut-and-Come-Again (CC) and Once-and-Done (OD) harvesting methods

4.3 Quality Analysis

The quality of lettuce was analyzed because the repeated harvest of the same plants in the CC method would damage the lettuce and potentially decrease the marketability. There was no difference in ratings of leaf color and leaf texture among the lettuce harvested by CC or OD method (data not shown). However, decay and poor visual ratings due to uncharacteristic reddish leaves and rib veins were observed in the CC method after 2nd harvest (Figure 2), while decay and visual quality ratings were consistently better for lettuce in the OD treatment from the 3rd harvest until the last harvest. Multiple harvesting from the same plants might have caused the lettuce to be stressed (Figures 2 and 3), and consequently resulted in the reduction of average yield (Table 3). Increase in red pigmentation could be due to an increase of anthocyanins in response to the stress (Gazula, Kleinhenz, Streeter, & Raymond, 2005). Rib discoloration was known to be one of the physiological disorders induced by heat stress in lettuce (Jenni, 2005). High quality ratings are essential to crop production for marketability. Although productivity was comparable or greater using the CC method compared to the productivity of OD method for the first four harvests, it is recommended to resow new lettuce seeds after two harvests. While some lettuce at the 3rd or 4th harvest was salvageable, it was time consuming to carefully harvest healthy looking lettuce. Furthermore, direct contact with decaying lettuce using tools and hands while harvesting can potentially spread the damage in other production areas and therefore the lettuce must be handled carefully to prevent further damage and/or the spreading of disease. Decay can render a crop unsalable and affect profitability of the lettuce crop. More importantly, mishandling during harvest or post-harvest could cause an epidemic of plant diseases or food contamination that may result in harmful food consumption to consumers and therefore, food safety testing is also required to guarantee food quality. However, no plant disease symptoms were observed in this study.

Results may also vary based on the time of year. This experiment was conducted in the summer when lettuce could regrow within a week due to warmer temperature and high light levels, however this warm environment would have increased the likely hood of decay. If this experiment were conducted in the spring or fall the growth rate may have been slower, but the reduction in quality using the CC harvest method may also have been ameliorated allowing for more marketable harvests increasing the average yield. With the CC method, lettuce was harvested as loose leaf, because lettuce was cut above the meristem without a base that holds the leaves together. Separated leaves tend to be sensitive to water deficit stress and cannot be stored as long. Therefore, these lettuce leaves need to be stored in a refrigerated storage, sold, or eaten quickly.

4.4 Estimated Lettuce Production Using Combination of CC and OD Methods

It required six weeks to harvest the first two rounds of lettuce with good quality from one planting using the CC method, compared to ten weeks of growing time from two separate plantings using the OD method. The CC harvest method provided several advantages over the OD method; the same quantity and quality of lettuce can be harvested in six weeks instead of ten weeks, the seed needs to be sown only once instead of twice, and the number of trays used for lettuce production also can be reduced by half. This result suggests that lettuce production could be increased by choosing the CC harvest method instead of replanting after every harvest. Considering the comparable average yields and quality of the first two harvests between CC and OD methods as well as significantly higher productivity in CC, the OD method combined with the CC method would increase total yield and reduce cost for lettuce production (Table 4). For example, a similar amount of lettuce was harvested by the CC method (1.8 kg; average of 203.3 g/tray × 9 trays) and OD method (2.0 kg; average of 227.0 g/tray × 9 trays) for the first two harvests. However, it would take ten weeks to grow lettuce for the OD method, instead of just six weeks for the CC method. When harvest time is considered, the productivity of CC method would be 67.8 g per tray per week based on our harvest result (total of two first harvest 406.7 g/tray for a six week growing period) compared to 45.4 g per tray per week using the OD method (total of two first harvest 454.0 g/tray for a ten week growing period). When two consecutive harvests using the CC method is repeated as one OD harvest, significantly increased total lettuce yield would have been obtained (Table 4). Therefore, the CC method used in our study will increase not only the total yield of lettuce, but also reduce the cost of production by reducing the number of the trays, seeds and the amount of labor. In addition, there was also a large saving of space when the square footage used for growing lettuce was compared between the CC and OD methods. While the OD method maintained harvestable lettuce for all six harvests it required six times the space. To get the best utilization of space without reduction in yield it would be recommendable to use the CC harvest method twice as it would save space without reduction of productivity. Depending on the size of growing operation it could be desirable. Not all crops can be harvested by the CC method in FTS. Other vegetables such as Swiss chard, herbs and cilantro would also be good candidates for FTS production using the CC method combined with the OD method (Buzby, Waterland, Semmens, & Lin, 2016a; Buzby et al., 2016b; Rakocy, Bailey, Shultz, & Thoman, 2004).

Table 4. Estimated average yield per tray and growing periods using Cut-and-Come-Again (CC) and Once-and-Done (OD) methods

Harvest Method[z]			
CC		OD	
Estimated Average Yield (g/tray)[y]	No. of Weeks[x]	Estimated Average Yield (g/tray)[w]	No. of Weeks[v]
406.8	6	454.0	10
813.4	12		
1220.1	18	908.0	20
1626.8	24		
2033.5	30	1362.0	30

Note. [z] CC, Cut-and-Come-Again; OD, Once-and-Done.

[y] The estimated yield was calculated by sum of first two harvest average yield based upon 180.9 g/tray for 1st harvest and 225.8 g/tray for 2nd harvest using the CC method in our study.

[x] Number of weeks was calculated by sum of first two harvest period using the CC method in our study.

[w] The estimated yield was calculated by sum of first two harvest average yield based upon 221.0 g/tray for 1st harvest and 233.0 g/tray for 2nd harvest using the OD method in our study.

[v] Number of weeks was calculated by sum of first two harvest period using the OD method in our study.

5. Conclusions

Our study showed that the CC method in an FTS had an advantage over the OD method in productivity because it allowed multiple harvests in a shorter cultivation time and had a benefit of increasing yield. Additionally, the production would be further increased by utilizing a combined harvest method of CC and OD, and the cost labor and materials could also be reduced.

Acknowledgements

We would like to thank Matthew Ferrell and Gene Jacobs for their assistance with maintaining the aquaculture system.

References

Adler, P. R., Harper, J. K., Takeda, F., Wade, E. M., & Summerfelt, S. T. (2000). Economic evaluation of hydroponics and other treatment options for phosphorus removal in aquaculture effluent. *HortScience, 35*, 993-999.

Blidariu, F., & Grozea, A. (2011). Increasing the economical efficiency and sustainability of indoor fish farming by means separation of aquaponics-review. *Animal Sci. and Biotech, 44*(2), 1-8.

Borrelli, K., Koenig, R. T., Jaeckel, B. M., & Miles, C. A. (2013). Yield of leafy greens in high tunnel winter production in the Northwest United States. *HortScience, 48*(2), 183-188.

Boyd, C. E., & Tucker, C. S. (2014). *Handbook for aquaculture water quality*. Auburn, AL: Craftmaster Printers, Inc.

Buzby, K. M., & Lin, L.-S. (2014). Scaling aquaponic systems: Balancing plant uptake with fish output. *Aquacultural Eng., 63*, 39-44. http://dx.doi.org/10.1016/j.aquaeng.2014.09.002

Buzby, K. M., Waterland, N. L., Semmens, K. J., & Lin, L.-S. (2016a). Evaluating aquaponic crops in a freshwater flow-through fish culture system. *Aquaculture, 460*, 15-24. http://dx.doi.org/10.1016/j.aquaculture.2016.03.046

Buzby, K. M., West, T. P., Waterland, N. L., & Lin, L.-S. (2016b). Remediation of flow-through trout raceway effluent via aquaponics. *North American Journal of Aquaculture* (In press).

Dyer, D. J. (2006). *Effectiveness of aquatic phytoremediation of nutrients via watercress (Nasturtium officinale), basil (Ocimum basilicum), dill (Anethum graveolens) and lettuce (Lactuca sativa) from effluent of a flow-through aquaculture operation* (MS Thesis, West Virginia University, Morgantown).

Gazula, A., Kleinhenz, M. D., Streeter, J. G., & Raymond, M. A. (2005). Temperature and cultivar effects on anthocyanin and chlorophyll b concentrations in three related lollo rosso lettuce cultivars. *HortScience, 40*(6), 1731-1733.

Hinshaw, J. M., & Fornshell, G. (2002). Effluent from raceways. In J. R. Tomasso (Ed.), *Aquaculture and the environment in the United States* (pp. 77-104). U.S. Aquaculture Society, Baton Rouge, LA.

Jenni, S. (2005). Rib discoloration: A physiological disorder induced by heat stress in crisphead lettuce. *HortScience, 40*(7), 2031-2035.

Kader, A. A., Lipton, W. J., & Morris, L. L. (1973). Systems for scoring quality of harvested lettuce. *HortScience, 8*(5), 408-409.

Love, D. C., Fry, J. P., Genello, L., Hill, E. S., Frederick, J. A., Li, X., & Semmens, K. (2014). An international survey of aquaponics practitioners. *Plos One, 9*(7), e102662. http://dx.doi.org/10.1371/journal.pone.0102662

Oziel, J., & Oziel, T. (2013). A basic overview of growing food in water along with fish. *Water Garden J., 28*(2), 6.

Rakocy, J. E., Bailey, D. S., Shultz, R. C., & Thoman, E. S. (2004). Aquaponic production of tilapia and basil: Comparing a batch and staggered cropping system. *Acta Hort. (ISHS), 648*, 63-69. http://dx.doi.org/10.17660/ActaHortic.2004.648.8

Rakocy, J. E., Masser, M. P., & Losordo, T. M. (2006). *Recirculating Aquaculture Tank Production Systems: Aquaponics – Integrating Fish and Plant Culture* (SRAC Publication No. 454, pp. 1-16). Southern Regional Aquaculture Center. Retrieved from https://srac.tamu.edu/viewFactSheets

Rakocy, J. E. (2007). Aquaponics: Vegetable hydroponics in recirculating systems. In M. B. Timmons & J. M. Ebeling (Eds.), *Recirculating aquaculture systems* (2nd ed., pp. 631-672). Cayuga Aqua Ventures, Ithaca, NY.

Rana, S., Bag, S. K., Golder, D., Mukherjee, S., Pradhan, C., & Jana, B. B. (2011). Reclamation of municipal domestic wastewater by aquaponics of tomato plants. *Ecol. Eng., 37*(6), 981-988. http://dx.doi.org/10.1016/j.ecoleng.2011.01.009

Snow, A., Anderson, B., & Wootton, B. (2012). Flow-through land-based aquaculture wastewater and its treatment in subsurface flow constructed wetlands. *Environ. Rev., 20*(1), 54-69. http://dx.doi.org/10.1139/a11-023

Soderberg, R. W. (1995). Flowing Water Fish Culture, CRC Press LLC, Boca Raton, FL

U.S. Department of Agriculture. (2013). Western Region SARE Research Innovations. *Organic winter production scheduling in unheated high tunnels in Colorado.* U.S. Dept. of Agr., Washington, D.C.

Viadero, R. C., Cunningham, J. H., Semmens, K. J., & Tierney, A. E. (2005). Effluent and production impacts of flow-through aquaculture operations in West Virginia. *Aquacultural Eng., 33*(4), 258-270. http://dx.doi.org/10.1016/j.aquaeng.2005.02.004

20

Physicochemical and Sensory Quality of Brown Sugar: Variables of Processing Study

Raphael Della Maggiore Orlandi[1], Marta Regina Verruma-Bernardi[1], Simone Daniela Sartorio[1] & Maria Teresa Mendes Ribeiro Borges[1]

[1] Department of Agroindustrial Technology and Rural Socieconomy, Federal University of São Carlos School (CCA), Araras, SP, Brazil

Correspondence: Raphael Della Maggiore Orlandi, Department of Agroindustrial Technology and Rural Socieconomy, Federal University of São Carlos School (CCA), Rodovia Anhanguera, km 174, Mail Box 153, Zip Code: 13600-970, Araras, SP, Brazil. E-mail: raphael.dmo@gmail.com

The research is financed by FAPESP (2012/20234-3).

Abstract

The lack of standardization in the processing of brown sugar reflects in its physicochemical and sensory quality and, consequently, harms the small producers and cottage industries in the products commercialization. In this context, this work aimed to study the influence of the variables – period of the year, variety of sugar cane, pH and final temperature of juice cooking – on the acquisition of a product that is acceptable according to the physicochemical and sensory requisites. The physicochemical parameters of the sugars that best classified the product presented juice neutral pH (7.0) and finalization temperature at 118 °C for both varieties, in the late period. In the sensory aspect, the sugars of the variety RB92579 found, in a general context, greater acceptance, being classified, also by the judges, as sugars of a darker appearance, smaller granules, less intense sweet aroma and flavor, and high solubility.

Keywords: temperature, pH, sugar cane variety

1. Introduction

The brown sugar, contrary to the refined one, does not go through a large number of chemical process; therefore, customers consider it as a more natural substitute compared to the white sugar.

The brown sugar is the raw, moist and dark sugar obtained after the sugar cane dehydration. Since it does not go through a refining process, it maintains the calcium and the iron, in addition to other vitamins and minerals.

According to Mendonça et al. (2000), the brown sugar aims the group of people with eating habits based on the minimization or absence of chemical products during the food processing.

This type of sugar is easily recognized by its color and flavor, which are similar to panela or even of the sugar cane juice. This characteristic is due to the fact that the brown sugar does not go through more elaborated processes of juice clarification.

Verruma-Bernardi et al. (2010) evaluated the sensorial characteristics and the preference for brown sugar brands, which showed some differences regarding their appearance and texture.

The the brown sugar color is one of the main attributes that influence purchase, according to Lopes and Borges (1998), and it can be modified according to the differences of temperature and juice pH during the product processing.

In another study made with brown sugar brands, Verruma-Bernardi et al. (2007) emphasize the lack of standardization in the product elaboration, which proves the sugars sensory variability and the occurrence of a high percentage of moisture and reduced sugars in the analyzed samples – aspects that would consequently affect their shelf life.

According to Durán-Rojas et al. (2012), the lack of standardization in the brown sugar production process harms its market positioning, what leads to the rejection of the product by the customer.

Studies by Mosquera et al. (2007) and Mujica et al. (2008) reinforce the importance of process standardization and the study of variables that influence the production of brown sugar. Among the variables they mentioned, the process temperature, pH of the juice, variety of sugarcane, climate, soil, mode of extraction and the juice cleaning. Mujica et al. (2008), demonstrate in their study the influence of temperature and sugarcane variety, they show the final pH differences, reducing sugars contents, moisture, color, and shelf life of the product.

The objective of this work was to produce brown sugars and evaluate the sensorial and physical-chemical quality, by combining 4 factors: time of year, variety of sugarcane, the juice pH and final process temperature.

The time of year (the age of sugarcane), is related to the quantity of Reducing Sugars present in the raw material. The different physiological properties of sugar cane, may influence the chemical attributes and consequently the quality of sugar.

The pH of the juice and the final process temperature, influence together the acquisition of a sensorially attractive and chemical properties that increase its quality and shelf life. Generally this proprieties are related with the moisture, amount of reducing sugars and color of the brown sugar.

2. Materials and Methods

2.1 Brown Sugar Processing

Brown sugars were processed with variation in pH (5.5 and 7.0), temperature (118 and 128 °C) and sugar cane (RB92579 and RB965917). Treatments were conducted in two periods of the year (July – medium, and October – late), numbering 16 treatments.

The brown sugars were produced in 10L metallic containers on workbench stoves. For the production, six liters of sugar cane juice were used, and pH was adjusted with the aid of limewater and a potentiometer, until the liming the juice to pH 7.0 (Figure 1).

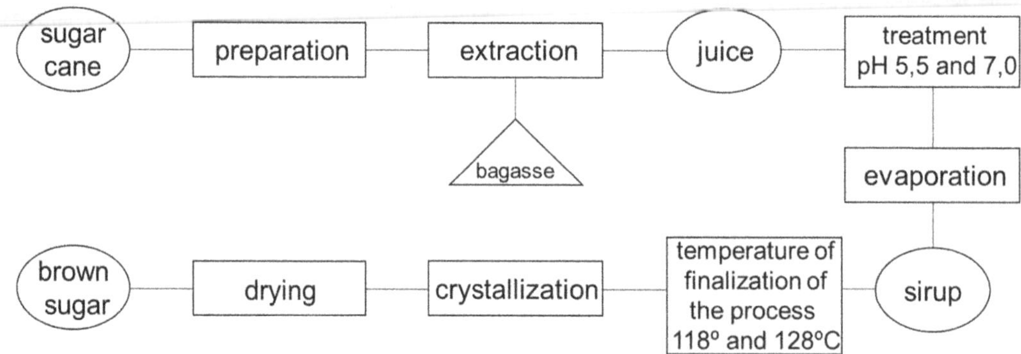

Figure 1. Flow chart of brown sugars production

2.2 Physicochemical Analyses

The physicochemical analyses were carried out, in triplicate, in the Laboratory of Analysis and Technological Simulation (LAST). The analyzed pyshicochemical parameters were moisture, which was analyzed using the gravimetric method by weight loss after drying at 100-105 °C, according to the Instituto Adolfo Lutz [IAL] (2005); polarization (Pol), according to the ICUMSA methods (2001); conductivity ashes, measured by the conductivity of the 5% (m/m) sugar solution (ICUMSA, 2011); spectrophotometric color, by the absorbency of the 1% (m/v) brown sugar solution in 420 nm wavelength; and reducing sugars (RS) and total reducing sugars (TRS), quantified via Lane-Eynon method (FERMENTEC, 2003).

2.3 Sensory Analysis

The work was approved by the UFSCar Ethics Committee, according to Certificate for Ethics Assessment (CAAE) n°17964513.3.0000.5504.

The sensory analysis was carried out in the Laboratory of Sensory Analysis of Centro de Ciências Agrárias (UFSCar). The samples of brown sugar (20 g) were served in coded transparent glasses together with mineral water. To prepare the samples, the sugars produced in the medium period were homogenized with the ones produced in the late period, both belonging to the same treatment.

Twenty-five participated of the ordering test of difference and preference (ABNT, 1994). The samples were presented and the judge was asked to classify them in ascending order of intensity for each of the following attributes: color (light – dark), granularity (thin – thick), cane aroma (weak – strong), sweet aroma (weak – strong), sweet flavor (weak – strong), bitter flavor (weak – strong), solubility of the sugar in the mouth (slow – fast) and dry texture (little – much). At last, the ordering test of preference (least preferred – most preferred).

The sensory attributes used in the ordering test were chosen through the utilization of the 8 samples in 4 sessions (2 samples at a time) by the Method Grid described by Kelly (1955), mentioned by Moskowitz (1983). The selection criteria attribute was the most mentioned, that is, when more than a half of the judges reported the same attribute.

2.4 Statistic Analyses

The statistic design was randomized in blocks (seasons) in factorial scheme 2 (variety) × 2 (pH) × 2 (temperature). The data were submitted to the analysis of variance (ANOVA) in its simplest form, and the Tukey's test was used as multiple comparisons test. The adopted significance level was 5%, and all the analyses were carried out using the R software, 3.0.1 version. The ordening test data were evaluated using the Friedman test (5% of probability), with the Newell and MacFarlane (1987).

3. Results and Discussion

3.1 Pyshicochemical Analyses

Table 1 presents the results of moisture, ashes, *Pol*, color, RS, and TRS of the varieties that were harvested in the different ripening periods, medium (July) and late (October), of the eight treatments.

Table 1. Average results of the physicochemical analyses of brown sugar samples in the medium period (July) and late period (October)

Variety	pH	Temperature (%)	Moisture (%)	Ashes (%)	Pol	Color	RS (%)	TRS (%)
Medium Period								
RB92579	5.5	118 °C	7.4	1.60	83.4	2803	8.7	93
RB92579	5.5	128 °C	3.3	1.70	80.1	4520	10.4	97
RB92579	7.0	118 °C	6.7	1.80	85.7	3495	6.7	99
RB92579	7.0	128 °C	3.2	1.80	84.5	4268	8.2	97
RB965917	5.5	118 °C	5.6	1.30	87.7	2245	7.1	97
RB965917	5.5	128 °C	3.9	1.30	85.4	2360	8.0	98
RB965917	7.0	118 °C	6.2	1.30	89.9	2253	4.0	96
RB965917	7.0	128 °C	3.6	1.30	89.4	2293	3.9	97
Later Season								
RB92579	5.5	118 °C	5.1	1.10	85.9	1160	6.3	100
RB92579	5.5	128 °C	3.9	1.10	83.0	1718	8.4	98
RB92579	7.0	118 °C	4.3	1.20	90.8	1460	2.1	99
RB92579	7.0	128 °C	2.9	1.20	87.6	1913	5.1	100
RB965917	5.5	118 °C	6.2	0.86	90.0	0895	4.7	100
RB965917	5.5	128 °C	3.8	0.70	83.6	1400	8.9	100
RB965917	7.0	118 °C	5.4	0.97	92.0	0905	2.4	100
RB965917	7.0	128 °C	2.3	0.97	89.9	1335	4.2	100

Note. RS: reducing sugars; TRS: total reducing sugars.

According to the results, in the ratio of pH and temperature for each treatment, there is a tendency in the physicochemical behavior of both attributes for the two periods.

Regarding the moisture, the increase in the process finalization temperature caused a decrease of the the sugars' moisture percentage. Also, there was a direct influence of the juice's acidity on the sugar's moisture, since the juices that were neutralized up to pH 7.0 presented less moisture when compared to the treatments with pH 5.0 at the same temperature. This phenomenon can be explained by the higher proportion of reducing sugars (RS) in the sugars whose juices were vaporized in their natural pH (5.5). The acid environment corroborates, together with the heating, the sucrose hydrolysis and, consequently, the generation of RS, which is hygroscopic.

Significant differences were verified in the parameters pH 7.0 (4.3^a); 5.5 (4.9^b) and Temperature 118 °C (5.9^a); 128 °C (3.4^b). According to Lopes and Borges (1998), to meet the specifications, a good-quality brown sugar must have a maximum of 5.0% of moisture, as it provides sensory and microbiological disadvantages when risen, such as the caking of sugar and the decrease in the product's shelf life due to the increase of water activity and, consequently, the easiness for microbial activity (P. A. Bobbio & F. O. Bobbio, 1995).

Regarding the ashes, the slight difference that was observed in some results can be attributed to the inorganic compounds that were incorporated with the addition of limewater for the liming of the juice's pH, since there was a small increase in the amountamount of ashes in the sugar that had the juices evaporated at pH 7.0, compared to its correspondent evaporated at a natural pH.

For the same atribute, significant differences of pH – 7.0 (1.3^a); 5.5 (1.2^b) and Variety – RB965917 (1.1^a); RB92579 (1.4^b) were noticed. According to the authors, to meet the specifications, a good-quality brown sugar must have a maximum of 2.4% of ashes. A large amount of ashes modifies the flavor of the product, giving it a bitter or salt flavor, what harms its acceptability.

The proportion of sucrose, represented by the Polatization (*Pol*) value, was significantly different for the attributes pH – 7.0 (88.7^a); 5.5 (84.9^b), Temperature – 118 °C (88.2^a); 128 °C (85.4^b) and Variety – RB965917 (88.5^a); RB92579 (85.1^b). The data also showed an increase in *Pol* for the samples from the late period, compared to the ones from the medium period, what was expected due to the sugar cane ripening.

As noticed ratio between pH and Temperature for each treatment, there was a decrease in *Pol* for the treatments of higher temperature and with the juice's natural pH (5.5), what explains again the chemical phenomenon of sucrose hydrolysis in reducing sugars. The variable Variety also influences on the proportion of sucrose, since the varieties present different phenotypic properties.

These results showed that the brown sugar Polarization is closely related to the variations of pH, temperature and variety of the cane that was used in the sugar production. To Lopes and Borges (1998), to meet the specifications, a good-quality brown sugar must have a minimum of 85°Z of Polarization.

Regarding the color, besides showing varietal influence, the higher the temperature is and the lower the juice pH is, it shows a tendency to the darkening of the sugar. Therefore, there were, statistically, significant differences of the attributes Temperature – 128 (1985.83^a); 118 (2475.78^b) and Variety – RB965917 (1710.6^a); RB92579 (2802.0^b).

The reducing sugars, originated in the sucrose hydrolysis due to the high temperatures, are one of the precursors of the brown sugar's color, since they generate compounds that result from the reaction of these sugars with amino acids (Maillard reaction).

Another importante factor in the color of the brown sugar is the structural modification, due to pH limings of natural pigments that are present in the sugar cane. The flavonoid group is the most critical one to the sugar processing and the responsible for up to 30% of the color of raw sugar with pH 7.0. The flavonoids can modify its coloration according to the environment's pH (ARAÚJO, 2001).

As mentioned, the values of Reducing Sugars (RS) increased in the treatments with higher temperature and lower pH, which are related to the sucrose's hydrolysis in glucose and fructose. The variables Variety and Period of the year are factors that also influence on this characteristic. Statistically, for the variable RS, there was a significant difference in the attributes pH – 5.5 (7.84^a); 7.0 (4.43^b), Temperature – 128 °C (7.09^a); 118 °C (5.19^b) and Variety – RB92579 (6.91^a); RB965917 (5.38^b). According to Lopes and Borges (1998), to meet the specifications, a good-quality brown sugar must have a maximum of 3.5% of reducing sugars.

For the variable TRS, the interection was noticed only in the factor Period of the year: the higher concentrations of TRS are found in the late period, due to the sugar cane ripening. According to the authors, to meet the specifications, a good-quality brown sugar must have a minimum of 90% of total reducing sugars. The factor Period of the year was significant for all the variables.

The study conducted by Mujica et al. (2008) states that the sugar cane variety had great importance in the quality of brown sugar, since it influenced the amount of reducing sugars, in the pH and color of the final product, concluding still, that is necessary to drying the product after crystallization, for the monitoring of humidity. The present study shows that it is possible to produce sugars with color, moisture content and reducing sugars quantity controlable, just using different varieties and combining appropriate process temperature and pH. Therefore, there would be no more needs for further processes to ensure the quality of the product.

3.2 Sensory Analyses

The results that were obtained in the ordering test for difference and preference show that the samples had significant differences ($p \leq 0.05$) among themselves in all the attributes, including global preference (Tables 2 and 3).

Regarding the attribute color, the samples differed among varieties, and, inside each variety, they did not present significant differences related to the treatments. Consequently, it is clear that the final product color was related to the variety of sugar cane used in the brown sugar production and that the modifications that were caused by the different treatments were not visibly noticed, but only by spectrophotometry.

For the attribute granularity, there was a significant difference among the varieties, since the brown sugar that was produced with the variety 2 (RB92579) distinguished itself with a thinner texture than the variety 1 (RB965917).

Regarding the sugar cane aroma, the samples of brown sugar $V_1T_{128}pH_{5,5}$ and $V_1T_{128}pH_{7,0}$ presented the highest sums and, consequently, higher intensity of the sugar cane aroma, in opposition to the samples $V_1T_{118}pH_{5,5}$, $V_2T_{118}pH_{5.5}$ and $V_2T_{128}pH_{5.5}$, which presented the lowest sums.

About the sweet aroma, the samples of the variety RB965917 presented the highest sums, being characterized as having the most pronounced sweet aroma, while the samples of the variety RB92579 presented the lowest sum. Thus, it is possible to consider the existence of a ratio between the sugar cane variety and the final product's sweet aroma.

Table 2. Sum of the orders of the sensory attributes of appearance and aroma to the studied brown sugars

Samples	Appearance		Aroma	
	Color	Granularity	Cane	Sweet
$V_1T_{118}pH_{5.5}$	44^{b}	193^{a}	89^{b}	118^{abc}
$V_1T_{128}pH_{5.5}$	48^{b}	160^{ab}	153^{a}	142^{ab}
$V_1T_{118}pH_{7.0}$	75^{b}	152^{ab}	125^{ab}	136^{ab}
$V_1T_{128}pH_{7.0}$	83^{b}	132^{bc}	151^{a}	163^{a}
$V_2T_{118}pH_{5.5}$	162^{a}	45^{d}	77^{b}	74^{c}
$V_2T_{128}pH_{5.5}$	171^{a}	64^{d}	91^{b}	78^{c}
$V_2T_{118}pH_{7.0}$	140^{a}	69^{d}	105^{ab}	95^{bc}
$V_2T_{128}pH_{7.0}$	177^{a}	85^{cd}	109^{ab}	92^{bc}

Note. V_1: variety RB965917; V_2: variety RB92579. Values followed by the same letters on the horizontal axis do not differ significantly ($p \geq 0.05$). Minor difference ≥ 53.

For the sweet flavor, the sample $V_1T_{128}pH_{7.0}$ presented the highest sum, differing significantly from the sample $V_2T_{118}pH_{5,5}$, which presented the lowest sum and, consequently, lower intensity of sweet flavor. Regarding this attribute, it is not possible to establish a ratio between variety and treatment, because the other samples do not differ among themselves statistically. For bitter flavor, the sample $V_1T_{118}pH_{7.0}$ presented the highest sum, differing from the samples $V_2T_{128}pH_{5.5}$ and $V_2T_{118}pH_{7.0}$, of lower sums. Nevertheless, the three samples do not differ significantly from the others.

Regarding the solubility, the sample $V_2T_{118}pH_{5.5}$ presented the highest sum, that is, it is the sample that was characterized with higher speed of solubility in the mouth, differing significantly from the samples of variety 1. The sample $V_1T_{118}pH_{5.5}$ presented, inside the variety 1, the lowest sum. It is possible to establish a ratio between solubility and texture of the sugar, because the variety 1 presented the samples with the lowest sum for solubility and the highest sum for granularity, while the samples of variety 2 presented an opposed behavior, with higher sum for solubility and lower sum for granularity.

For dry texture, there was a slight variation. The sample $V_1T_{118}pH_{5.5}$ presented the highest sum, differing significantly from the sample $V_2T_{118}pH_{7.0}$, which presented the lowest sum. Both of them do not differ significantly from the other samples.

Regarding the global preference, the samples of variety 2 presented the highest sum, differing significantly only from the sample $V_1T_{118}pH_{5.5}$, inside the variety 1, which presented the lowest sum, being, therefore, the least favorite.

Table 3. Sum of the orders of the sensory attributes of brown sugars

Samples	Flavor		Texture		Preference Global
	Sweet	Bitter	Solubility	Dry	
$V_1T_{118}pH_{5.5}$	122^{ab}	119^{ab}	58^{c}	139^{a}	69^{b}
$V_1T_{128}pH_{5.5}$	130^{ab}	126^{ab}	98^{bc}	124^{ab}	112^{ab}
$V_1T_{118}pH_{7.0}$	117^{ab}	150^{a}	96^{bc}	106^{ab}	91^{ab}
$V_1T_{128}pH_{7.0}$	161^{a}	115^{ab}	96^{bc}	122^{ab}	108^{ab}
$V_2T_{118}pH_{5.5}$	62^{c}	111^{ab}	160^{a}	102^{ab}	139^{a}
$V_2T_{128}pH_{5.5}$	96^{bc}	95^{b}	141^{ab}	111^{ab}	135^{a}
$V_2T_{118}pH_{7.0}$	109^{abc}	83^{b}	129^{ab}	85^{b}	122^{a}
$V_2T_{128}pH_{7.0}$	98^{bc}	98^{ab}	122^{ab}	111^{ab}	124^{a}

Note. V_1: variety RB965917; V_2: variety RB92579. Values followed by the same letters on the horizontal axis do not differ significantly ($p \geq 0.05$). Minor difference ≥ 53.

4. Conclusion

➢ The parameters that were studied – pH, temperature of finalization, period of the year and variety – are important in the quality control of brown sugar production.

➢ The sugars with the best physicochemical parameters presented neutral juice pH (7.0) and temperature of finalization at 118 °C in the late period for both varieties.

➢ The sugars that were produced with the ripest sugar cane (late period) presented physicochemical characteristics that increase their shelf life, since they presented a smaller amount of RS and lower moisture, in a general way.

➢ The results that were obtained in the ordering analysis show that the treatments differed significantly ($p \leq 0.05$) among themselves in almost all the attributes, including global preference, what proves the heterogeneity among the samples.

➢ The variety presented great sensory influence on the attributes color, granularity, sweet aroma, sweet flavor, and solubility, in which there were at least two samples that differed significantly compared to the other variety. In this case, the variety RB965917 (V_1) can be considered as the variety of a clearer brown sugar, with bigger granules, more intense sweet aroma and flavor and low solubility.

➢ It can also be concluded that the choice of the variety of sugar cane used plays a key role to obtain a final product of higher quality, since the data of global preference with higher sum were concentrated in the group of samples of brown sugar of the variety 2 (RB92579), what can mean that the judges prefer a sugar of a darker appearance, smaller granules, less intense sweet aroma and flavorm and high solubility.

References

ABNT (Associação Brasileira de Normas Técnicas). (1994). Teste de ordenação em análise sensorial. *NBR 13170*. Rio de Janeiro.

Araújo, E. R., Borges, M. T. M. B., Ceccato-Antonini, S. R., & Verruma-Bernardi, M. R. (2011). Qualidade de açúcares mascavo produzidos em um assentamento da reforma agrária. *Alimentos e Nutrição, 22*(4), 617-621.

Araújo, J. M. A. (2001). *Química de Alimentos* (2nd ed.). Viçosa: Teoria e Prática.

Bobbio P. A., & Bobbio, F. O. (1995). *Química do Processamento de Alimentos* (2nd ed.). São Paulo: Varela Ltda.

Durán Rojas, E., Pérez, R., Cardoso, W., & Pérez, O. (2012). A Colorimetria e aceitação de açúcar mascavo. *Temas Agrários, 17*(2), 30-42.

Fermentec. (2003). *Métodos analíticos para o controle da produção de açúcar e álcool* (3rd ed.). Piracicaba: Fermentec.

IAL (Instituto Adolfo Lutz). (2005). *Métodos físico-químicos para análise de alimentos* (4th ed.). Brasília: ANVISA.

ICUMSA. (2011). International Commission for Uniform Methods of Sugar Analysis. *ICUMSA methods book.* England: Icumsa.

Lopes, C. H., & Borges, M. T. M. R. (1998). *Produção de açúcar mascavorapadura e melado de cana* (1st ed.). Rio Grande do Sul: Capacitação Tecnológica para a Cadeia Agroindustrial.

Mezaroba, S., Meneguetti, C. C., & Groff, A. M. (2010). Processo de produção do açúcar de cana e os possíveis reaproveitamentos dos subprodutos e resíduos resultantes do sistema. *IV Encontro de Engenharia de Produção Agroindustrial (FECICLAM).* Campo Mourão/PR, Brasil.

Moskowitz, H. R. (1983). Product testing and sensory evaluation of foods. *Marketing and R & D Approaches* (p. 605). Westport: Food and Nutrition Press.

Mosquera, S. A., Carrera, J. E., & Villada, H. S. (2007). Variables que afectan la calidad de la panela procesada en el departamento del cauca. *Facultad Ciencias Agropecuarias, 5*(1), 17-27.

Mujica, M. V., Guerra, M., & Soto, N. (2008). Efecto de la variedad, lavado de la caña y temperatura de punteo sobre la calidad de la panela granulada. *Interciencia, 33*(8), 598-602.

Newell, G. J., & Macfarlane, J. D. (1987). Expanded tables for multiple comparison procedures in the analysis of ranked data. *Journal of Food Science, 52*(6), 1721-1725. https://doi.org/10.1111/j.1365-2621.1987.tb05913.x

Verruma-Bernardi, M. R., Borges, M. T. M. R., Lopes, C. H., Della-Modesta, R. C., & Ceccato-Antonini, S. R. (2007). Avaliação Microbiológica, Físico-Química e Sensorial de Açúcares Mascavos Comercializados na Cidade de São Carlos, SP. *Brazilian Journal of Food Technology, 10*(3), 205-211.

Verruma-Bernardi, M. R., Silva, T. G. E. R., Borges, M. T. M. R., Lopes, C. H., & Deliza, R. (2010). Avaliação sensorial de açúcar mascavo. *Brazilian Journal of Food Technology, 14*(1), 29-38.

21

Year-Round Lettuce (*Lactuca sativa* L.) Production in a Flow-Through Aquaponic System

Gaylynn E. Johnson[1], Karen M. Buzby[2], Kenneth J. Semmens[3,4] & Nicole L. Waterland[1]

[1] Division of Plant and Soil Sciences, West Virginia University, Morgantown, WV, USA

[2] Department of Civil and Environmental Engineering, West Virginia University, Morgantown, WV, USA

[3] Division of Animal and Nutritional Sciences, West Virginia University, Morgantown, WV, USA

[4] Aquaculture Research Center, Kentucky State University, Frankfort, KY, USA

Correspondence: Nicole L. Waterland, Division of Plant and Soil Sciences, West Virginia University, Morgantown, WV, 26506, USA. E-mail: nicole.waterland@mail.wvu.edu

The research is financed by the U.S. Department of Agriculture Special Grant Agreement 2010-34386-21745 and West Virginia University Agriculture and Forestry Experimental Station (Hatch Grant WVA00640). Scientific Article No. 3292 of the West Virginia Agricultural and Forestry Experiment Station, Morgantown.

Abstract

Aquaponics is the combination of hydroponics and aquaculture that sustainably produces both animal and plant food products. Soluble nutrients are released into water by the fish providing nutrition for plant growth. Lettuce (*Lactuca sativa* L.) is one of the most popular vegetables grown in aquaponic systems. In this experiment, the feasibility of year-round lettuce production utilizing a cold water flow-through aquaponic system (FTS) growing trout (*Oncorhynchus mykiss*) in a high tunnel was evaluated. A high tunnel is a greenhouse-like facility constructed with polyethylene covering a metal frame which extends the growing season and protects the crop from cold temperatures. The average night air temperature inside the high tunnel during winter in Wardensville, WV was 2.9±3.4 °C and it helped extend the growing period into the fall and winter. Results from this pilot scale experiment showed the potential for year-round lettuce production in an FTS. Average yield (fresh harvest weight per tray) in the spring season was the highest, while productivity (average yield per week) during the summer season was higher than that in spring. During the extended growing seasons (fall and winter), more than a quarter (30.6%) of the total lettuce production was obtained. The yield per unit area (7.4 kg m^{-2}) from our pilot study was significantly higher than that from the reported average field production (3.1 kg m^{-2}) in the U.S. except California and Arizona where year-round production of lettuce occurs. To compensate for lower lettuce yields during cold seasons, high value crops requiring less nutrients and tolerant to the colder environment may be considered.

Keywords: aquaculture, flow-through aquaponics, high tunnel, lettuce, trout

1. Introduction

Aquaponics refers to any agricultural system that integrates aquaculture (rearing fish) with hydroponics (producing plants without soil). Soluble nutrients are released into water by the fish providing nutrition for plant growth while plants remove compounds such as ammonia that may impact growth and health of the fish. Traditionally, aquaponics has combined recirculating aquaculture system (RAS) with plant production. Nutrient availability in RAS has been shown to be similar to that in nutrient rich hydroponic systems (Rakocy, Masser, & Losordo, 2006). Recently, Buzby et al. (2016) demonstrated that a cold water flow-through aquaculture production system may also be used to produce aquaponic crops. Flow-through aquaculture systems differ from RAS in that water is not reused or recycled and nutrients do not accumulate over time. Nutrient concentrations in the flow-through aquaculture system are substantially lower than that in RAS (Buzby & Lin, 2014). In addition, water temperature is dependent on water source and tends to be cooler in spring fed flow-through aquaculture systems than that in RAS (Losordo, Masser, & Rakocy, 1992).

The plant production component in an aquaponic system utilizes hydroponic technology. Hydroponic systems employ different arrangements to provide nutrients to the plants' roots. Among them, raft or deep water culture (DWC), nutrient film technique (NFT) and media-filled beds (MFB) are commonly practiced methods in hydroponics (Love, Uhl, & Genelloa, 2015; Love et al., 2014; Resh, 2013; Saavas, 2002). In a DWC system, plants are grown in the polystyrene trays or rafts that float on top of the water. Roots of plants grow into the water to take up the nutrients. DWC systems require a large volume of water in order for the root system to be submerged to the nutrient solution. In our system DWC was used in a high tunnel.

A high tunnel is a tunnel-shaped enclosed structure constructed with polyethylene, polycarbonate, plastic or fabric over a metal or wooden frame which is easier and more economical to construct compared to building a greenhouse. The use of a high tunnel not only protects the crop from harsh environments, but also extends the growing season to produce cold-tolerant crops such as lettuce (Lee, Liao, & Lo, 2015). Cold-tolerant crops are able to grow during winter months for both a longer cropping season and a winter market (Borrelli, Koenig, Jaeckel, & Miles, 2013). Crop production in high tunnels can exploit market conditions where supplies are limited and prices are high (Castoldi, Bechini, & Ferrante, 2011). Production of several vegetables and herbs including basil (Adler, Harper, Takeda, Wade, & Summerfelt, 2003; Rakocy, Bailey, Shultz, & Thoman, 2004), cucumber and herbs (Savidov, Hutchings, & Rakocy, 2007), and lettuce and tomato (Rakocy, Hargreaves, & Bailey, 1993) have been reported in aquaponic systems. Recently, performance of a wide variety of food crops including lettuce, herbs, Asian greens and vegetables was evaluated in a flow-through aquaponic system (Buzby et al., 2016).

Lettuce is one of the most popular crops grown aquaponically. Lettuce is genetically diverse and comes with variety of shapes, colors and textures (Kim, Moon, Tou, Mou, & Waterland, 2016). Lettuces rank first in the United States in total value of production of fresh market vegetables (United States Department of Agriculture, 2016). National consumption, on average, exceeded 23 lbs per person annually between 1980 and 2010 (United States Department of Agriculture, 2014). Lettuce production is limited by temperature. If air temperature exceeds 30/16 °C day/night, the crop will develop tipburn, produce premature flower stalks (bolting), and form loose heads (Fukuda et al., 2011). Thus, lettuce is a good candidate for a cold water flow-through aquaponic system (FTS).

Aquaponic production of crops has gained popularity and crop production in RAS is well documented (Losordo et al., 1992; Ido, 2016; Martins et al., 2010; Rakocy et al., 2004; Rakocy, 2007). However, there is limited information regarding crop production, especially year-round lettuce production in an FTS. In this study, the feasibility of year-round aquaponic lettuce production using effluent from an FTS in a high tunnel was evaluated. The cultivar 'Red Sails' was chosen because it resists bolting and grows well in cold environments (Park Seed, 2016).

2. Materials and Methods

2.1 Facility

Experiments were conducted in a 7.9 m × 14.6 m × 3.7 m tall high tunnel covered in 6 mil (0.15 mm thick) clear greenhouse plastic located in Wardensville, West Virginia (394′32″N, 7835′40″W). Ventilation in the high tunnel was provided by rolling up the side walls approximately 1.4 m and removing the end walls. A shade cloth cover was put on the high tunnel on June 29th and removed October 1st to reduce light intensity and moderate air temperature inside the high tunnel. The end and side walls were closed in mid-October. Three channels (13.7 m × 2.7 m) were constructed of dry stacked concrete block on a fine gravel base and then covered with a 45 mil, heavy-duty, black plastic pond liner (Firestone Pondgard, ethylene propylene diene monomer (EPDM), Model PG 5000; Indianapolis, IN, USA) liner. The block was stacked such that the channel was 30.5 cm deep.

2.2 Growth Conditions

The high tunnel maintained the inside temperature above outside temperatures during the entire year. This was especially apparent during the late fall through winter season (Figure 1). Average night temperatures inside the high tunnel remained above the freezing point, which allowed lettuce to grow without being damaged during the winter months. Water temperature of the spring fed system was relatively constant through the year (12.1-13.8 °C) and represented a heat source during the cold winter season.

Effluent from rainbow trout (*Oncorhynchus mykiss*) reared in a spring fed flow-through system in a nearby structure was used as the sole water and nutrient source in this study. Trout effluent was pumped to the high tunnel and distributed through a PVC manifold. Flow of effluent into each channel was independently controlled with a ball valve. Water depth was maintained at 23 cm with a standpipe drain.

'Red Sails' lettuce (*Lactuca sativa* L. var. *crispa* 'Red Sails') were grown in this experiment from seed purchased from Johnny's Selected Seeds (Winslow, ME, USA). Lettuce plants were grown in 128-cell styrofoam Speedling (67.6 cm × 34.6 cm; Speedling, Inc., Ruskin, FL) trays and vermiculite (Therm-O-Rock East Inc., Grade 3A; New Eagle, PA, USA) was used as the growing medium. Seeds were sown directly into the trays using a vacuum seeder. After seeds were sown, the trays were placed directly into the channels. Nine trays (3 × 3) as a block were placed at the top of the channel next to the influent. Initially, the remainder of the channel was filled with empty trays. An additional block of trays was sown and placed closest to the inlet next to the previous block of trays and placement of trays was repeated weekly until the channel was full. Each channel held maximum of 45 trays (five blocks of nine trays). Lettuce production was evaluated for an entire year from December 1, 2011 through December 21, 2012.

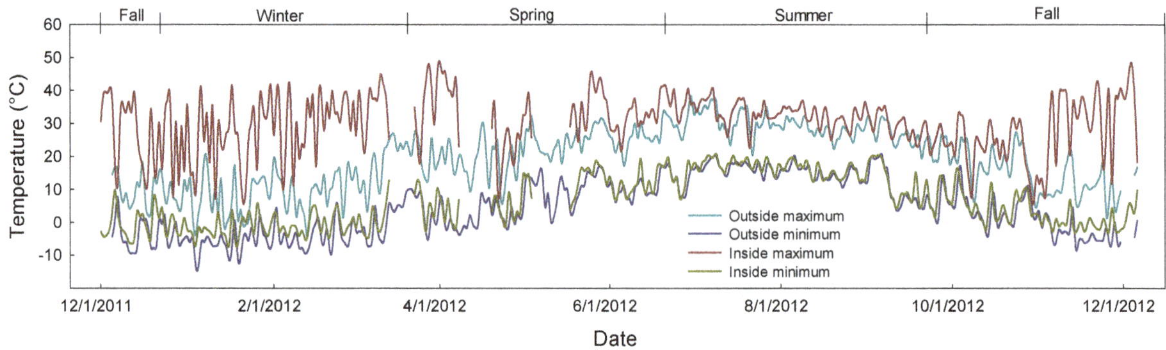

Figure 1. Daily maximum and minimum temperatures outside and inside of the high tunnel

Note. Maximum and minimum temperatures outside the high tunnel were obtained from National Oceanic and Atmospheric Administration (Wardensville in Hardy County).

2.3 Stand Establishment, Average Yield and Productivity

Stand establishment (SE), the ability of the seedlings to survive after germination, was determined two weeks after seeds were sown. The SE was defined as the percentage of visible seedlings at least 5 mm tall in a tray. Lettuce was considered ready to harvest when it reached an average height of 12.7-15.2 cm as measured from the surface of media to the top of the tallest leaf (United States Department of Agriculture, 2013). The trays of lettuce were harvested using an electric fillet knife which was drawn across the top of the tray severing the shoot from the roots. All lettuce in a block of nine trays were harvested at the same time and considered one harvest. Total fresh weight of all lettuce combined from nine trays was defined as harvest biomass. Average yield was determined as the average fresh harvest weight per tray. Seasonal yield was determined as the total fresh weight harvested in each season. Productivity was determined on a seasonal basis as the average yield per week.

2.4 Water Quality

Water samples were collected from the end of each channel in acid washed plastic bottles every two weeks. A 2 L bottle was used for TSS and a 250 ml bottle for the nutrient analyses. As soon as water samples were collected, they were immediately placed on ice. Upon transfer to the lab, samples were stored at 4 °C until they were analyzed. The samples were analyzed according to methods delineated by APHA (1995) for total ammonia nitrogen (TAN) (4500-NH_3 colorimetric method), nitrite (4500-NO_2) colorimetric method), nitrate (4500-NO_3 cadmium reduction colorimetric method), phosphate (4500-P ascorbic acid method) and total suspended solids (TSS) (2540D). Samples were analyzed within 24 hours of collection and then frozen. Nitrite concentrations never exceeded 0.02 mg L^{-1}, were often less than 0.01 mg L^{-1}, and they are not reported. Concentration of nitrate was determined within 30 days of collection.

2.5 Statistical Analysis

Temperature, light intensity and crop production data were analyzed by season. The seasons were defined as: spring (March 20-June 19), summer (June 20-September 21), fall (September 22-December 19) and winter (December 20-March 19).

Experimental design was randomized complete design. Stand establishment (SE), average yield and productivity values are the means of three replications (n = 3). Analysis of variance (ANOVA) was performed with PROC

GLM (generalized linear model) by SAS version 9.3 (SAS Institute, Inc., Cary, NC). Differences were determined using Tukey's significance test at $P \leq 0.05$.

3. Results

3.1 Environmental Conditions

The experiment was conducted in a high tunnel where air temperature inside the high tunnel was expected to be higher than outside, especially during the colder seasons (Table 1 and Figure 1). During the cold seasons (fall and winter), the average night temperature inside high tunnel was above the freezing point, allowing lettuce growth throughout the year. Once there was no danger of frost, the sides of the high tunnel were rolled up and the plastic was removed from the end walls. Additionally, shade cloth was placed over the high tunnel during the summer months. Both measures prevented air temperatures within the high tunnel from greatly exceeding the outside temperatures. Maintaining passive ventilation during the spring and summer and closing the high tunnel during fall and winter reduced the seasonal difference in daytime air temperatures (Figure 1).

Light intensity, measured as photosynthetic photon flux density (PPFD), was significantly higher in the spring than the other seasons. Installation of the shade cloth reduced light intensity during the summer months to levels similar to fall and winter (Table 1). Daily light integral, the integrated light intensity, was significantly greater during the spring than the other seasons. Water temperature was measured three times in December (12.1±0.05 °C), January (13.1±0.05 °C) and February (13.8±0.05 °C).

Table 1. The seasonal average of day and night temperatures, photosynthetic photon flux density (PPFD), and daily light integral (DLI) in the high tunnel

Season	Temperature (°C)		PPFD (μmol m^{-2} s^{-1})[z]	DLI (mol m^{-1} d^{-1})[y]
	Day	Night		
Spring	24.5±8.7	11.4±5.6	652.8a	31.8a
Summer	27.1±5.9	17.4±4.1	345.6b	17.0b
Fall	18.4±9.7	5.5±5.5	346.6b	12.3b
Winter	18.3±11.4	2.6±4.5	383.5b	15.4b

Note. [z]Photosynthetic photon flux density (PPFD) is a measure of the number of photons in the 400-700nm range of the visible light spectrum that fall on a square meter of target area per second.

[y]Daily light integral (DLI) is a measure of the amount of photosynthetically active radiation photons received each day in a square meter.

Different letter down the column indicate significant difference by Tukey's significance test at $P \leq 0.05$.

3.2 Stand Establishment, Average Yield and Productivity

Stand establishment was not significantly different among summer, fall and winter seasons. In spring, however, stand establishment was generally low for each sowing including very poor germination for the 11th harvest (49.7%). After the shade cloth was installed, SE was improved. Overall, stand establishment averaged 83.5±11.6% for the year (Table 2).

Lettuce was harvested on 30 occasions; ten harvests each in spring and summer, seven harvests in fall and three harvests in winter (Table 2). The number of harvests was dependent on the growth rate of the lettuce. Growth rates were highest in summer when it took 4.6 weeks to achieve harvestable size and lowest in winter when it took 9 weeks.

Total annual yield was 644.0 kg with significant differences among all four seasons ($P \leq 0.001$). The highest yields occurred in the spring with an average yield of nearly 700 g per tray and a seasonal yield of 250.4 kg. Yield in summer was also higher than that in colder seasons with an average yield of 622.0 g per tray and a seasonal yield of 196.5 kg. Yields in fall and winter were lower with an average yields of 520.8 g and 435.6 g per tray, respectively. Average yield per harvest was 25.0, 19.7 and 21.4 kg for spring, summer and fall, respectively, while 15.7 kg of lettuce was harvested during winter. Nearly 70% of lettuce production (446.9 kg) was obtained from the spring and summer seasons.

Because there was a significant difference in growing period ($P \leq 0.001$) among seasons, productivity (g produced per week per tray) was determined. While yield was the greatest in spring, productivity was highest in summer (139.2 g/wk/tray) and increased more than 30% compared to that in spring. Productivity was higher in spring (105.5 g/wk/tray) than fall (85.3 g/wk/tray) and winter (48.7 g/wk/tray). Productivity during summer was nearly three times greater than in winter.

Table 2. Seasonal standard establishment, number of harvest times, yield, productivity of lettuce.

Season	Standard establishment (%)	No. of harvest times	Seasonal yield (kg)	% of total yield	Yield/harvest (kg)	Average yield (g/tray)	Growing period (weeks)	Productivity (g/tray/week)
Spring	78.4b[z]	10	250.4	38.9	25.0	699.5a	6.6 b	105.5b
Summer	86.7a	10	196.5	30.5	19.7	622.0b	4.6 c	139.2a
Fall	85.1a	7	150.0	23.3	21.4	520.8c	6.6 b	85.3c
Winter	86.6a	3	47.0	7.3	15.7	435.6d	9.0 a	48.7d
Significance	***	N/A	N/A	N/A	N/A	***	***	***

Note. [z]Values are the means of three replications (n = 3).

Different letters down the column indicate significant difference by Tukey's significance test at $P \leq 0.05$.

NA, not applicable.

*** Significant at $P \leq 0.001$.

3.3 Water Quality

The average influent nutrient and TSS concentrations were 0.34 mg L^{-1}, 0.30 mg L^{-1}, 0.17 mg L^{-1} and 1.95 mg L^{-1} for TAN, NO_3^{-1}, phosphate (PO_4^{-3}) and TSS, respectively. There were significant seasonal differences in influent and effluent concentration for all nutrients and the influent TSS ($P \leq 0.01$) (Table 3). Influent TAN concentrations ranged between 0.22 and 0.40 mg L^{-1} with the highest concentrations in summer. The concentration of TAN in effluent during the summer season was the highest (0.35 mg L^{-1}) and the lowest in fall (0.21 mg L^{-1}). The seasonal differences in the quantity of TAN removal were significant ($P \leq 0.001$). During spring and summer removal of TAN was significantly higher than that in fall and winter.

Influent nitrate concentrations ranged between 0.29 and 0.41 mg L^{-1} and the highest concentrations of nitrate was detected in winter (0.41 mg L^{-1}), while there was no difference among other seasons. Removal of nitrate was low all year ranging from -0.02 to 0.02 mg L^{-1} with no significant difference among seasons. Influent phosphate concentrations ranged between 0.14 and 0.17 mg L^{-1} and generally were low during fall and winter compared to spring and summer. There was minimal removal during spring and summer and no detectable level of removal was observed during the fall and winter. Influent TSS concentration during summer (2.25 mg L^{-1}) was higher than that in fall (1.55 mg L^{-1}) that was the lowest among the four seasons. There were significant differences in influent concentration among seasons, however, amount of TSS removed was not significantly different.

Table 3. Seasonal nutrient removal of total ammonia nitrogen (TAN), nitrate (NO_3^{-1}), and phosphate (PO_4^{-3}) and total soluble solids (TSS) in flow-through aquaponic system by lettuce[z]

Season	TAN (mg L^{-1})			NO_3^{-1} (mg L^{-1})			PO_4^{-3} (mg L^{-1})			TSS (mg L^{-1})		
	Influent	Effluent	Removal[y]	Influent	Effluent	Removal	Influent	Effluent	Removal	Influent	Effluent	Removal
Spring	0.34b	0.30ab	0.04ab	0.30b	0.29b	0.01	0.17a	0.16a	0.01a	1.95ab	1.2	0.75
Summer	0.40a	0.35a	0.05a	0.29b	0.27b	0.02	0.17ab	0.16a	0.01ab	2.25a	1.25	1.00
Fall	0.22d	0.21c	0.01bc	0.30b	0.28b	0.02	0.14c	0.14b	0.00c	1.55b	1.11	0.44
Winter	0.28c	0.28b	0.00c	0.41a	0.43a	-0.02	0.15bc	0.15ab	0.00bc	1.79ab	1.27	0.52
Significance	***	***	***	***	***	NS	***	**	***	*	NS	NS

Note. [z]Values are the means of three replications (n = 3).

[y]Removal concentration was calculated by subtracting effluent concentration from influent concentration.

Different letters down the column indicate significant difference by Tukey's significance test at $P \leq 0.05$.

NS, not significant.

*, **, *** Significant at $P \leq 0.01$, 0.05 or 0.001, respectively.

4. Discussion

4.1 Stand Establishment, Yield and Productivity

The seeds of 'Red Sails' lettuce were able to germinate and establish well throughout the year including during the cold seasons in FTS (Table 2). However, it took longer until harvest during the cool seasons (fall and winter) and fewer harvests were made compared to the warm seasons (spring and summer). In other systems lettuce seeds are usually sown in a germination area and seedlings were transplanted to the aquaponic facility. In our study seeds were directly sown into the tray. Direct sowing method can be beneficial because it saves labor, time and space that are associated with germination and transplanting, however, it could cause lower SE, because 100% seed germination can't always be guaranteed and more often than not there will be empty cells in a tray. Empty cells can't be refilled after seeds were directly sown into the tray and it could result in lower SE as observed during the spring season in our study.

Seasonal average yield of 'Red Sails' lettuce was varied. Average yield in spring was the greatest followed by summer, fall and winter in descending order. In spring average yield was higher than in summer likely due to cooler temperature and higher light intensity (Table 1). Nutrients in the aquaculture wastewater are unlikely to be a major factor influencing seasonal average yield. Although the concentration of TAN during the summer was higher than that in the spring, removal of TAN was not different between the spring and summer (Table 3). Additionally, there was no difference in nitrate, phosphate and TSS concentration in either influent or effluent. During the fall and winter seasons, the average yields were lower than warmer seasons, most likely due to lower night temperatures and the low concentration of TAN (both seasons), nitrate (winter only), and potentially phosphate. It should be noted that in summer two harvests were discarded because lettuce appeared to be damaged by stress and became unmarketable. Summer yield would have been higher if there were no such incident. The exact cause for the damage was not known.

Seasonal yields of lettuce were variable across season with the greatest yield in spring higher than in summer. In spring, temperatures were moderate, and light intensity and DLI were the highest while in summer temperatures were higher and light intensity was comparable to fall and winter due to the shade cloth. Particularly, SE was lower in spring than that in other seasons, but the seasonal yield and average yield in the spring harvests were highest, indicating that SE alone may not be a definite indicator of the crop production. Additionally, higher intensity of light during the spring season might have negatively affected SE. The lower yield during cold seasons, although SE was similar to that in warm seasons, might have been caused by not only cold air temperatures, but also limited supply of nutrients (Table 3).

During the summer months, lettuce is not generally grown in the field because elevated temperatures and long day photoperiod stimulate bolting and reduce lettuce production (Simko, Hayes, Mou, & McCreight, 2014). For the opposite reason in the cold season, lettuce is not grown in the field because the air and soil temperatures are too low for crop production in most temperate regions. Generally, water keeps relatively constant temperature compared to air. In our FTS, spring water provided steady root-zone temperature for lettuce production year-round. Root-zone temperature significantly affects plant production (Gosselin & Trudel, 1985). Moderate root-zone temperature might have mitigated the adverse effect of either elevated and decreased air temperature during the summer and winter, respectively, and helped lettuce to grow in a cold water FTS.

Growing period until harvest varied among seasons. During the summer, lettuce was ready to be harvested in less than five weeks (Table 2) compared to longer than six weeks in spring and fall. Elevated temperature and/or adequate light intensity, considering similar amounts of nitrogen and phosphate supply or removal in FTS during the spring and summer, could have accelerated plant growth and consequently shortened harvest time in the summer. Additionally, lower night time temperatures in the spring (11.6 °C) than the summer (17.6 °C) might have contributed to a longer growing period in the spring and resulted in lower productivity, although with same reason average yield were higher in spring. During the extended growing seasons (fall and winter), harvest time was longer and fewer harvests were made compared to the warm seasons (spring and summer), potentially due to significantly lower night temperature and lower level of nutrient caused by slow growth rate or low number of fish in the aquaculture system.

During the warm seasons (spring and summer), the temperature both inside and outside the high tunnel was often higher than the optimum temperature range (16-20 °C) for lettuce production. Lettuce is prone to heat stress and often bolts, i.e. produces flowers prematurely. Bolting occurs when lettuce is grown under elevated temperature and high light intensity, and it diminishes the marketability of the lettuce (Zhao & Carey, 2009). However, bolting was not observed during the year-round production in a cold water FTS. Shading during the summer season, in conjunction with cool water temperatures may have helped prevent bolting by reducing the temperature and light intensity inside the high tunnel.

4.2 Water Quality

The only nutrient source supporting lettuce production was the nutrients coming from the aquaculture facility. The aquaculture effluent nutrient concentrations were low and frequently there was little difference in concentration between the effluent entering the channels and that leaving them. Nonetheless, this system was able to produce lettuce at comparable rates to soil based systems. This may be due in part to the high delivery rates of effluent to the hydroponic channels. At a flow rate of 76 liter min^{-1} an average of 25.8 mg L^{-1}, 22.8 mg L^{-1} and 12.9 mg L^{-1} of TAN, nitrate and phosphate, respectively, was delivered to the top of the channel. Due to the conveyor production system used, the smallest plants received the most concentrated nutrients yet would have the lowest uptake due to low biomass.

Nutrient availability may have been greater than the measured concentrations entering the channels due to decomposition and mineralization of accumulated solids (uneaten fish food and feces) in the bottom of the channel. No attempt was made to remove the solids comprised of vermiculite and particulate organic material that accumulated in the channel. The aquaculture facility does have quiescent zones at the end of each raceway section, however they do not trap all of the solids. The entrained solids deposit mostly in the upper third of the channel and represent the largest particles. The deposition is apparent from the reduction in TSS concentrations at the end of the channel. Accumulated solids then decompose creating an unmeasured nutrient source.

TAN removal was higher than nitrate or phosphate removal and varied seasonally (Table 3). The highest removal rates occurred during spring and summer when the lettuce was growing the fastest. TAN removal was substantially greater than nitrate removal. Buzby and Lin (2014) determined that lettuce grown in this cool water flow-through aquaponic system has a strong preference for TAN. Xu et al. (1992) determined that ammonium was the preferred nitrogen source when N concentrations were low while nitrate was preferred when N concentrations were high. Low water temperatures may also have played a role in the preference for TAN. Multiple studies, reviewed by von Wirén et al. (1997) have demonstrated that low temperatures generally increased the reliance of plants on ammonium as a mineral N source.

Lower removal rates for PO_4^{-3} may be related to the N:P ratio which was 6.6. Koerselman and Meuleman (1996) determined that for most plants an N:P ratio of less than 14 is indicative of N limitation. Buzby and Lin (2014) determined that this system was strongly N limited and that additions of ammonia increased PO_4^{-3} removal rates. It is unclear whether the slightly higher TAN concentrations in summer increased PO_4^{-3} removal or if it was due to increased growth rates of lettuce.

Concentrations of all the soluble nutrients in the flow-through raceway effluent were substantially lower than nutrient concentrations found in either hydroponic culture or in effluent from recirculating fish culture. Despite the low nutrient concentrations, no nutrient deficiencies were observed and all plants appeared healthy, even those at the end of the channel where nutrient concentrations would be the lowest. This may be due to the flow-through nature of the system where plants were supplied with a continual nutrient supply. In flow-through culture, fresh effluent, with its full complement of nutrients derived from both the source water and fish culture, is delivered to the plant growing channel. In a recirculating system nutrients are removed with each pass through the system and those that are not replaced by inputs from fish culture can become limited, which may cause an imbalance of nutrients leading to nutrient deficiencies (Rakocy et al., 2006).

4.3 Lettuce Production

In our study, lettuce was produced in 115.9 m^2 high tunnel (26' × 48' or 1,268 ft^2) using three of four available channels and it produced 5.6 kg m^{-2} (644 kg/115.9 m^2) of lettuce. If all four channels were used, the estimated total yield per m^2, by adding 33% yield to the total production, would have been 7.4 kg m^{-2}, which is significantly higher than average field lettuce production in the U.S. in 2012 (3.1 kg m^{-2}, based upon 40.3, 27.3 and 35.6 t ha^{-1} for iceberg, leaf and romaine lettuce production, respectively) (Simko et al., 2014; Galinato & Miles, 2013), and the soil based high tunnel in West Virginia (4.9 kg m^{-2}) (personal communication, L.W. Jett). Combined yield during the spring and summer in FTS was 446.9 kg or 3.9 kg m^{-2}, which was higher than field lettuce production per area. Lettuce production during the extended seasons (197.0 kg or 1.7 kg m^{-2}) could make up almost 54.4% of the yearly field production.

In a 128-cell tray (67.6 cm × 34.6 cm) with annual average SE of 83.5%, 107 lettuce plants would be expected to grow in a tray or 166 plants per m^2 (based upon 107 lettuces per tray, 45 trays in a channel, four channels in the facility of 115.9 m^2). The density of lettuce grown in a tray was about 26 times higher than the crop densities in the field production (6.4 plants m^{-2} based upon 26,000 heads per acre) (Galinato & Miles, 2013) and 3 times higher than the soil based high tunnel (53.8 plants m^{-2} based upon 0.2 ft^2 per plant in West Virginia) (personal communication, L.W. Jett). High density lettuce production requires high level of nutrients. In an FTS, the content of nutrients was generally limited and supplied at a lower level, although there was a moderate increase of TAN in the summer and nutrient was constantly supplied at a low level. Insufficient nutrients in the wastewater in an FTS might have limited lettuce growth. The average fresh weight of a single lettuce plant grown in a tray was assumed to be about 44.6 g m^{-2} throughout the year, based upon the annual yield 7.4 kg m^{-2} from estimated 166 lettuces per m^2, was significantly low compared to lettuce grown in the field (484.4 g m^{-2}; 3.1 kg per 6.4 heads m^{-2}) at maturity. Depending on the configuration of a tray manufactured for a hydroponic system, up to 128 lettuces can grow in one tray. Using a tray with smaller cells, it is recommended to produce baby or spring mix lettuces instead of fully grown loose leaf or whole head lettuce. There is no defined size for baby or spring mix vegetables. Usually, baby lettuces are lettuces with four to six true leaves and spring mix refers to mixture of baby sized leafy vegetables. A higher number of lettuce can be produced for a shorter growing period in a tray with smaller cells. Bulk production of loose leaf lettuce or baby lettuce for salad processing, food service or value added products would be a better choice if the scale of lettuce production and marketing strategy fit such operation (Simko et al., 2014). It is a growing trend that farmers do not grow head lettuce to their full mature size because lettuce often needs to be immediately packaged in a fixed size of a carton for 24 heads at the production site (Simko et al., 2014). In addition, multiple research studies indicated that premature lettuces such as microgreen and baby greens contained more nutritious phytochemicals and demand for miniature sized lettuce is increasing (Kim et al., 2016; Pinto, Almeida, Aguiar, & Ferreira, 2015; Xiao et al., 2015; Ascensión, Luna, Selma, Tudela, Abad, & Gil, 2012; Martínez-Sánchez et al., 2012).

5. Conclusions

The present study has demonstrated that year-round lettuce production using an FTS inside a high tunnel was not only feasible, but also better than the field or soil based high tunnel production based upon yield per area. However, yield during the winter months was low and needs to be improved. To further increase the yield of lettuce during winter, low tunnels within a high tunnel and solar powered supplemental lighting systems such as low energy light emitting diode (LED) lighting system can be installed to enhance low light conditions and increase photosynthetic activity during cold seasons to a certain extent. Depending on the goals of the grower, a selection of different plant species that are more tolerant to a cold growing environment such as cole crops (a group of vegetables in the Brassica family) could be another option for consideration. Conversely, some growers may wish to shut down an aquaponic operation during the winter months.

Acknowledgements

We would like to thank Matthew Ferrell and Gene Jacobs for their assistance with maintaining the aquaculture system.

References

Adler, P. R., Harper, J. K., Takeda, F., Wade, E. M., & Summerfelt, S. T. (2000). Economic evaluation of hydroponics and other treatment options for phosphorus removal in aquaculture effluent. *HortScience, 35*, 993-999.

APHA (American Public Health Administration). (1995). *Standard methods for the examination of water and wastewater* (19th ed.). American Public Health Administration, Washington, D.C.

Ascensión, M.-S., Luna, M. C., Selma, M. V., Tudela, J. A., Abad, J., & Gil, M. I. (2012). Baby-leaf and multi-leaf of green and red lettuces are suitable raw materials for the fresh-cut industry. *Postharvest Biology and Technology, 63*, 1-10. https://doi.org/10.1016/j.postharvbio.2011.07.010

Borrelli, K., Koenig, R. T., Jaeckel, B. M., & Miles, C. A. (2013). Yield of leafy greens in high tunnel winter production in the Northwest United States. *HortScience, 48*(2), 183-188.

Buzby, K. M., & Lin, L. (2014). Scaling aquaponic systems: Balancing plant uptake with fish output. *Aquacultural Eng., 63*, 39-44. https://doi.org/10.1016/j.aquaeng.2014.09.002

Buzby, K. M., Waterland, N. L., Semmens, K. J., & Lin, L.-S. (2016). Evaluating aquaponic crops in a freshwater flow-through fish culture system. *Aquaculture, 460*, 15-24. https://doi.org/10.1016/j.aquaculture.2016.03.046

Castoldi, N., Bechini, L., & Ferrante, A. (2011). Fossil energy usage for the production of baby leaves. *Energy, 36*, 86-93. https://doi.org/10.1016/j.energy.2010.11.004

Fukuda, M., Matsuo, S., Kikuchi, K., Kawazu, Y., Fujiyama, R., & Honda, I. (2011). Isolation and functional characterization of the FLOWERING LOCUS T homolog, the *LsFT* gene, in lettuce. *Journal of Plant Physiology, 168*, 1602-1607. https://doi.org/10.1016/j.jplph.2011.02.004

Galinato, S. P., & Miles, C. A. (2013). Economic profitability of growing lettuce and tomato in western Washington under high tunnel and open-filed production systems. *HortTechnology, 23*(4), 453-460.

Gosselin, A., & Trudel, M.-J. (1985). Influence of root-zone temperature on growth, development and yield of cucumber plants cv. Toska. *Plant and Soil, 85*, 327-336. https://doi.org/10.1007/BF02220188

Ido, S. (2016). Growth models of gilthead sea bream (*Sparus aurata* L.) for aquaculture: A review. *Aquacultural Engineering, 70*, 15-32. https://doi.org/10.1016/j.aquaeng.2015.12.001

Jett, L. W. (2016). *Personal communication.* April 8, 2016.

Kim, M. J., Moon, Y., Tou, J. C., Mou, B., & Waterland, N. L. (2016). Nutritional value, bioactive compounds and health benefits of lettuce (*Lactuca sativa* L.). *J. of Food Composition and Analysis, 49*, 19-34. https://doi.org/10.1016/j.jfca.2016.03.004

Lee, A.-C., Liao, F.-H., & Lo, H.-F. (2015). Temperature, daylength, and cultivar interact to affect the growth and yield of lettuce grown in high tunnels in subtropical regions. *HortScience, 50*(10), 1412-1418.

Losordo, T. M., Masser, M., & Rakocy, J. (1992). *Recirculating aquaculture tank production systems: An overview of critical considerations* (No. 451). Southern Regional Aquaculture Center, SRAC Publication. Retrieved from http://university.uog.edu/cals/people/pubs/aquac/451recag.pdf

Love, D. C., Fry, J. P., Genello, L., Hill, E. S., Frederick, J. A., Li, X., & Semmens, K. (2014). An international survey of aquaponics practitioners. *Plos One, 9*(7), e102662. https://doi.org/10.1371/journal.pone.0102662

Love, D. C., Uhl, M. S., & Genelloa, L. (2015). Energy and water use of a small-scale raft aquaponics system in Baltimore, Maryland, United States. *Aquacultural Engineering, 68*, 19-27. https://doi.org/10.1016/j.aquaeng.2015.07.003

Martínez-Sánchez, A., Lunaa, M. C., Selma, M. V., Tudela, J. A., Abad, J., & Gil, M. I. (2012). Baby-leaf and multi-leaf of green and red lettuces are suitable raw materials for the fresh-cut industry. *Postharvest Biol. and Tech., 63*, 1-10. https://doi.org/10.1016/j.postharvbio.2011.07.010

Martins, C. I. M., Eding, E. H., Verdegem, M. C. J., Heinsbroek, L. T. N., Schneider, O., Blancheton, J. P., … Verreth, J. A. J. (2010). New developments in recirculating aquaculture systems in Europe: A perspective on environmental sustainability. *Aquacult. Eng., 43*(3), 83-93. https://doi.org/10.1016/j.aquaeng.2010.09.002

National Oceanic and Atmospheric Administration (Wardensville in Hardy County). (2016). Retrieved April 1, 2016, from https://www.ncdc.noaa.gov/cdo-web/search

Park Seed. (2016). Retrieved May 9, 2016, from http://parkseed.com/red-sails-lettuce-seeds/p/05153-PK-P1/

Pinto, E., Almeida, A. A., Aguiar, A. A., & Ferreira, I. M. P. L. V. O. (2015). Comparison between the mineral profile and nitrate content of microgreens and mature lettuces. *Journal of Food Composition and Analysis, 37*, 38-43. https://doi.org/10.1016/j.jfca.2014.06.018

Rakocy, J. E. (2007). Aquaponics: Vegetable hydroponics in recirculating systems. In M. B. Timmons & J. M. Ebeling (Eds.), *Recirculating aquaculture systems* (2nd ed., pp. 631-672). Cayuga Aqua Ventures, Ithaca, NY.

Rakocy, J. E., Bailey, D. S., Shultz, R. C., & Thoman, E. S. (2004). Aquaponic production of tilapia and basil: Comparing a batch and staggered cropping system. *Acta Hort. (ISHS), 648*, 63-69. https://doi.org/10.17660/ActaHortic.2004.648.8

Rakocy, J. E., Hargreaves, J. A., Bailey, D. S. (1993). *Nutrient accumulation in a recirculating aquaculture system integrated with hydroponic vegetable production.* Paper Presented at the American Society of Agricultural engineers Conference, Spokane WA, June 21-23, 1993.

Rakocy, J. E., Masser, M. P., & Losordo, T. M. (2006). *Recirculating aquaculture tank production systems: Aquaponics–integrating fish and plant culture* (No. 454, pp. 1-16). Southern Regional Aquaculture Center, SRAC Publication. Retrieved from https://srac.tamu.edu/

Resh, H. M. (2013). *Hydroponic Food Production* (7th ed.). CRC Press, Boca Raton, FL.

Saavas, D. (2002). Nutrient solution recycling. In D. Saavas & H. Passam (Eds.), *Hydroponic production of vegetables and ornamentals* (pp. 299-343). Embryo Publications, Athens, Greece.

Savidov, N., Hutchings, E., & Rakocy, J. E. (2007). Fish and plant production in a recirculating aquaponic system: an new approach to sustainable agriculture in Canada. *Acta Hortic., 742*, 209-213. https://doi.org/10.17660/ActaHortic.2007.742.28

Silva, E. C., Maluf, W. R., Leal, N. R., & Gomes, L. A. A. (1999). Inheritance of bolting tendency in lettuce *Lactuca sativa* L. *Euphytica, 109*, 1-7. https://doi.org/10.1023/A:1003698117689

Simko, I., Hayes, R. J., Mou, B., & McCreight, J. D. (2014). Lettuce and Spinach. In S. Smith, B. Diers, J. Specht, & B. Carver (Eds.), *Yield gains in major U.S. field crops* (pp. 53-85). American Society of Agronomy, Madison, WI. https://doi.org/10.2135/cssaspecpub33.c4

U.S. Department of Agriculture. (2013). Western Region SARE Research Innovations. *Organic winter production scheduling in unheated high tunnels in Colorado*. U.S. Dept. of Agr., Washington, D.C.

U.S. Department of Agriculture. (2014). National Statistics Service. *2012 Census of Aquaculture*. U.S. Dept. of Agr., Washington, D.C.

U.S. Department of Agriculture. (2016). National Agricultural Statistics Service. *Vegetables 2015 summary*. U.S. Dept. of Agr., Washington, D.C. Retreived from http://usda.mannlib.cornell.edu/usda/current/VegeSumm/VegeSumm-02-04-2016.pdf

von Wirén, N., Gazzarrini, S., & Frommer, W. B. (1997). Regulation of mineral nitrogen uptake in plants. *Plant and Soil, 196*, 191-199. https://doi.org/10.1023/A:1004241722172

Xiao, Z., Lester, G. E., Park, E., Saftner, R. A., Luo, Y., & Wang, Q. (2015). Evaluation and correlation of sensory attributes and chemical compositions of emerging fresh produce: Microgreens. *Postharvst Biology and Technology, 110*, 140-148. https://doi.org/10.1016/j.postharvbio.2015.07.021

Xu, Q. F., Tsai, C. L., & Tsai, C. Y. (1992). Interaction of potassium with the form and amount of nitrogen nutrition on growth and nitrogen uptake of maize. *J. of Plant Nutri., 15*, 23-33. https://doi.org/10.1080/01904169209364299

Zhao, X., & Carey, E. E. (2009). Summer production of lettuce, and microclimate in high tunnel and open field plots in Kansas. *HortTechnology, 19*(1), 113-119.

Permissions

All chapters in this book were first published in JAS, by Canadian Center of Science and Education; hereby published with permission under the Creative Commons Attribution License or equivalent. Every chapter published in this book has been scrutinized by our experts. Their significance has been extensively debated. The topics covered herein carry significant findings which will fuel the growth of the discipline. They may even be implemented as practical applications or may be referred to as a beginning point for another development.

The contributors of this book come from diverse backgrounds, making this book a truly international effort. This book will bring forth new frontiers with its revolutionizing research information and detailed analysis of the nascent developments around the world.

We would like to thank all the contributing authors for lending their expertise to make the book truly unique. They have played a crucial role in the development of this book. Without their invaluable contributions this book wouldn't have been possible. They have made vital efforts to compile up to date information on the varied aspects of this subject to make this book a valuable addition to the collection of many professionals and students.

This book was conceptualized with the vision of imparting up-to-date information and advanced data in this field. To ensure the same, a matchless editorial board was set up. Every individual on the board went through rigorous rounds of assessment to prove their worth. After which they invested a large part of their time researching and compiling the most relevant data for our readers.

The editorial board has been involved in producing this book since its inception. They have spent rigorous hours researching and exploring the diverse topics which have resulted in the successful publishing of this book. They have passed on their knowledge of decades through this book. To expedite this challenging task, the publisher supported the team at every step. A small team of assistant editors was also appointed to further simplify the editing procedure and attain best results for the readers.

Apart from the editorial board, the designing team has also invested a significant amount of their time in understanding the subject and creating the most relevant covers. They scrutinized every image to scout for the most suitable representation of the subject and create an appropriate cover for the book.

The publishing team has been an ardent support to the editorial, designing and production team. Their endless efforts to recruit the best for this project, has resulted in the accomplishment of this book. They are a veteran in the field of academics and their pool of knowledge is as vast as their experience in printing. Their expertise and guidance has proved useful at every step. Their uncompromising quality standards have made this book an exceptional effort. Their encouragement from time to time has been an inspiration for everyone.

The publisher and the editorial board hope that this book will prove to be a valuable piece of knowledge for researchers, students, practitioners and scholars across the globe.

List of Contributors

Purnomo
Laboratory of Plant Systematic, Faculty of Biology, Universitas Gadjah Mada, Indonesia

Budi Setiadi Daryono
Laboratory of Genetic, Faculty of Biology, Universitas Gadjah Mada, Indonesia

Hironobu Shiwachi
Graduate School of Agriculture, Tokyo University of Agriculture, Tokyo, Japan

Mohammad Omrani, Mohammad Nabi Shahiki Tash and Ahmad Akbari
School of Management & Economics, University of Sistan and Baluchestan, Zahedan, Iran

Nada Sedky
Zewail City of Science and Technology, Egypt

Nagwa Elnwishy
Zewail City of Science and Technology, Egypt
Biotechnology Research Center, Suez Canal University, Egypt

Emad Rashwan
Department of Agronomy, Faculty of Agriculture, Tanta University, Egypt

Ahmed Mousa
Fiber Crops Research Section, Field Crops Research Institute, Egypt

Ayman EL-Sabagh
Department of Agronomy, Faculty of Agriculture, Kafrelsheikh University, Egypt

Celaleddin Barutçular
Department of Field Crops, Faculty of Agriculture, Cukurova University, Turkey

Valdir Alves and Tamiel Khan B. Jacobson
Programa de Pós Graduação em Meio Ambiente e Desenvolvimento Rural, Faculdade UnB Planaltina, Universidade de Brasília, Planaltina, DF, Brazil
Programa de Pós Graduação em Agroecossistemas, Centro de Ciências Agrárias, Universidade Federal de Santa Catarina, Florianópolis, SC, Brazil

Clarilton Edzard D. Cardoso Ribas
Programa de Pós Graduação em Agroecossistemas, Centro de Ciências Agrárias, Universidade Federal de Santa Catarina, Florianópolis, SC, Brazil

Fernando F. Goulart
Análise e Modelagem de Sistemas Ambientais/Centro de Sensoriamento Remoto, Dept. Cartografia, Instituto de Geociências, Universidade Federal de Minas Gerais, Belo Horizonte, MG, Brazil

Reinaldo J. de Miranda Filho
Faculdade UnB Planatina, Universidade de Brasília, RADIS/Fup, Planaltina, DF, Brazil

P. I. Ifejika
National Institute for Freshwater Fisheries Research, New Bussa, Niger State, Nigeria

Verónica Lango-Reynoso, Juan L. Reta-Mendiola, Felipe Gallardo López and Alberto Asiain-Hoyos
Colegio de Postgraduados, Campus Veracruz, Manlio Fabio Altamirano, Veracruz, México

Katia A. Figueroa-Rodríguez
Colegio de Postgraduados, Campus Córdoba, Amatlán de los Reyes, Veracruz, México

Fabiola Lango-Reynoso
Instituto Tecnológico de Boca del Río, Boca del Río, Veracruz, México

Firew Elias and Delelegn Woyessa
Department of Biology, Jimma University, Ethiopia

Diriba Muleta
Environmental Biotechnology Unit, Addis Ababa University, Ethiopia

Jean Dominique Gumirakiza
Department of Agriculture, Western Kentucky University, USA

Amber Daniels
School of Law, University of Louisville, USA

Bibhuti Bhusan Dalei and Manoj Kumar Meena
AICRP on Niger, Regional Research & Technology Transfer Station (OUAT), Semiliguda, Koraput, Odisha, India

Bibhuti Bhusan Sahoo, Lalatendu Nayak, Pravamayee Acharya and Niranjan Senapati
Regional Research & Technology Transfer Station (OUAT), Semiliguda, Koraput, Odisha, India

Amit Phonglosa
Regional Research & Technology Transfer Sub-Station (OUAT), Umarkot, Nabarangapur, Odisha, India

Daniel Bomfim Manera and Daniel Ribeiro Menezes
Universidade Federal do Vale do São Francisco (Univasf), Petrolina/PE, Brazil

Tadeu Vinhas Voltolini and Gherman Garcia Leal de Araújo
Embrapa Semiárido, Petrolina/PE, Brazil

Kuniaki Sato, Toshiyuki Wakatsuki and Tsugiyuki Masunaga
Faculty of Life and Environmental Sciences, Shimane University, Matsue, Japan

Adha Fatmah Siregar
Faculty of Life and Environmental Sciences, Shimane University, Matsue, Japan Indonesian Soil Research Institute, Bogor, Indonesia

Ibrahim Adamy Sipahutar, Husnain and Heri Wibowo
Indonesian Soil Research Institute, Bogor, Indonesia

Madegwa Yvonne, Onwonga Richard and Karuku George
Department of Land Resource Management and Agricultural Technology, University of Nairobi, Nairobi, Kenya

Shibairo Solomon
Department of Plant Science and Crop Protection, University of Nairobi, Nairobi, Kenya

Luma S. Albanna and Nidá M. Salem
Department of Plant Protection, Faculty of Agriculture, the University of Jordan, Amman, Jordan

Akl M. Awwad
Nanotechnology Laboratory, Royal Scientific Society, Amman, Jordan

Jesús Montoya-Mendoza, María del Refugio Castañeda-Chávez, Fabiola Lango-Reynoso and Salvador Rojas-Castañeda
Instituto Tecnológico de Boca del Río, División de Estudios de Posgrado e Investigación, Veracruz, México

Khalil Allali
Département d'Economie Rurale, Ecole Nationale d'Agriculture de Meknes, Meknes-El Menzeh, Morocco

Boubaker Dhehibi and Aden Aw-Hassan
Sustainable Intensification and Resilient Production Systems Program (SIRPSP), International Center for Agricultural Research in th+s (ICARDA), Amman, Jordan

Shinan N. Kassam
Sustainable Intensification and Resilient Production Systems Program (SIRPSP), International Center for Agricultural Research in the Dry Areas (ICARDA), Cairo, Egypt

Jun Jing, Xinfeng Xiao, Yu Zhou, Xiaoyu Wang, Shangqiao Chen and Bin Bao
Department of Marine Pharmacology, College of Food Science and Technology, Shanghai Ocean University, Shanghai, China

Wenhui Wu
Department of Marine Pharmacology, College of Food Science and Technology, Shanghai Ocean University, Shanghai, China
Huaihai Institute of Technology, Lianyungang, China

Shujun Wang
Huaihai Institute of Technology, Lianyungang, China

Yongxu Cheng and Xugan Wu
Key Laboratory of Exploration and Utilization of Aquatic Genetic Resources, Shanghai Ocean University, Shanghai, China

Hosein Mohammadi and Solmaz Nojavan
Agricultural Economics Department, Agriculture Faculty, Ferdowsi University of Mashhad, Iran

Abdolhamid Moarefi Mohammadi
Department of Economics, University of Isfahan, Isfahan, Iran

Gaylynn E. Johnson and Nicole L. Waterland
Division of Plant and Soil Sciences, West Virginia University, Morgantown, WV, USA

Karen M. Buzby
Department of Civil and Environmental Engineering, West Virginia University, Morgantown, WV, USA

Kenneth J. Semmens
Division of Animal and Nutritional Sciences, West Virginia University, Morgantown, WV, USA
Aquaculture Research Center, Kentucky State University, Frankfort, KY, USA

Raphael Della Maggiore Orlandi, Marta Regina Verruma-Bernardi, Simone Daniela Sartorio and Maria Teresa Mendes Ribeiro Borges
Department of Agroindustrial Technology and Rural Socieconomy, Federal University of São Carlos School (CCA), Araras, SP, Brazil

Gaylynn E. Johnson and Nicole L. Waterland
Division of Plant and Soil Sciences, West Virginia University, Morgantown, WV, USA

Karen M. Buzby
Department of Civil and Environmental Engineering, West Virginia University, Morgantown, WV, USA

Kenneth J. Semmens
Division of Animal and Nutritional Sciences, West Virginia University, Morgantown, WV, USA
Aquaculture Research Center, Kentucky State University, Frankfort, KY, USA

Index

O

P

R

S

T

V

W

www.ingramcontent.com/pod-product-compliance
Lightning Source LLC
Chambersburg PA
CBHW080624200326
41458CB00013B/4492

* 9 7 8 1 6 3 2 3 9 9 6 5 6 *